MOLECULAR AND CELLULAR ASPECTS OF
MICROBIAL EVOLUTION

SYMPOSIA OF THE SOCIETY FOR GENERAL MICROBIOLOGY*

1. THE NATURE OF THE BACTERIAL SURFACE
2. THE NATURE OF VIRUS MULTIPLICATION
3. ADAPTATION IN MICRO-ORGANISMS
4. AUTOTROPHIC MICRO-ORGANISMS
5. MECHANISMS OF MICROBIAL PATHOGENICITY
6. BACTERIAL ANATOMY
7. MICROBIAL ECOLOGY
8. THE STRATEGY OF CHEMOTHERAPY
9. VIRUS GROWTH AND VARIATION
10. MICROBIAL GENETICS
11. MICROBIAL REACTION TO ENVIRONMENT
12. MICROBIAL CLASSIFICATION
13. SYMBIOTIC ASSOCIATIONS
14. MICROBIAL BEHAVIOUR, 'IN VIVO' AND 'IN VITRO'
15. FUNCTION AND STRUCTURE IN MICRO-ORGANISMS
16. BIOCHEMICAL STUDIES OF ANTIMICROBIAL DRUGS
17. AIRBORNE MICROBES
18. THE MOLECULAR BIOLOGY OF VIRUSES
19. MICROBIAL GROWTH
20. ORGANIZATION AND CONTROL IN PROKARYOTIC AND EUKARYOTIC CELLS
21. MICROBES AND BIOLOGICAL PRODUCTIVITY
22. MICROBIAL PATHOGENICITY IN MAN AND ANIMALS
23. MICROBIAL DIFFERENTIATION
24. EVOLUTION IN THE MICROBIAL WORLD
25. CONTROL PROCESSES IN VIRUS MULTIPLICATION
26. THE SURVIVAL OF VEGETATIVE MICROBES
27. MICROBIAL ENERGETICS
28. RELATIONS BETWEEN STRUCTURE AND FUNCTION IN THE PROKARYOTIC CELL
29. MICROBIAL TECHNOLOGY: CURRENT STATE, FUTURE PROSPECTS
30. THE EUKARYOTIC MICROBIAL CELL
31. GENETICS AS A TOOL IN MICROBIOLOGY

* Published by the Cambridge University Press, except for the first Symposium, which was published by Blackwell's Scientific Publications Limited.

MOLECULAR AND CELLULAR ASPECTS OF MICROBIAL EVOLUTION

EDITED BY

M. J. CARLILE, J. F. COLLINS AND
B. E. B. MOSELEY

THIRTYSECOND SYMPOSIUM OF THE SOCIETY
FOR GENERAL MICROBIOLOGY
HELD AT
THE UNIVERSITY OF EDINBURGH
SEPTEMBER 1981

Published for the Society for General Microbiology
CAMBRIDGE UNIVERSITY PRESS
CAMBRIDGE
LONDON   NEW YORK   NEW ROCHELLE
MELBOURNE   SYDNEY

Published by the Press Syndicate of the University of Cambridge
The Pitt Building, Trumpington Street, Cambridge CB2 1RP
32 East 57th Street, New York, NY 10022, USA
296 Beaconsfield Parade, Middle Park, Melbourne 3206, Australia

© The Society for General Microbiology Limited 1981

First published 1981

Printed in Great Britain at The Pitman Press, Bath

ISBN 0 521 24108 1

British Library Cataloguing in Publication Data

Society for General Microbiology. *Symposium*
(*32nd: 1981: University of Edinburgh*)
Molecular and cellular aspects of microbial
evolution. – (Society for General Microbiology
symposia)
1. Microbiology – Congresses
I. Carlile, M. J.   II. Collins, J. F.
III. Moseley, B. E. B.   IV. Series
567   QR1   80–42172
ISBN 0 521 24108 1

CONTRIBUTORS

ABELSON, J., Department of Chemistry, University of California, San Diego, La Jolla, California 92093, USA

BAUMBERG, S., Department of Genetics, University of Leeds, Leeds LS2 9JT, UK

CAMMACK, R., Department of Plant Sciences, University of London King's College, 68 Half Moon Lane, London SE24 9JF, UK

CAVALIER-SMITH, T., Department of Biophysics, University of London King's College, 26–29 Drury Lane, London WC2B 5RL, UK

CORDINGLEY, J. S., MRC Biochemical Parasitology Unit, The Molteno Institute, University of Cambridge, Downing Street, Cambridge CB2 3EE, UK

CULLUM, J., Max-Planck-Institut für Züchtungsforschung, 5 Köln 30 (Vogelsang), German Federal Republic

DAWES, IAN W., Department of Microbiology, University of Edinburgh, Edinburgh, EH9 3JG, UK

DEVOS, R., Laboratory of Molecular Biology, State University of Ghent, Ledeganckstraat 35, B-9000, Ghent, Belgium

FANG, R. X., Laboratory of Molecular Biology, State University of Ghent, Ledeganckstraat 35, B-9000, Ghent, Belgium

FIERS, W., Laboratory of Molecular Biology, State University of Ghent, Ledeganckstraat 35, B-9000, Ghent, Belgium

FINCHAM, J. R., Department of Genetics, University of Edinburgh, Edinburgh EH9 3JN, UK

GARLAND, P. B., Biochemistry Department, Medical Sciences Institute, University of Dundee, Dundee DD1 4HN, Scotland, UK

HUYLEBROECK, D., Laboratory of Molecular Biology, State University of Ghent, Ledeganckstraat 35, B-9000, Ghent, Belgium

JOHNSON, J. D., Department of Chemistry, University of California, San Diego, La Jolla, California 92093, USA

JOHNSON, P. F., Department of Chemistry, University of California, San Diego, La Jolla, California 92093, USA

KNAPP, G., Department of Chemistry, University of California, San Diego, La Jolla, California 92093, USA

KREBS, H., Metabolic Research Laboratory, Nuffield Department of Clinical Medicine, Radcliffe Infirmary, Oxford OX2 6HE, UK

MIN JOU, W., Laboratory of Molecular Biology, State University of Ghent, Ledeganckstraat 35, B-9000, Ghent, Belgium

OGDEN, R. C., Department of Chemistry, University of California, San Diego, La Jolla, California 92093, USA

PEEBLES, C. L., Department of Chemistry, University of California, San Diego, La Jolla, California 92093, USA

RAO, K. K., Department of Plant Sciences, University of London King's College, 68 Half Moon Lane, London SE24 9JF, UK

SAEDLER, H., Max-Planck-Institut für Züchtungsforschung, 5 Köln 30 (Vogelsang), German Federal Republic

STRACKEBRANDT, E., Technical University Munich, Arcisstr. 21, 8000 Munich 2, German Federal Republic

TURNER, M. J., MRC Biochemical Parasitology Unit, The Molteno Institute, University of Cambridge, Downing Street, Cambridge CB2 3EE, UK

VERHOEYEN, M., Laboratory of Molecular Biology, State University of Ghent, Ledeganckstraat 35, B-9000 Ghent, Belgium

WOESE, C. R., Department of Genetics and Development, College of Liberal Arts and Sciences, University of Illinois, 515 Morrill Hall, Urbana, IL 61801, USA

CONTENTS

Editors' preface   *page* ix

ERKO STACKEBRANDT AND CARL R. WOESE
   The evolution of prokaryotes   1

T. CAVALIER-SMITH
   The origin and early evolution of the eukaryotic cell   33

IAN W. DAWES
   Sporulation in evolution   85

JOHN CULLUM AND HEINZ SAEDLER
   DNA rearrangements and evolution   131

J. ABELSON, G. KNAPP, C. L. PEEBLES, R. C. OGDEN, P. F. JOHNSON AND J. D. JOHNSON
   tRNA splicing in yeast   151

K. K. RAO AND R. CAMMACK
   The evolution of ferredoxin and superoxide dismutase in microorganisms   175

HANS KREBS
   The evolution of metabolic pathways   215

SIMON BAUMBERG
   The evolution of metabolic regulation   229

PETER B. GARLAND
   The evolution of membrane-bound bioenergetic systems: the development of vectorial oxidoreductions   273

W. MIN JOU, M. VERHOEYEN, R.-X. FANG, R. DEVOS, D. HUYLEBROECK AND W. FIERS
   Shift and drift in influenza viruses   285

M. J. TURNER AND J. S. CORDINGLEY
Evolution of antigenic variation in the salivarian trypanosomes   313

J. R. S. FINCHAM
Summarizing remarks   349

Index   355

EDITORS' PREFACE

On the occasion of this Edinburgh Symposium, the Society for General Microbiology has returned to the theme of evolution, to which the Twenty-fourth Symposium in 1974, entitled 'Evolution in the Microbial World', was devoted. Evolution forms a natural meeting point for many areas of research which are of interest and importance to the Society's members, and we hope that this will recommend the Symposium to a wide audience.

Since 1974, progress in all fields has been rapid, and with the improvements in methodology pertinent to the study of microbial evolution, particularly of analysis at the molecular level, this book may be regarded as more a sequel than a companion to the previous volume.

Within such a unifying theme, however, there is a diversity of approach possible, and this is well illustrated here. For example, evolution is viewed both in an historical perspective and as a process which is still occurring in Nature. The contrast between these two approaches is very marked, the former being necessarily speculative and inferential, while the latter is essentially documentary; yet both are based upon the most recent experimental techniques.

The contributions have been chosen to cover a wide range of topics, including evolution of the major groups of microorganisms, aspects of their development, and analyses of the ways in which metabolic pathways and their necessary and sophisticated controls may have evolved. In these areas, we are made aware of the increasing role of studies on the sequences and properties of nucleic acids, which form the focus of another set of contributions. There is no doubt that nucleic acid sequence studies will be a rich source of evolutionary paradigms for a long time to come, and will play a unique role in linking or discriminating between groups of living organisms. One of the pleasures of being an Editor for this volume has been to sense the momentum in this field, as the manuscripts have arrived from our contributors, each one quickening the pace such that, in this instance, we may justifiably claim that the whole is greater than the sum of its parts.

This volume can also be regarded as a prelude of things to come, and nothing in it is necessarily final. As C. H. Waddington (in *Towards a Theoretical Biology*, ed. C. H. Waddington. Edinburgh University Press, 1968, p. 108) said, 'After all, evolution has had a long time to cook up some really clever tricks'. We are confident that the Society will find it appropriate, from time to time in the future, to select the theme of evolution for its symposia, and we in our turn look forward to reading the next volume in this series.

Finally, we would like to thank all the authors and those members of the SGM Council and the Cambridge University Press who have been involved in the successful production of this volume.

M. J. CARLILE
J. F. COLLINS
B. E. B. MOSELEY

THE EVOLUTION OF PROKARYOTES

ERKO STACKEBRANDT* AND CARL R. WOESE†

*Department of Microbiology, Technical University Munich,
Arcisstr. 21, 8000 Munich 2, German Federal Republic
†Department of Genetics and Development, College of Liberal Arts
and Sciences, University of Illinois, 515 Morrill Hall, Urbana, IL
61801, USA

INTRODUCTION

The microbiologist has sought for a century to establish the natural relationships among the myriad bacterial species. This has been largely a frustrating task because of the simplicity of their morphologies and other characteristics. In the higher forms where morphologies are indeed complex, morphological convergence is for the most part ruled out, and morphology is then a reliable phylogenetic indicator. However, distinctions involving spherical, rod, and spiral shapes, etc., are clearly not sufficient either reliably to group bacteria phylogenetically or necessarily to exclude species from groups so defined. The caveat concerning the use of these simple characters in attempting to determine the natural relationships among bacteria has been pronounced many times. Yet, to this day, morphological characters have been heavily relied upon for classification of bacteria simply because no better criteria existed until recently. It is obvious, therefore, that what bacterial classification we have (say up through the eighth edition of Bergey's Manual (1923–1974)) is probably not in very good accord with the natural relationships that exist among organisms.

Genetic sequence is an historical record (Bryson & Vogel, 1965; Zuckerkandl & Pauling, 1965a, b). Comparative analysis of genetic sequence can then be used to establish genealogical relationships among organisms. There are many approaches now available that directly or indirectly reflect genetic sequence to one degree or another, including the ultimate one of exact sequence determination. Although of limited use, one of the best techniques in principle is DNA–DNA hybridization. The method is relatively simple, rapid, and inexpensive. More importantly, the method gives an averaged measure for the entire genome, and so, unlike most other methods, it is necessarily representative of the whole organism. Unfortunately, this method permits detection of only the closest genealogical relationships among bacteria, failing above the intrageneric level (Johnson, 1973; Steigerwalt, Fanning, Fife-Ashbury & Brenner, 1976) and this makes it of limited utility.

Most other methods that compare genetic sequence reflect one or a few genes only. Given that individual bacterial genes are often subject to lateral (interspecific) transfer, there is then a serious question as to whether a method based on a single gene reflects true bacterial phylogeny or merely the phylogeny of that gene *per se*. In fact, if lateral transfer of bacterial genes were extensive enough, there could be no such thing as a phylogeny representative of the whole bacterium. Fortunately this seems not to be the case. A prediction of the lateral transfer hypothesis is that independent genes (or gene clusters) will not exhibit the same pattern of interspecific transfer. In other words, a set of phylogenetic relationships determined with one gene would not be the same as a set determined with genes unrelated to that gene. As we will see below, phylogenies of an extensive group of bacteria determined by the use of two independent genes, give practically identical trees.

What is the optimal system for making phylogenetic measurements among bacteria? There are a number of requirements. First, the system must not be subject to appreciable lateral transfer. (This would rule out antibiotic resistance factors, nitrogen fixation, and so on.) Second, the system should be universally distributed. Third, the system must exhibit functional constancy; i.e., one does not want *selected* mutations (as opposed to neutral single or multiple mutations) to distort the measurement, to give the appearance of saltatory evolution. Fourth, the gene(s) involved has (have) to provide a sufficiently slow 'clock', i.e., change in sequence slowly enough that the largest of genealogical distances can be detected (which is definitely not the case, for example, with DNA–DNA hybridization methods). And finally, the system has to be an experimentally feasible one.

We decided more than a decade ago that the 16S ribosomal RNA was well suited to the purpose of measuring genealogical relationships among bacteria, and hopefully for constructing the universal phylogenetic tree as well. The molecule was universally distributed. It was easily isolated. It appeared to be highly constant in function (as ribosome reconstitution experiments demonstrated) (Nomura, Traub & Bechmann, 1968; Higo, Held, Kaham & Nomura, 1973). Parts of it, at least, seemed to change very slowly with time, as rRNA–DNA hybridization studies had shown (Pace

& Campbell, 1971; Moore & McCarthy, 1967) and, although the molecule was too large to sequence, it was possible, using the then current nucleic-acid-sequencing technology, to sequence large enough fragments of it to make feasible a comparative analysis of its primary structure. Moreover, the molecule was large (about 1540 residues) which seems to give it a useful, but more subtle, advantage. Smaller molecules, e.g., cytochrome *c* and the 5S rRNA, exhibit saltatory evolutionary behaviour when the structure of one of their 'domains', loosely speaking, changes; we have noted several examples in the 5S rRNA in which the sequence of one of its four helical elements appears to change drastically for a given organism (Woese *et al.*, 1976*a*). These saltatory changes, undoubtedly involving some strongly selected mutations, distort the phylogenetic picture. For a large molecule like 16S rRNA, which has about fifty helical elements (Woese *et al.*, 1980*b*) a drastic 'redesigning' of any one of them would have far less effect on the apparent phylogenetic distance measure than would be the case for a small molecule like 5S rRNA, making the former the more accurate phylogenetic indicator.

The choice of 16S rRNA for measuring phylogenetic relationships among bacteria has proved to be a good one. Not only can the molecule span the greatest phylogenetic distances (Woese & Fox, 1977*a*), because parts of its sequence change only slowly with time, but it can be used to measure close phylogenetic relationships as well (Zablen *et al.*, 1975*a*; Stackebrandt *et al.*, 1980*b*) because other parts of its sequence change relatively rapidly with time. The main limitations of the technique are its expense and the fact that it is too slow to be used on thousands of bacterial species. Nevertheless, compared to the amount of time and effort that have gone into attempts to establish bacterial phylogenetic relationships in the past, the expense and time used for this one seem slight. Another potential drawback to the method in some cases, the necessity to incorporate moderately high levels of ^{32}P into the RNA of the growing cell, which is not always possible due to medium composition, pool levels, or radiation damage to cells, has now been alleviated by the development of *in vitro* labelling methods for the rRNA digest (E. Stackebrandt, W. Ludwig, K. H. Schleifer & H. J. Gross, 1981). The 16S rRNA cataloguing approach, as our method is known, should be used for establishing the major phylogenetic units down to the level of what is conventionally seen as a genus, e.g., the more 'recent' genera like *Escherichia*,

Arthrobacter, and so on. Within these recent genera, the bulk of the species can then more rapidly and less expensively be interrelated by the technique of DNA–DNA hybridization.

THE METHOD

The cataloguing of 16S rRNA is performed in one of two related ways. In the original method, a bacterial culture, typically 10–50 ml, is labelled in exponential phase growth with $^{32}PO_4$, and the 16S rRNA isolated by standard techniques, e.g., phenol extraction and polyacrylamide gel electrophoretic separation. Ideally 100 μg or more of purified, labelled 16S rRNA at a specific activity of roughly 1 μCi/μg is required. The RNA is then digested by T1 ribonuclease, to produce a set of oligonucleotides, usually up to 15–20 residues in length, each ending in a G residue and preceded by some number (from zero up) of non-G residues. These oligonucleotides are resolved from one another by the two dimensional paper electrophoretic method originally developed by Sanger and coworkers (Sanger, Brownlee & Barrell, 1965) and modified by us (Uchida *et al.*, 1974; Woese, Luehrsen, Pribula & Fox, 1976). This employs a cellulose acetate first dimension at pH 3.5 followed by transfer to DEAE cellulose, which is then run in 0.1 M buffer at a pH of about 2.3. (For a typical fingerprint, see Uchida *et al.*, 1974 and Balch *et al.* 1979.) The individual oligonucleotides are then sequenced by a combination of endonuclease digestion procedures, to produce finally a list, or catalogue, of sequences that is characteristic of the organism in question.

Alternatively, unlabelled 16S rRNA is digested with T1 ribonuclease, the terminal phosphate(s) removed, and a ^{32}P labelled phosphate enzymatically placed on the 5' end of each oligonucleotide (Stackebrandt, Ludwig, Schleifer & Gross, 1980*b*). Separation of the oligonucleotides is then performed as described above or, more usually, the second dimension DEAE paper is replaced by a DEAE cellulose thin-layer plate, which is developed in a relatively high-salt-buffer system. Sequencing of the individual oligonucleotides in this case is somewhat different, since only the 5' terminus is labelled, and involves a two-dimensional method of separating partial digests of the oligonucleotide (Silberklang, Gillum & RajBhandary, 1979).

Oligonucleotide catalogues of organisms are then generally analysed by a binary method; each catalogue is compared individually to every other catalogue, and the oligonucleotides, of six residues or larger, common to any two catalogues are scored to produce a 'S_{AB} value' characteristic of that pair of organisms. (The function S_{AB} is defined as twice the total number of residues in sequences common to a pair of catalogues, divided by the total number of residues in all sequences in the two catalogues, consideration being confined, as stated, to hexamers and larger. These S_{AB} values range from 1.0 for identical RNAs to about 0.03 for randomly related sequences of 1 500 nucleotides length.) A table of S_{AB} values for any given set of organisms is then analysed by standard clustering procedures (average linkage among merged groups) to produce a dendrogram (Fox, Pechman & Woese, 1977).

The dendrogram so produced is a reasonable approximation to the true phylogenetic relationships provided that the 'mutational clocks' in all organisms are isochronic; in other words, all organisms introduce mutations into rRNA at the same rate. As we shall see, this is the case for the bacteria with a small number of interesting exceptions.

THE PRIMARY PHYLOGENETIC DIVISIONS

The ribosomal RNA cataloguing method is able to detect the most distant phylogenetic relationships. The lowest S_{AB} values, observed among the primary groups, are in the range of 0.10, which corresponds roughly to 50–60 residues in common oligonucleotides (hexamers and larger). The method can therefore be used to identify and define what should be called the 'primary kingdoms', or 'urkingdoms', those major phylogenetic units that directly stem from the common ancestor of all extant life (Woese & Fox, 1977a). The term 'primary kingdom' is used to distinguish these fundamental phylogenetic units from the classically defined 'kingdoms', such as animals, plants, etc. The classical eukaryotic kingdoms are not related to the primary kingdoms; the former are phylogenetic groupings within the eukaryotic domain of organization, whereas the primary kingdoms are defined for the underlying prokaryotic domain. (A more complete discussion of this question of phylogenetic units and domains, or levels, of biological organization can be found in Woese & Fox (1977a).)

Ribosomal RNA cataloguing shows that there exist at least three primary kingdoms, or three primary lines of descent. This initial discovery came as a considerable shock to the scientific community, for biologists had for some time accepted that the dichotomy eukaryote–prokaryote somehow defined not only mutually exclusive categories of cell types, but in addition, mutually exclusive phylogenetic categories (Chatton, 1937; Allsopp, 1969; Margulis, 1970; Murray, 1974). In other words, all organisms were seen as belonging either to a prokaryotic or to an eukaryotic line of descent. As it turns out, there exist at least two distinct bacterial lines of descent, lines that are no more related to one another than either of them is to the 'eukaryotic line of descent' (see below).

The three primary kingdoms then are these (Woese & Fox, 1977*a*; Fox *et al.*, 1980): (I) A grouping that includes the vast majority of recognized bacteria. To date, this primary kingdom can also be defined as those organisms that do, or whose ancestors did, possess the muramic-acid type of cell wall. The true mycoplasmas, which come from an ancestry within the classical Gram-positive bacteria, are included herein (see below), and both the chloroplast and (plant) mitochondrion trace their ancestry back to this kingdom. This primary kingdom has been called the *true bacteria* or *eubacteria*. ('Eubacteria' is a term that has been used in many contexts, and microbiologists may be reluctant to see it in yet another.) (II) The second primary kingdom is known as the archaebacteria. These organisms are bacterial in size and simplicity, morphological and genetic, but differ from true bacteria in the details of most, if not all, of their organization at the molecular level. At present, the archaebacterial group is known to contain only three (rather bizarre) phenotypes, the methanogens, the extreme halophiles, and certain extreme thermoacidophiles. (III) The third primary kingdom may not exist in the strictest sense of the word. At present it is known to be represented only by the 18S rRNA of the eukaryotic cell; a prokaryotic example of the group has yet to be found. It is tempting to say that such should be called the eukaryotic line of descent, or the eukaryotic primary kingdom. However, the matter of eukaryotic origins is complicated and not yet well understood. The mitochondrion and chloroplast are of true bacterial ancestry (Bonen & Doolittle, 1975; Zablen, Kissil, Woese & Buetow, 1975*b*; Bonen, Cunningham, Gray & Doolittle, 1977). One of the eukaryotic ribosomal proteins seems to be of archaebacterial ancestry (Matheson, Möller, Amons & Yaguchi, 1980). The eukaryotic 18S rRNA, as we have just seen, seems of an ancestry neither archaebacterial nor eubacterial. Therefore, the eukaryotic cell is a phylogenetic chimera; how radical a chimera, i.e., how many gene or gene cluster 'capture' events are involved we have no idea. At least until the genealogies for many of the eukaryotic gene families are traced into the prokaryotic realm, i.e., the various primary kingdoms they represent are identified, it is not useful to speak of an ancestral eukaryotic line of descent; there may be no single line of descent that accounts for enough of the eukaryotic gene families to be called *the* eukaryotic line of descent. What we do now know, however, is that there exist two primary kingdoms of bacteria, the true bacteria and the archaebacteria, and that these two lines of descent, along with others yet to be defined at the prokaryotic level, are variously represented in the genetic chimera that is the eukaryote. And for the present, we should leave it at that. The eukaryotes will not be discussed further in this chapter.

THE PHYLOGENETIC STRUCTURE OF THE TRUE BACTERIA

Conventionally, the true bacteria are divided into Gram-positive and Gram-negative groups, with some uncertainty and debate surrounding the cyanobacteria and mycoplasmas. As we will see, this grouping is only partially in accord with the phylogenetic structure of the true bacteria. Fig. 1 is an overview of the phylogeny of the true bacteria, as seen in terms of the rRNA cataloguing method.

The Gram-positive eubacteria

With the exception of the *Micrococcus radiodurans* group, whose members possess atypical walls and other features (Brooks *et al.*, 1980), the Gram-positive bacteria form a phylogenetically coherent unit, albeit a deep, i.e., ancient, one. By and large, the conventional separation of low G + C DNA-content organisms within the group from high G + C ones is seen to hold (Figs 2 and 3). In other words, the actinomycete–coryneform type of phenotype is phylogenetically distinct from the clostridium–bacillus–streptococcus type. However, in detail, there are some surprises and considerable rearrangement of the traditional groupings.

The high G + C Gram-positive bacteria are seen to structure

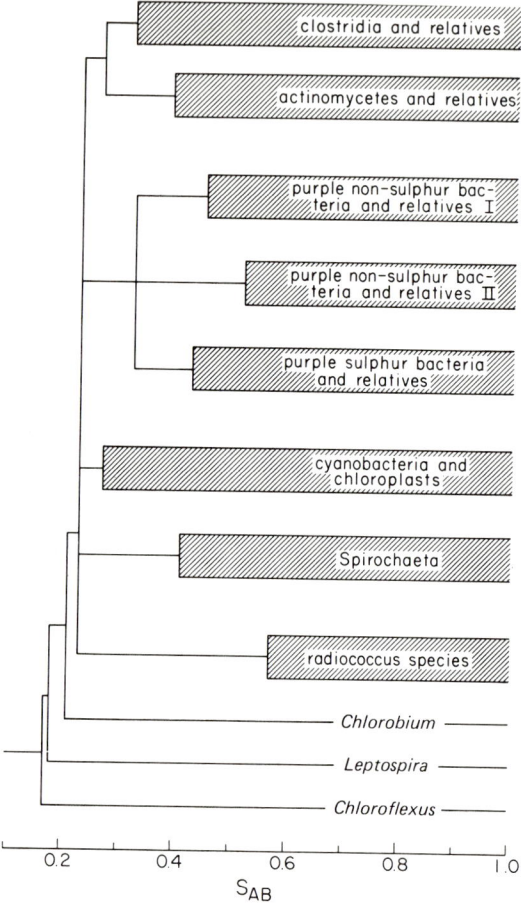

Fig. 1. Dendrogram of relationships among the true bacteria. Organisms forming the clusters of clostridia, actinomycetes, cyanobacteria, and purple non-sulphur bacteria are listed in Figs. 2 and 4–7 (Fox et al., 1980; C. R. Woese, P. Blanz & C. M. Hahn, unpublished). Purple sulphur group: *Aerobacter aerogenes, Aeromonas hydrophila, Chromatium vinosum, Escherichia coli, Oceanospirillum maris, O. minutulum, Pasteurella multocida, Photobacterium phosphoreum, Proteus mirabilis, Pseudomonas aeruginosa, P. alcaligenes, P. fluorescens, P. pseudoalcaligenes, P. putida, P. stutzeri, P. syringae, Serpens flexibilis, Serratia marcescens, Thiocapsa pfennigii, Vibrio marinus, Yersinia pestis.* The *Spirochaeta* cluster contains *S. aurantia, S. halophila, S. litoralis, S. stenostrepta.* The cluster of the radio-resistant micrococci is defined by *M. radiodurans, M. radiophilus, M. roseus* UWO 294 (University of Western Ontario).

phylogenetically in the following way (Fig. 2). The members of *Arthrobacter* form a major subunit along with *Cellulomonas*, the plant pathogen coryneforms, *Microbacterium* and certain other genera. Peripherally these are related to members of the genus *Actinomyces*. It is surprising that the genuine species of *Micrococcus*

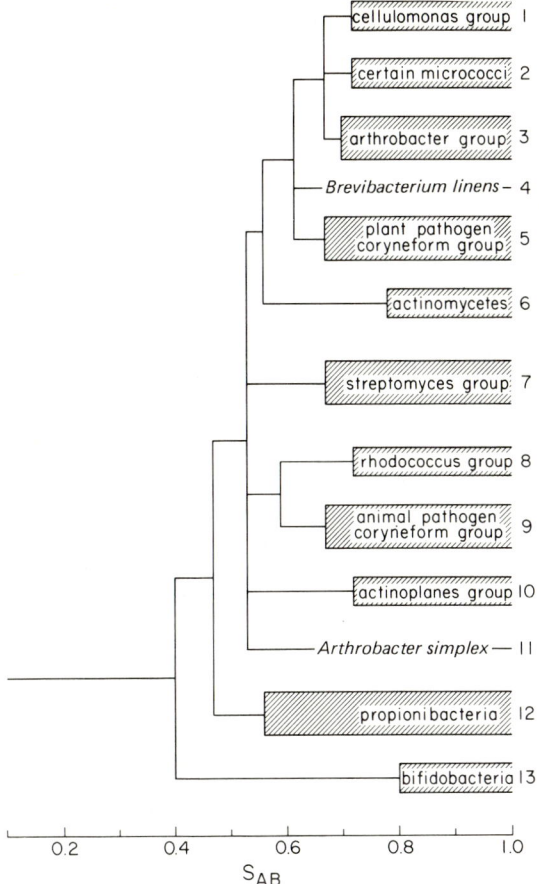

Fig. 2. Dendrogram of relationships among Gram-positive bacteria with a high DNA G + C content (≥55 Mol%). *Cellulomonas*: *Cellulomonas cartae*, *Cell. flavigena*, *Nocardia cellulans*, *Oerskovia turbata*. *Micrococcus*: *Micrococcus sedentarius*, *M. nishinomiyaensis*. *Arthrobacter*: *Arthrobacter globiformis*, *A. oxidans*, *A. atrocyaneus*, *M. luteus*, *M. lylae*, *M. roseus*, *M. varians*. Plant pathogen coryneforms: *Microbacterium lacticum*, *Corynebacterium betae*, *C. mediolanum*. Actinomycetes: *Actinomyces bovis*, *Act. viscosus*. Streptomyces: *Actinomadura dassonvillei*, *Chainia antibiotica*, *Elythrosporangium brasiliense*, *Kitasatoa kauaiensis*, *Microellobosporia cinerea*, *Streptomyces griseus*, *Streptosporangium roseum*, *Streptoverticillium baldacchi*. Rhodococcus group: *C. fascians*, *Nocardia corallina*, *N. calcarea*. Animal pathogen coryneforms: *C. diphtheriae*, *A. variabilis*, *C. glutamicum*. Actinoplanes: *Actinoplanes philippinensis*, *Ampulariella regularis*, *Dactylosporangium aurantiacum*, *Micromonospora chalcea*. Propionibacteria: *P. freudenreichii*, *P. acnes*. Bifidobacteria: *Bif. bifidum*, *Bif. breve*.

also fit into this grouping, and more surprising that these cannot be phylogenetically separated from species of *Arthrobacter*. It appears then that *Micrococcus* is not a phylogenetically valid genus. Rather the micrococci seem to have arisen as degenerate forms of the arthrobacteria, locked into the coccoid stage of the arthrobacterial

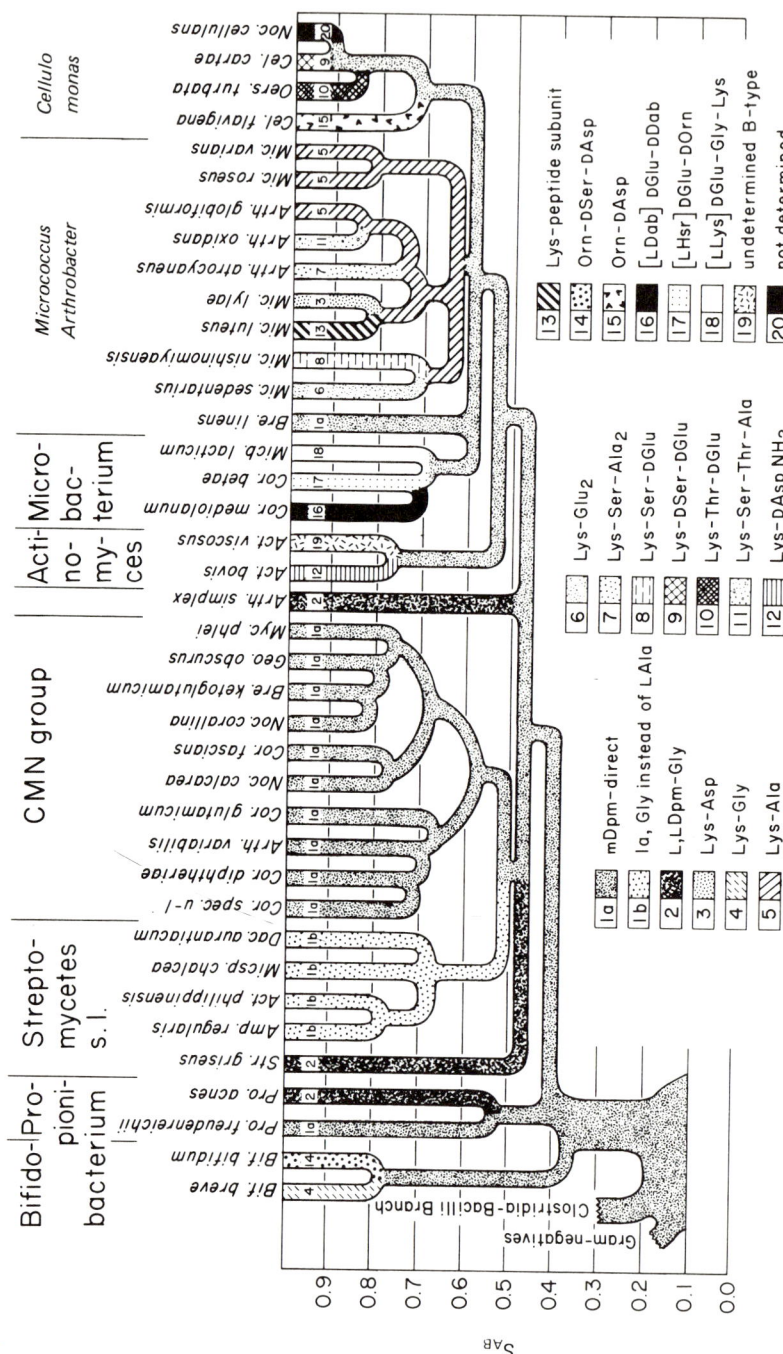

Fig. 3. Distribution of peptidoglycan types within the RNA clusters of actinomycetes and relatives. The abbreviation of the types are listed in Schleifer & Kandler (1972). Type 20 was recently found to be the same as type 9 (Stackebrandt, Häringer & Schleifer, 1980a). The figure was kindly provided by Professor O. Kandler.

life cycle (Stackebrandt & Woese, 1979; Stackebrandt, Lewis & Woese, 1980b).

Streptomyces forms a distinct subunit within the general group, along with *Elythrosporangium, Kitasatoa, Microellobosporia, Chainia, Streptoverticillium* and *Streptosporangium* (Stackebrandt, et al., unpublished). The genera *Mycobacterium, Nocardia,* and certain members of *Corynebacterium* form a unit. *C. diphtheriae, C. glutamicum* and *Arthrobacter variabilis* also form a group, perhaps peripherally related to the preceding one. *Dactylosporangium, Micromonospora, Actinoplanes, Ampulariella,* and *Amorphosporangium* cluster separately. *Arthrobacter simplex* appears to stand more or less alone among the previously described groupings.

It is notable that the phylogenetic depth to the high G + C Gram-positive unit is provided by the anaerobic and microaerophilic representatives, the propionibacteria and the bifidobacteria. Note also the phylogenetic depth (low S_{AB} value) within the propionibacterial phenotype, which suggests that it is an ancient group.

There is quite reasonable agreement between the phylogenetic structure of the high G + C Gram-positive bacteria and various of their other properties, e.g., the distribution of mycolic acids and isoprenoid quinones (Minnikin, Goodfellow & Collins, 1978) as well as cytochromes (Faller, 1980). An interesting correlation exists also between cell wall structure, which is so variable among these organisms, and the proposed phylogeny; this can be seen in Fig. 3.

It should be noted, however, that the traditional tendency to classify the morphologically simpler coryneforms apart from the more complex Actinomycetes (Rogosa, Cummins, Lelliott & Keddie, 1974) does not hold. Note in particular phylogenetic juxtapositions such as the morphologically simple cellulomonads together with the morphologically complex representatives of *Oerskovia* (Stackebrandt, Häringer & Schleifer, 1980a), or the simple corynebacteria and mycobacteria with organisms producing a multilocular thallus like *Geodermatophilus* (Stackebrandt & Woese, unpublished).

Certain organisms presently classified with the actinomycetes and their coryneform relatives are seen not to belong there. Members of the genus *Eubacterium,* which also are not high G + C organisms, group phylogenetically with the clostridia and relatives (Fox et al., 1980). Indeed the former appear to be no more than non-spore-forming clostridia. *Thermoactinomyces,* an actinomycete atypical in possessing heat-resistant, dipicolinic acid-containing spores (Cross, Davies & Walfer, 1971), does not belong with the actinomycetes, but indeed with the aerobic endospore formers, i.e., *Bacillus*. *Kurthia*, which biochemically resembles the lactobacilli, groups phylogenetically with them (Stackebrandt, unpublished).

The low G + C Gram-positive bacteria are seen to be structured largely in terms of the clostridia. Far from being simply a genus, *Clostridium* appears to be a major phylogenetic unit, deeper (more ancient) even than the entire group of high G + C Gram-positive bacteria (Fig. 4). The clostridial group seems to have spawned some interesting offshoots: a number of anaerobic Gram-positive cocci, e.g., *Sarcina ventriculae*, seem to come from clostridia that have lost spore-forming capacity and shape, i.e., they have rounded up. In fact, note that as was the case with high G + C Gram-positive bacteria, the coccoid members of the group, e.g., *Peptococcus*, *Peptostreptococcus* and *Ruminococcus*, generally possess close relatives that are morphologically more complex.

Bacillus, *Lactobacillus*, *Streptococcus* and a few other genera seem to form a unit, microaerophilic to aerobic in phenotype, that is relatively coherent phylogenetically and is a sub-branch of the clostridia. *Staphylococcus*, *Planococcus*, and *Sporosarcina* can be considered morphologically-degenerate forms of *Bacillus*.

Perhaps the most surprising offshoot of the clostridia are the true mycoplasmas, i.e., the genera *Mycoplasma* and *Acholeplasma* and presumably, therefore, *Anaeroplasma* and *Ureaplasma* (Woese, Maniloff & Zablen, 1980a). Two clostridial species, *C. ramosum* and *C. innocuum*, are specific relatives of the mycoplasmas. The latter therefore seem to have arisen by a highly degenerate evolution from a clostridial ancestry involving loss of spore formation, shape, wall, most of the genome and consequently many functions. The phylogenetic data strongly suggest, but do not definitively demonstrate (see discussion by Woese *et al.*, 1980a) that *Acholeplasma* and *Mycoplasma* do not share a common ancestor that was itself a mycoplasma. In other words, the mycoplasmal state may have arisen more than once from clostridial ancestry.

It is tempting to try to envisage the ancestral phenotype of the Gram-positive true bacteria. It must have been anaerobic, perhaps rod-shaped, at least not coccoid. It must have involved the unique Gram-positive form of the eubacterial wall. (Seemingly the Gram-positive wall derived from something like the Gram-negative one, not the reverse.) Various types of spores and antibiotic production

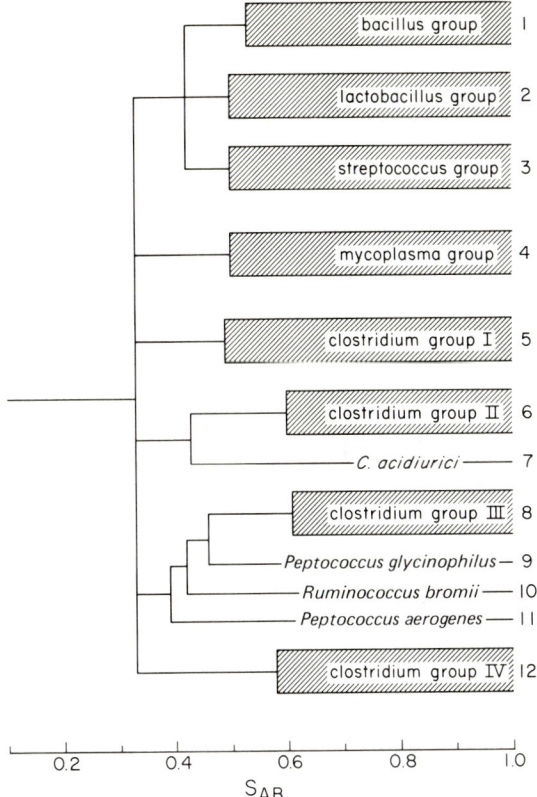

Fig. 4. Dendrogram of relationships among Gram-positive bacteria with a low DNA G + C content (≥50 Mol%). *Bacillus* group: *B. acidocaldarius, B. alvei, B. badius, B. brevis, B. cereus, B. coagulans, B. fastidiosus, B. firmus, B. insulitus, B. megaterium, B. pasteurianum, B. polymyxa, B. psychrophilus, B. pumilus, B. sphaericus, B. stearothermophilus,* Peptococcus saccharolyticus, Planococcus citreus, Sporolactobacillus inulinus, Sporosarcina urea, Staphylococcus aureus, S. epidermidis, S. haemolyticus, S. sciuri, S. xylosus, Thermoactinomyces vulgaris. *Lactobacillus* group: Kurthia zopfii, Lactobacillus acidophilus, L. brevis, L. casei, L. fermentum, L. helveticus, L. lactis, L. plantarum, L. ruminis, L. viridescens, Leuconostoc mesenteroides, Pediococcus pentosaceus. *Streptococcus* group: Streptococcus faecalis, S. lactis, S. salivarius. *Mycoplasma* group: Acholeplasma laidlawii, Clostridium innocuum, C. ramosum, Mycoplasma capricolum, M. galli, Spiroplasma citri. *Clostridium* group I: *C. butyricum, C. scatalogenes, C. pasteurianum,* plus *Clostridium* group I, according to Johnson & Francis (1975). *Clostridium* group II: *C. litus-eburense,* plus group II, according to Johnson & Francis (1975), *Eubacterium tenue, Peptostreptococcus anaerobius.* *Clostridium* group III: *C. sphenoides, C. aminovalericum.* *Clostridium* group IV: *C. barkeri, Acetobacterium woodii, Eubacterium limosum.*

are common to the group and so too may have been an ancestral characteristic.

The purple photosynthetic bacteria and relatives

Customarily the Gram-negative bacteria are seen as a phylogenetically coherent unit. This is not the case. The group actually comprises a number of distinct phylogenetic units, each comparable to (as ancient as) the Gram-positive bacterial group. Moreover, the phylogenetic structure of the Gram-negative bacteria *vis à vis* the photosynthetic bacteria is again not according to the conventional prejudices.

The most extensively characterized division of Gram-negative bacteria is the grouping that appears to centre around the purple photosynthetic bacterial phenotype. Fig. 5 shows that the purple photosynthetic bacteria fall into three major phylogenetic groups; two distinct groups of purple non-sulphur bacteria and one of purple sulphur bacteria, though only two species of the latter have been characterized to date. The admixture of generic names in Fig. 5 demonstrates that the presently accepted classification of purple non-sulphur bacteria has little phylogenetic correspondence. The fact that there are two distinct groups of purple non-sulphur photosynthetic bacteria indicates that if this phenotype arose from the purple sulphur one, as has been suggested (Almassy & Dickerson, 1978), then it did so more than once. (Since so few purple sulphur species have been characterized in terms of rRNA cataloguing to date, we do not yet have a feeling for how ancient (deep) that phenotype is.)

The groupings of the purple non-sulphur photosynthetic bacteria generated by ribosomal RNA typing are virtually identical to those generated by comparative analysis of cytochrome *c* sequences. This effectively rules out interspecific transfer of genes as the cause of either phylogeny (Ambler *et al.*, 1979; Dickerson, 1980; Woese, Gibson & Fox, 1980).

Several rather surprising results come out of this phylogeny (Fig. 5). One is that each of the three phylogenetic groupings defined by the purple photosynthetic bacteria contains a variety of non-photosynthetic organisms representing well-known genera (Gibson *et al.*, 1979; Fox *et al.*, 1980; C. R. Woese, P. Blanz & C. M. Hahn, unpublished). The group defined by *Chromatium* contains the enterics, vibrios, the fluorescent group of pseudomonads and others. The group defined by *Rhodopseudomonas sphaeroides*, *Rps. palustris*, *Rhodospirillum rubrum* and others, additionally contains the rhizobacteria, *Paracoccus denitrificans*, *Aquaspirillum itersonii* and others. And the group defined by *Rps. gelatinosa* and *R. tenue* harbours species of *Alcaligenes*, *Pseudomonas*, *Chromobacterium*, *Sphaerotilus*, and so on (Gibson *et al.*, 1979; C. R. Woese, P. Blanz

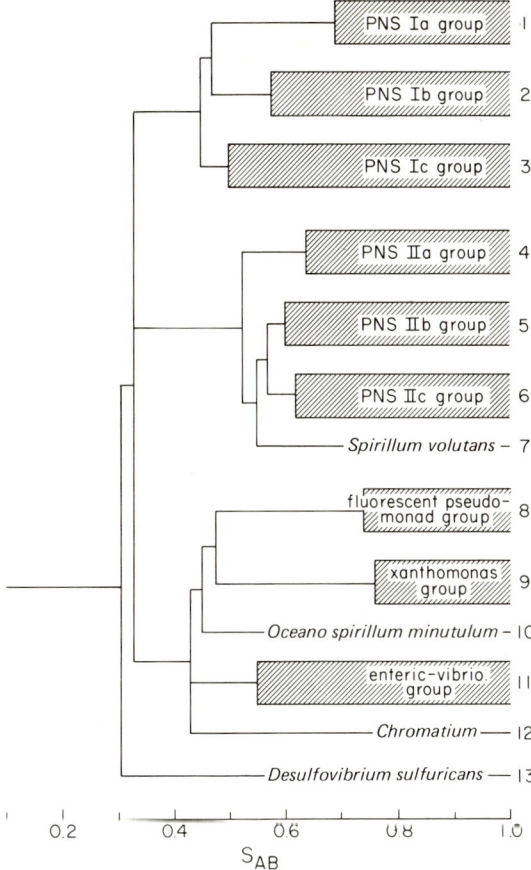

Fig. 5. Dendrogram of relationships among purple photosynthetic bacteria and their non-photosynthetic relatives. PNS (Purple non-sulphur) group Ia: *Rhodopseudomonas sphaeroides, Rps. capsulata, Paracoccus denitrificans*. PNS group Ib: *Rps. viridis, Rps. palustris, Rhizobium leguminosarum, Agrobacterium tumefaciens, Rhodomicrobium vaniellii, Pseudomonas diminutia*. PNS group Ic: *Rhodospirillum rubrum, Aquaspirillum itersonii, Azospirillum braziliense*. PNS group IIa: *Rps. gelatinosa, Rhodospirillum tenue, Sphaerotilus natans, Pseudomonas testosteroni, P. acidovorans, Comamonas terrigena, Aquaspirillum gracile*. PNS group IIb: *Chromobacterium lividum, P. cepacia, Alcaligenes faecalis, A. eutrophus*. PNS group IIc: *Chromobacterium violaceum, Aquaspirillum serpens*. Xanthomonas group: *X. campestris, P. maltophilia, Lysobacter enzymogenes*. Enteric–Vibrio group: *Escherichia coli, Yersinia pestis, Proteus mirabilis, Aeromonas hydrophila, Pasteurella multocida, Benekea harveyi, Photobacterium fischeri*.

& C. M. Hahn, unpublished). In other words, it appears as though a host of non-photosynthetic Gram-negative bacteria have arisen from a photosynthetic ancestry. Prejudice would have us believe that this is not the case, i.e., that photosynthetic bacteria are phylogenetically rather distinct from the vast majority of non-photosynthetic bacteria.

A second surprising outcome of this portion of the study has been that many, perhaps most, of the genera or other groupings now defined as Gram-negative bacteria are not phylogenetically coherent; neither are they phylogenetically comprehensive, i.e., species are ruled out of the genera, as now defined, that should not be. A prime example is the genus *Pseudomonas* (Doudoroff & Palleroni, 1974). The so-called fluorescent group of pseudomonads is not specifically related to the remaining organisms in the genus *Pseudomonas* (C. R. Woese, P. Blanz & C. M. Hahn, unpublished). Moreoever, the fluorescent group does encompass organisms, e.g., *Azotobacter vinelandii* and *Serpens flexibilis* that are by existing definition ruled out. The second and third subgroups of the pseudomonads (see Bergey's Manual (Buchanan & Gibbons, 1974)) appear coherent and related to one another to the extent they have been investigated by rRNA cataloguing, except that many non-pseudomonads (as now defined) are closely related to these pseudomonads (C. R. Woese, P. Blanz & C. M. Hahn, unpublished). Finally, the two species that constitute the fourth group of pseudomonads are unrelated to one another, one of them, *P. diminutia*, having *Rps. palustris* and others as specific relatives, the other, *P. maltophilia*, being related to *Lysobacter* and others (C. R. Woese, P. Blanz & C. M. Hahn, unpublished). Palleroni, Kunisawa, Contopoulou & Doudoroff (1973), using rRNA–DNA hybridization initially found the five groups of pseudomonads defined here. What their study did not do, however, was reveal the relationships among the groups and the fact that each was not really a pseudomonad group, but a phenotypically much broader group.

In a similar way the spiral bacteria, the genera *Spirillum*, *Aquaspirillum*, *Azospirillum* and *Oceanospirillum* show no specific relationship to one another and lack internal phylogenetic coherence as well (C. R. Woese, P. Blanz & C. M. Hahn, unpublished).

The cyanobacteria

The only other group of Gram-negative bacteria to receive moderately extensive characterization is the cyanobacteria (Doolittle *et al.*, 1975; Bonen *et al.*, 1977; Bonen & Doolittle, 1976, 1978). Fig. 6 shows the cyanobacteria to be a moderately deep phylogenetic unit. The full depth of the unit is provided by the chloroplast examples from the higher plants, and *Euglena* (Bonen & Doolittle, 1975; Schwarz & Kössel, 1980; Dyer & Woese, unpublished). The

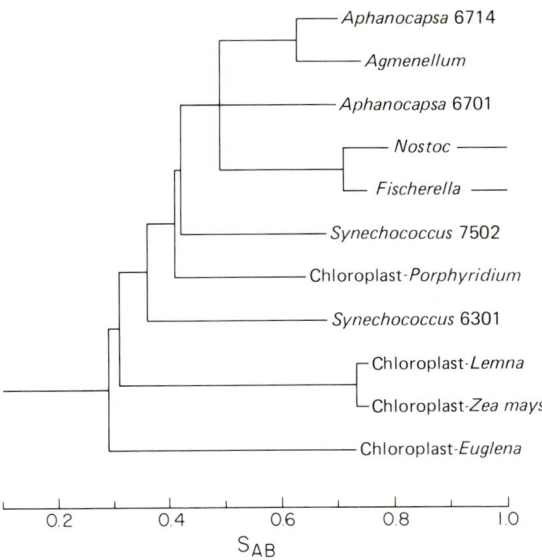

Fig. 6. Dendrogram of relationships among Cyanobacteria and chloroplasts from plants and algae.

cyanobacterial group *per se,* surprisingly, is not as deep as one might be led to expect from the usual interpretation given to microfossil and stromatolite evidence. What this suggests is that the ancestral phenotype of the group may not have been exactly the cyanobacterial one, but rather a related one, e.g., that contained both chlorophylls *a* and *b*. (Organisms of the genus *Prochloron* (Lewin, 1977) as well as chloroplasts of the higher plants, possess both of these chlorophylls. A phylogenetic characterization of some of the other algal chloroplasts, e.g., from brown algae, should prove interesting in this regard.)

Note that the morphologically more complex, advanced cyanobacteria encompass only a relatively small portion of the cyanobacterial tree.

Other Gram-negative eubacteria

The remaining major divisions of the true bacteria that have so far been detected but remain to be explored (Fox *et al.,* 1980) are a group, perhaps two groups, defined by the green photosynthetic bacteria, *Chlorobium* and *Chloroflexus,* one defined by the genus *Spirochaeta,* the group defined by *Leptospira* (apparently unrelated to *Spirochaeta*), that defined by the radioresistant Gram-positive cocci, and certain others, e.g., defined by preliminary studies on *Myxococcus, Flavobacterium,* and an unnamed anaerobic isolate from the Great Salt Lake (Woese *et al.*, unpublished).

It appears, unfortunately, that the primary branching which gave rise to the major branches (divisions) of the true bacteria happened over so short a time, once the common ancestral eubacterial phenotype evolved, that their exact order of branching cannot be determined. (It will take something of the order of full sequencing of 16S rRNA from many of the major divisions before the matter of exact order of major branching can approach resolution; if not by primary structure comparisons, then by comparison of the details of secondary structure.)

THE ARCHAEBACTERIA AND THEIR PHYLOGENY

Whereas the deepest phylogenetic splits within the true bacterial kingdom correspond to S_{AB} values in the 0.20–0.25 range, that between the true bacteria and archaebacteria corresponds, as stated above, to S_{AB} values of 0.10, a factor of two lower! The phylogenetic tree for the archaebacterial urkingdom is shown in Fig. 7. Although the group is nowhere near as extensively represented as are the eubacteria, minimal S_{AB} values within the archaebacteria are at least as low as those seen for the true bacteria. Unfortunately, it is not possible to say whether archaebacterial 'clocks' run at the same rate as do eubacterial clocks (see discussion below) so relative time statements cannot be adduced in this case from relative S_{AB} values.

The archaebacteria fall into a number of major groups (Balch *et al.*, 1979; Fox *et al.*, 1977, 1980; Magrum, Luehrsen & Woese, 1978). Perhaps the deepest division involves the thermoacidophiles, *Sulfolobus* and *Thermoplasma* (which appear not to be specifically related to one another) on the one hand, and the methanogens and extreme halophiles on the other. Such a division is consistent with the fact that the two types of RNA polymerase found in the archaebacteria distribute accordingly (Sturm *et al.*, 1980).

As judged by the depth of the groupings, the methanogenic phenotype is indeed an ancient one. It appears to be older than any of the major eubacterial phenotypes (again with the caveat about inferring time from S_{AB} values). Three major groups of metha-

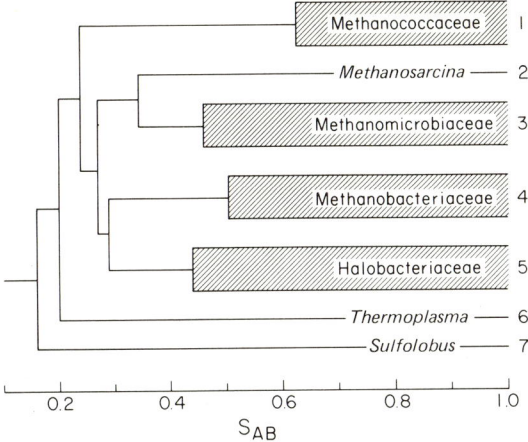

Fig. 7. Dendrogram of relationships among archaebacteria. *Methanococcaceae: M. vanielii, M. voltae. Methanomicrobiaceae: Methanomicrobium mobile, Methanogenium cariaci, M. marisnigri, Methanospirillum hungatei. Methanobacteriaceae: Methanobacterium formicicum, M. bryantii, M. thermoautotrophicum, Methanobrevibacter arboriphilus, M. ruminantium, M. smithii. Halobacteriaceae: Halobacterium halobium, H. volcanii, Halococcus morrhuae.*

nogens are seen. This phylogenetic partitioning correlates very well with various phenotypic characteristics of the organisms (Balch *et al.*, 1979).

To the extent that it has been characterized, the extreme halophile does not appear all that ancient a phenotype. This is consistent with its being aerobic. (S_{AB} values for the group are roughly comparable to those for a well-established group of eubacterial aerobes, e.g., the genus *Bacillus* (Fox *et al.*, 1980).) Two things should be noted about the phylogeny of the extreme halophiles. For one, most of the recognized species of *Halobacterium*, i.e., *H. halobium, H. cutirubrum* and *H. salinarium*, have identical rRNA oligonucleotide catalogues, and so should be considered as strains or other variants of the same species. More important, the halophilic phenotype appears to have arisen specifically from one of the branches of the methanogens. Data of other sorts, other qualities of the organisms, are needed to substantiate this conjecture.

The ways in which archaebacteria differ from true bacteria and eukaryotes are indeed striking. One particularly notable way is their walls. Unlike the true bacteria, whose walls all contain murein, the archaebacteria exhibit a great fundamental variety of wall types. Some exhibit a 'pseudomurein' structure (Kandler & Hippe, 1977; Kandler, 1979; König & Kandler, 1979*a, b*), others a polysaccharide sacculus (Steber & Schleifer, 1975), others a basically proteinaceous wall (Weiss, 1974; Jones, Bowers & Stadtman, 1977), and, of course, *Thermoplasma* is wall-less. Yet none of them exhibit the murein-wall type of the true bacteria.

More telling even is archaebacterial membrane structure. Archaebacterial lipids are not of the usual ester-linked, straight-chain variety found in true bacteria and eukaryotes. Rather, they are glycerol *ethers* (and so non-saponifiable) involving branched, mainly phytane side-groups (Kates, 1972; Langworthy, 1977; Tornabene & Langworthy, 1978; Tornabene, Langworthy, Holzer & Oro, 1979). In the thermoacidophiles two such glycerol ethers are often found covalently linked through the ends of their side-chains producing C_{40} side-chains linked at both ends to glycerols, resulting in membranes that are chemically, not just physically, bonded bilayers.

Although the same genetic code appears to be employed by archaebacteria as by other living forms, the details of the translation process are as unique in archaebacteria as they are in true bacteria or eucaryotes. For example, the transfer RNAs of archaebacteria contain no ribothymidine in the so-called common arm, as true bacteria and eukaryotes do; neither do these tRNAs contain dihydro-uridine or 7-Me guanosine (with the exception of one species of methanogen, *Methanosarcina*) (R. Gupta & C. R. Woese, unpublished). The initiator tRNA of *Halobacterium* carries methionine, not *N*-formyl methionine as in the true bacteria (RajBhandary, 1978). Significant ribosomal differences between true and archaebacteria are also seen (Woese, Magrum & Fox, 1978).

Archaebacteria, most particularly methanogens, are noted for the unusual coenzymes they possess (Balch *et al.*, 1979). This bespeaks some novelty, if not radical differences, in archaebacterial intermediary metabolism. Regulatory mechanisms for archaebacteria as well as their genome organization have not been studied. One can confidently predict that these will be as unique as are other aspects of archaebacteria.

The archaebacterial RNA polymerases (see above) have a different subunit structure from the eubacterial RNA polymerases (Zillig, Stetter & Janekovic, 1979; Zillig *et al.*, 1980; Sturm *et al.*, 1980).

The general conclusion that seems to be emerging with regard to the differences among archaebacteria, true bacteria, and eukaryotes is that all are identical in the basic aspects of their basic processes, yet all differ from one another in the details of these processes.

GENERAL CONSIDERATIONS

The ultimate goal of phylogenetic studies is not merely to create a natural classification for organisms but to create a framework in which the course of their evolution can be rationalized. Although bacterial phylogeny at its present state falls short of this, the studies discussed above constitute a significant advance in two senses; first, they reveal shortcomings and misconceptions in earlier attempts to adduce bacterial phylogeny and the course of early evolution, and second, they let us glimpse for the first time the true outline of the universal phylogenetic tree.

Phylogenetically valid characters

The fact that current bacterial classification is rather far from being phylogenetically valid is now clear. The heavy emphasis placed, in many cases necessarily, on morphological characters in classifying bacteria is seen to be misleading. Many microbiologists were undoubtedly ready to admit this already for spherical shape as a phylogenetic character. Of the families and genera defined for walled spherical bacteria, most if not all are not valid phylogenetic units. Perusal of the figures in this chapter shows not only that *Micrococcaceae* is not a family, but all three of its (current) component genera, *Micrococcus, Staphylococcus,* and *Planococcus,* are not valid either (Stackebrandt & Woese, 1979). (The case is marginal for *Staphylococcus,* and depends upon whether or not *Bacillus* is considered merely a genus.) The same goes for spherical genera such as *Paracoccus* (a close relative of *Rps. capsulata*) and *Sarcina* (a member of the *C. pasteurianum* subgroup of clostridia). Even the wall-less highly unique, spherical organisms, the mycoplasmas, must be demoted from lofty phylogenetic status, as a separate Class or Division, to members of one of the sublines of the clostridia.

How many of us, though, were prepared to see the more complex bacterial morphological characters evaporate as phylogenetic indicators? Spiral shape has little or no phylogenetic significance (C. R. Woese, P. Blanz & C. M. Hahn, unpublished). *Serpens flexibilis* joins certain pseudomonads; *Rhodospirillum tenue* is related to

Alcaligenes species among others; and so on. A budding mode of division defines only a low level unit phylogenetically (Gibson *et al.*, 1979), and does not group all such organisms (J. Gibson & C. R. Woese, unpublished). We have also seen how even the complex morphologies exhibited by the actinomycetes and relatives are unreliable phylogenetic indicators (Stackebrandt, Häringer & Schleifer, 1980*a*). Our findings have been ample demonstration of van Niel's warnings against using morphology as a guide to the natural relationships among bacteria (Van Niel, 1946).

Some biochemical characteristics are also deceptive phylogenetic indicators. This seems particularly true in the case of the pseudomonads, distinctions among which, and grouping of which, were often made on the basis of what might be called the 'superficial aspects' of their biochemistry, e.g., the ability to metabolize this or that esoteric carbon source.

Among the phylogenetically more reliable characters appear to be spores and spore-associated structures, and cell-wall compositions. Even in the latter case, however, confusion can be introduced at times by evolutionary convergence (in the high G + C Gram-positive group) (Schleifer & Kandler, 1972).

Photosynthesis and bacterial origins

The biologist, in particular the microbiologist, tends to accept the 'Oparin prejudice' concerning the origin of bacteria (Oparin, 1938). The general argument is that in prebiotic times the ocean was a soup of organic chemicals, an ideal growth medium for, say, a clostridium. Therefore the first organisms to arise were *anaerobic extreme heterotrophs*, requiring all their amino acids, nucleotides, etc.; they were non-photosynthetic as well. Only as the primeval stores of organic compounds in the oceans became depleted did these early fermenters evolve toward autotrophy (one step, i.e., enzyme, at a time (Horowitz, 1945)) and finally toward phototrophy as all energy-rich compounds became exhausted. This view translates into the idea that the first *bacteria* were also heterotrophic, non-photosynthetic anaerobes; fermentation was the primeval bacterial metabolism. Statements and diagrams to this effect abound in textbooks that touch on the evolution of bacteria. From the hypothetical fermentating ancestor most of the lines of bacterial descent arose; one or a few of these then developed phototrophy and/or autotrophy. For this reason photosynthetic lines of descent should be largely separate from the non-photosynthetic ones.

The findings reviewed here at the very least bring this view into question. Within the eubacterial kingdom many of the non-photosynthetic species have clearly come from photosynthetic ancestry. Moreover, three or four of the major divisions of eubacteria are basically photosynthetic. This constitutes a strong suggestion that the common ancestor of all eubacteria was a photosynthetic organism.

The case is not clear for the archaebacteria. Photosynthetic phenotypes are not widespread and there is no indication of where the halobacterial photosynthetic mechanism could have come from. The biosynthesis of rhodopsin includes an oxygen-dependent step, which would only evolve after the advent of oxygen. In any case we no longer feel that one should take for granted the idea that the universal common ancestor was a fermentative organism; it could just as well have been photosynthetic and/or autotrophic.

Phenotypic and genotypic measures of evolution

It is a striking finding that only among the anaerobic phenotypes does one encounter any great phylogenetic depth, i.e., low minimal S_{AB} values. The case of the methanogens is most striking, as is that for the clostridia (see above). Aerobic groups, like the genus *Bacillus*, or the aerobic section of the high $G + C$ Gram-positive eubacteria all exhibit considerably higher minimal S_{AB} values (about 0.50). These results imply that early on the earth's atmosphere was anaerobic. They also imply that evolution to an aerobic metabolism is a relatively easy thing to accomplish; there seem to be a number of independent examples among the purple photosynthetic bacteria alone where aerobic metabolism has arisen from a photosynthetic phenotype. (That it can also arise from a non-photosynthetic phenotype is seen by the evolution of a clostridial subline into *Bacillus*.)

It is evident in the above that the classically defined taxonomic categories, the genera and families, do not correspond to fixed (minimal) S_{AB} values. What are now called genera range all the way from shallow genealogical structures (high minimal S_{AB}) such as *Escherichia*, to substantial genera such as *Bacillus* (minimal S_{AB} in the range of 0.55), to groupings like *Clostridium*, a 'genus' whose phylogenetic depth exceeds that of 'major groupings' like the actinomycetes and their relatives. It is evident that two measures of evolution are being used here; these might be called the phenotypic versus the genotypic measure of evolution. The phenotypic measure is the classical one; it is the one we use in speaking of evolutionary *progression,* e.g., from fish to amphibians to mammals. For the most part it reflects ecological change, or the working out of some 'evolutionary potential' in a line of descent. Phenotypic change is sporadic and saltatory. Genotypic change, on the other hand, reflects the rate of mutation, and so is clock-like. Thus it is not surprising to find a genus like *Clostridium,* which has remained ecologically stable for eons, exhibiting a much lower minimal S_{AB} value than a more recent genus like *Bacillus.* It is not a question of whether the phenotypic or the genotypic measure is the correct one. Both are characteristic of the evolutionary process. It is proper that a genus be defined phenotypically, in some terms that reflect the degree of commonality and the spread in the phenotype. It is also proper that the age of the genus be designated.

Evolutionary rates

The degree to which the dendrograms generated by the S_{AB} analysis approximate true phylogenetic relationships is a function of the relative rates at which the various organisms evolve (actually the rates at which mutations become fixed in ribosomal RNA cistrons). The best approximation occurs when all organisms in a considered set are isochronic. Organisms with 'fast clocks' appear as branchings that are deeper than their true branchings (Woese *et al.,* 1980a). For any given group of organisms it is possible to ascertain whether or not the organisms therein are isochronic. Take for example the genus *Bacillus.* If all the species considered are isochronic, then the phylogenetic distance of every one of them (measured by S_{AB} value) from an organism or organisms known to be outside the group, e.g., *E. coli,* will be statistically the same. A *Bacillus* species with a faster 'clock' would then exhibit a lower S_{AB} value than the others with the outside reference organism, and so on. This analysis, using an outside reference organism(s), can be carried out for any group of organisms except the largest; there is no outside reference organism that enables one to assess whether mutational clocks in two different primary kingdoms run at the same or different rates.

Table 1 shows S_{AB} values for a group of *Bacillus* species with several outside reference organisms. It can be seen that the majority

Table 1. S_{AB} values for Bacillus species with various reference organisms

	Escherichia coli	Chromatium vinosum	Spirochaeta aurantia	Aphanocapsa 6714	Clostridium aminovalericum	Micrococcus luteus
B. cereus	0.24	0.29	0.28	0.31	0.36	0.36
B. megaterium	0.27	0.32	0.29	0.29	0.36	0.37
B. pumilus	0.26	0.31	0.29	0.29	0.34	0.38
B. subtilis	0.26	0.32	0.29	0.29	0.34	0.38
B. firmus	0.25	0.31	0.28	0.28	0.36	0.36
B. badius	0.28	0.33	0.31	0.31	0.39	0.40
B. pasteurii	0.24	0.29	0.27	0.28	0.33[a]	0.34
Sporosarcina	0.26	0.31	0.28	0.29	0.34	0.35
B. sphaericus	0.26	0.31	0.27	0.30	0.33	0.33
B. psychrophilus	0.26	0.33	0.28	0.27	0.32[b]	0.34
B. insolitus	0.26	0.32	0.26[a]	0.26	0.33[a]	0.35
B. coagulans	0.23[a]	0.30	0.26[a]	0.24[b]	0.36	0.31
B. stearothermophilus	0.29	0.35	0.30	0.31	0.36	0.35
B. brevis 8185	0.28	0.32	0.27	0.29	0.37	0.31
B. brevis 12991	0.26	0.31	0.29	0.27	0.35	0.31
B. polymxa	0.25	0.29	0.32	0.32	0.33[a]	0.32
B. alvei	0.23[a]	0.28[a]	0.28	0.31	0.35	0.33
Sporolactobacillus	0.22[b]	0.26[b]	0.24[b]	0.25[a]	0.33[a]	0.29[a]
B. acidocaldarius	0.26	0.34	0.27	0.25[a]	0.31[b]	0.28[b]

[a] next lowest value with reference organism.
[b] lowest value with reference organism.

of Bacillus species exhibit the same distance to the reference organisms, and so are isochronic. A few, however, exhibit significantly lower S_{AB} values to the reference organisms, and so, for some reason, must possess rRNA mutational clocks that run at a relatively rapid pace. In some instances, e.g., B. acidocaldarius (within the Gram-positives) where the physical environment of the ribosome is changed, this could merely reflect selection for mutations that alter the design of the ribosome. In other instances a general increase in the organism's mutation rate may be the cause. It is interesting that B. stearothermophilus, growing at elevated temperatures, does not exhibit a faster mutational clock.

By and large the true bacteria all appear isochronic. The reason for this is unclear. However, a similar condition holds for eukaryotes (Wilson, Carlson & White, 1977). Possibly factors such as the degree of complexity of the organism and energy throughput in the niche are involved here, as is another ill-defined parameter, extent of adaptation of organism to its niche.

Although we will not discuss in detail the exceptions to isochronicity, it is interesting to note that the majority of the exceptions are spherical bacteria (which appear to be morphologically degenerate forms of more morphologically complex bacteria). The most extreme case of fast mutational clocks among free-living bacteria are the mycoplasmas (Woese et al., 1980a). Their smaller genome sizes might rationalize the effect, i.e., for the same mutation rate per genome, they would have a higher mutation rate per base pair than do normal bacteria. (For a more detailed discussion of those relative evolutionary rates see Woese et al. (1980a).)

THE UNIVERSAL COMMON ANCESTOR

It is worth giving some thought to the implications of the above findings with regard to the common ancestor of all extant life. Fossil evidence strongly suggests that the major lines of eubacterial descent, i.e., the photosynthetic bacteria, were established 3.5 to 3.8 billion years ago, within a billion years of the earth's formation (Schidlowski, Appel, Eichmann & Junge, 1978). Thus the common ancestor of all eubacteria should have existed at an even earlier time, and the universal common ancestor in a still more remote era. This time scale demands that less than a billion years, perhaps *far less,* was spent in getting to the universal common ancestor of all extant life and from that form to the common ancestors of the various primary kingdoms. Evolution since the advent of the primary kingdoms has covered a far longer time than evolution before their advent. Yet the major evolutionary changes, those reflected in striking qualitative and quantitative (S_{AB} value) differences among the three major lines of descent, occurred during that relatively short period before the first known sedimentary rocks were laid down, 3.8 billion years ago.

The evolutionary changes that occurred in that early period have a qualitatively different feel about them from the later changes. Not only do they seem to occur more rapidly than later changes, but they appear to be of a more drastic character, affecting cell-wall type, lipid composition, modifications of the translation apparatus, RNA polymerase subunit structure, and so on. Therefore, one is easily led to believe that the evolution that transformed the universal common ancestor into the ancestor of the true bacteria, for example, was of a different quality from the evolution that transforms the latter into any of its progeny phenotypes. In other words, it does not seem that the universal common ancestor was a prokaryote; it was an organism far simpler, more capable of rapid and drastic evolutionary change (Woese & Fox, 1977a).

This case has also been argued from a different perspective (Woese, 1970; Woese & Fox, 1977b). If the versions of the translation apparatus characteristic of the three major lines of descent differ in such features as the patterns of modification of bases in rRNA and tRNA, it is quite possible that the universal common ancestral version of the translation apparatus did not possess these features, i.e., the latter apparatus may have been more rudimentary than its counterparts in the three major lines of descent. If so, what we are seeing in the evolution of the universal ancestor into the three primary lines of descent is primarily an evolution of the translation apparatus through its final stages, and this underlies the other drastic evolutionary changes in the cell that then occur. The changes in the translation apparatus almost certainly involve increases in precision and other control of that process (that is the only reasonable explanation for the existence of much of the translation apparatus, e.g., modified nucleotides). If translation in the universal ancestor was less accurate than today's versions of the process, the ancestor should not have possessed proteins such as those we encounter today in prokaryotes and eukaryotes (Woese, 1970; Woese & Fox, 1977b). Its proteins should have been smaller and, more important, possess less biological specificity and less accuracy in their functions. A result of this, of course, would be less accurate replication and maintenance of genetic information, and so a far smaller genome. It is for this reason, i.e., that at the time of the universal ancestor the genotype–phenotype link was still in the process of evolutionary refinement, that the universal ancestor has been named the *Progenote* (Woese & Fox, 1977b).

If this line of argument is correct, and future experiments offer the possibility of strongly substantiating it, then the universal phylogenetic tree traces life back past the prokaryotic stage, to a far simpler form called the Progenote, which opens the door to a new set of concepts, questions, and experiments.

REFERENCES

ALLSOPP, A. (1969). Phylogenetic relationships of the procaryotes and the origin of the eucaryotic cell. *New Phytologist,* **68,** 591–612.

ALMASSY, R. M. & DICKERSON, R. E. (1978). *Pseudomonas* cytochrome c_{551} at 2.0 Å resolution: enlargement of the cytochrome c family. *Proceedings of the National Academy of Sciences, USA,* **75,** 2674–8.

AMBLER, R. P., DANIEL, M., HERMOSO, J., MEYER, T. E., BARTSCH, R. G. & KAMEN, M. D. (1979). Cytochrome c sequence variation among the recognised species of purple nonsulphur photosynthetic bacteria. *Nature,* **278,** 659–60.

BALCH, W. E., FOX, G. E., MAGRUM, L. J., WOESE, C. R. & WOLFE, R. S. (1979). Methanogens: Reevaluation of a unique biological group. *Microbiological Reviews*, **43**, 260–96.

BERGEY's Manual of Determinative Bacteriology. 1st Edn (1923); 2nd Edn (1926); 3rd Edn (1930); 4th Edn (1934); 5th Edn (1939); 6th Edn (1948). The Williams and Wilkins Co., Baltimore. 7th Edn (1957) Ballière, Tindall & Cox, London. 8th Edn (1974) The Williams and Wilkins Co., Baltimore.

BONEN, L., CUNNINGHAM, R. S., GRAY, M. W. & DOOLITTLE, W. F. (1977). Wheat embryo mitochondrial 18S ribosomal RNA: evidence for its procaryotic nature. *Nucleic Acid Research*, **4**, 663–71.

BONEN, L. & DOOLITTLE, W. F. (1975). On the prokaryotic nature of red algal chloroplasts. *Proceedings of the National Academy of Sciences, USA*, **72**, 2310–14.

BONEN, L. & DOOLITTLE, W. F. (1976). Partial sequences of 16S rRNA and the phylogeny of blue-green algae and chloroplasts. *Nature*, **261**, 669–73.

BONEN, L. & DOOLITTLE, W. F. (1978). Ribosomal RNA homologies and the evolution of the filamentous blue-green bacteria. *Journal of Molecular Evolution*, **10**, 283–91.

BROOKS, B. W., MURRAY, R. G. E., JOHNSON, J. L., STACKEBRANDT, E., WOESE, C. R. & FOX, G. E. (1980). A study of the red-pigmented micrococci as a basis for taxonomy. *International Journal of Systematic Bacteriology*, (in press).

CHATTON, E. (1937). Titres et Traveaux Scientifiques. Sete, Sottano.

CROSS, T., DAVIES, F. & WALFER, P. D. (1971). *Thermoactinomyces vulgaris*. In *Spore Research*, Ed. A. N. Barker, G. W. Gould & J. Wolf. Academic Press, London.

DICKERSON, R. E. (1980). Evolution and gene transfer in purple photosynthetic bacteria. *Nature*, **283**, 210–12.

DOOLITTLE, W. F., WOESE, C. R., SOGIN, M. L., BONEN, L. & STAHL, D. (1975). Sequence studies on 16S ribosomal RNA from a blue-green alga. *Journal of Molecular Evolution*, **4**, 307–15.

DOUDOROFF, M. & PALLERONI, N. J. (1974). Pseudomonas. In *Bergey's Manual of Determinative Bacteriology, 8th Edn*, Ed. R. E. Buchanan & N. E. Gibbons, pp. 217–43. The Williams and Wilkins Co., Baltimore.

FALLER, A. (1980). Untersuchungen zum Cytochrom-Muster bei Gram-positiven Kokken and Coryneformen Bakterien. Ph.D. Thesis, Technical University, Munich, Federal Republic of Germany.

FOX, G. E., MAGRUM, L. J., BALCH, W. E., WOLFE, R. S. & WOESE, C. R. (1977). Classification of methanogenic bacteria by 16S ribosomal RNA characterization. *Proceedings of the National Academy of Sciences, USA*, **74**, 4537–41.

FOX, G. E., PECHMAN, K. J. & WOESE, C. R. (1977). Comparative cataloging of 16S ribosomal ribonucleic acid: Molecular approach to procaryotic systematics. *International Journal of Systematic Bacteriology*, **27**, 44–57.

FOX, G. E., STACKEBRANDT, E., HESPELL, R. B., GIBSON, J., MANILOFF, J., DYER, T., WOLFE, R. S., BALCH, W., TANNER, R., MAGRUM, L. J., ZABLEN, L. B., BLAKEMORE, R., GUPTA, R., LUEHRSEN, K. R., BONEN, L., LEWIS, B. J., CHEN, K. N. & WOESE, C. R. (1980). The phylogeny of prokaryotes. *Science*, **209**, 457–63.

GIBSON, J., STACKEBRANDT, E., ZABLEN, L. B., GUPTA, R. & WOESE, C. R. (1979). A genealogical analysis of the purple photosynthetic bacteria. *Current Microbiology*, **3**, 59–66.

HIGO, K., HELD, W., KAHAM, L. & NOMURA, M. (1973). Functional correspondence between 30S ribosomal proteins of *Escherichia coli* and *Bacillus stearothermophilus*. *Proceedings of the National Academy of Sciences, USA*, **70**, 944–8.

HOROWITZ, N. H. (1945). On the evolution of biochemical synthesis. *Proceedings of the National Academy of Sciences, USA,* **31,** 153–7.
JOHNSON, J. L. (1973). Use of nucleic-acid homologies in the taxonomy of anaerobic bacteria. *International Journal of Systematic Bacteriology,* **23,** 308–15.
JOHNSON, J. L. & FRANCIS, B. S. (1975). Taxonomy of the Clostridia: ribosomal ribonucleic acid homologies among the species. *Journal of General Microbiology,* **88,** 229–44.
JONES, J. B., BOWERS, B. & STADTMAN, T. C. (1977). *Methanococcus vanielii.* Ultrastructure and sensitivity to detergents and antibiotics. *Journal of Bacteriology,* **130,** 1357–63.
KANDLER, O. (1979). Zellwandstruktur bei Methan-Bakterien. Zur Evolution der Prokaryonten. *Naturwissenschaften,* **66,** 95–105.
KANDLER, O. & HIPPE, H. (1977). Lack of peptidoglycan in the cell wall of *Methanosarcina barkeri. Archives of Microbiology,* **113,** 57–60.
KATES, M. (1972). *Ether lipids, chemistry and biology,* Ed. F. L. Snyder. Academic Press, New York.
KÖNIG, H. & KANDLER, O. (1979*a*). N-Acetyltalosaminuronic acid, a constituent of the pseudomurein of the genus *Methanobacterium. Archives of Microbiology,* **123,** 295–9.
KÖNIG, H. & KANDLER, O. (1979*b*). The amino acid sequence of the peptide moiety of the pseudomurein from *Methanobacterium thermoautotrophicum. Archives of Microbiology,* **121,** 271–5.
LANGWORTHY, T. A. (1977). Long-chain diglycerol tetraethers from *Thermoplasma acidophilum. Biochimica et Biophysica Acta,* **487,** 37–50.
LEWIN, R. A. (1977). *Prochloron,* type genus of the *Prochlorophyta. Phycologia,* **16,** 217.
MAGRUM, L. J., LUEHRSEN, K. R. & WOESE, C. R. (1978). Are extreme halophiles actually 'bacteria'? *Journal of Molecular Evolution,* **11,** 1–8.
MARGULIS, L. (1970). *Origin of Eukaryotic Cells.* Yale University Press, New Haven.
MATHESON, A. T., MOLLER, W., AMONS, R. & YAGUCHI, M. (1980). Comparative studies on the structure of ribosomal proteins, with emphasis on the alanine-rich, acidic ribosomal 'A'-protein. In *Ribosomes: Structure, function, and genetics,* Ed. G. Chamblis, G. R. Craven, J. Davies, K. Davis, L. Kahan & M. Nomura, pp. 297–333. University Park Press, Baltimore.
MINNIKIN, D. E., GOODFELLOW, M. & COLLINS, M. D. (1978). Lipid composition in the classification and identification of coryneform and related taxa. In *Coryneform bacteria,* Ed. E. Bousfield & A. G. Callely. Academic Press, London.
MOORE, R. L. & MCCARTHY, B. (1967). Comparative study of ribosomal ribonucleic acid cistrons in enterobacteria and myxobacteria. *Journal of Bacteriology,* **94,** 1066–74.
MURRAY, R. G. E. (1974). A place for bacteria in the living world. In *Bergey's Manual of Determinative Bacteriology,* 8th edn, Ed. R. E. Buchanan & N. E. Gibbons, pp. 4–9. The Williams and Wilkins Co., Baltimore.
NOMURA, M., TRAUB, P. & BECHMANN, H. (1968). Hybrid 30S ribosomal particles reconstituted from components of different bacterial origins. *Nature,* **219,** 793–9.
OPARIN, A. I. (1938). *The Origin of Life.* Macmillan, New York.
PACE, B. & CAMPBELL, L. L. (1971). Homology of ribosomal ribonucleic acid of diverse bacterial species with *Escherichia coli* and *Bacillus stearothermophilus. Journal of Bacteriology,* **107,** 543–7.
PALLERONI, N. J., KUNISAWA, R., CONTOPOULOU, R. & DOUDOROFF, M. (1973).

Nucleic acid homologies in the genus *Pseudomonas*. *International Journal of Systematic Bacteriology*, **23**, 333–9.

RAJBHANDARY, U. L., unpublished, as cited in: Heckman, J. E., Hecher, L. I., Schwartzbach, S. D., Barnett, W. E., Baumstark, B. & RajBhandary, U. L. (1978). Structure and Function of initiator methionine tRNA from the mitochondria of *Neurospora crassa*. *Cell*, **13**, 83–95.

ROGOSA, M., CUMMINS, C. S., LELLIOTT, R. A. & KEDDIE, R. M. (1974). Coryneform Group of Bacteria. In *Bergey's Manual of Determinative Bacteriology. 8th Edn*, Ed. R. E. Buchanan & N. E. Gibbons, pp. 599–632. The Williams and Wilkins Co., Baltimore.

SANGER, F. J., BROWNLEE, G. G. & BARRELL, B. G. (1965). A two-dimensional fractionation procedure for radioactive nucleotides. *Journal of Molecular Biology*, **13**, 373–98.

SCHIDLOWSKI, M., APPEL, P. W. U., EICHMANN, R. & JUNGE, C. E. (1978). Carbon isotope geochemistry of the 3.7×10^9-yr-old Isua sediments, West Greenland: implications for the Archaean carbon and oxygen cycles. *Geochimica and Cosmochimica Acta*, **43**, 189–99.

SCHLEIFER, K. H. & KANDLER, O. (1972). The peptidoglycan types of bacterial cell walls and their taxonomic implications. *Bacteriological Reviews*, **36**, 407–77.

SCHWARZ, A. & KÖSSEL, H. (1980). The primary structure of 16S rDNA from *Zea mays* chloroplast is homologous to *E. coli* 16S rRNA. *Nature*, **283**, 739–42.

SILBERKLANG, M., GILLUM, A. M. & RAJBHANDARY, U. L. (1979). Use of in vitro ^{32}P labelling in the sequence analysis of nonradioactive tRNAs. In *Methods of Enzymology*, **590**, 58–109.

STACKEBRANDT, E., HÄRINGER, M. & SCHLEIFER, K. H. (1980a). Molecular genetic evidence for the transfer of *Oerskovia* species into the genus *Cellulomonas*. *Archives of Microbiology*, **127**, 179–85.

STACKEBRANDT, E., LEWIS, B. J. & WOESE, C. R. (1980b). The phylogenetic structure of the coryneform group of bacteria. *Zentralblatt für Bakteriologie, I. Abteilung Originale*, **C1**, 137–49.

STACKEBRANDT, E., LUDWIG, W., SCHLEIFER, K. H. & GROSS, H. J. (1981). Rapid cataloguing of ribonuclease T_1 resistant oligonucleotides from ribosomal RNAs for phylogenetic studies. *Journal of Molecular Evolution*, in press.

STACKEBRANDT, E. & WOESE, C. R. (1979). A phylogenetic dissection of the family Micrococcaceae. *Current Microbiology*, **2**, 317–22.

STEBER, J. & SCHLEIFER, K. H. (1975). *Halococcus morrhuae*: a sulfated heteropolysaccharide as the structural component of the bacterial cell wall. *Archives of Microbiology*, **105**, 173–7.

STEIGERWALT, A. G., FANNING, G. R., FIFE-ASHBURY, M. A. & BRENNER, D. J. (1976). DNA relatedness among species of *Enterobacter* and *Serratia*. *Canadian Journal of Microbiology*, **22**, 121–37.

STURM, S., SCHÖNEFELD, U., ZILLIG, W., JANEKOVIC, D. & STETTER, K. O. (1980). Structure and function of the DNA dependent RNA polymerase of the Archaebacterium *Thermoplasma acidophilum*. *Zentralblatt für Bakteriologie I. Abteilung, Originale*, **C1**, 12–25.

TORNABENE, T. G. & LANGWORTHY, T. A. (1978). Diphytanyl and dibiphytanyl glycerol ether lipids of methanogenic archaebacteria. *Science*, **203**, 51–3.

TORNABENE, T. G., LANGWORTHY, T. A., HOLZER, G. & ORO, J. (1979). Squalenes, phytanes and other isoprenoids as major neutral lipids of methanogenic and thermoacidophilic 'Archaebacteria'. *Journal of Molecular Evolution*, **13**, 73–83.

UCHIDA, T., BONEN, L., SCHAUP, H. W., LEWIS, B. J., ZABLEN, L. & WOESE, C. R. (1974). The use of ribonuclease U_2 in RNA sequence determination. *Journal of Molecular Evolution*, **3**, 63–77.

Van Niel, C. B. (1946). The classification and natural relationships of bacteria. *Cold Spring Harbor Symposium on Quantitative Biology,* **11,** 285–301.
Weiss, R. L. (1974). Subunit cell wall of *Sulfolobus acidocaldarius. Journal of Bacteriology,* **118,** 275–84.
Wilson, A. C., Carlson, S. S. & White, T. J. (1977). Biochemical evolution. *Annual Review of Biochemistry,* **46,** 573–639.
Woese, C. R. (1970). Organization and control in procaryotic and eucaryotic cells. 20th Symposium of the Society for General Microbiology, Ed. H. P. Charles & B. C. J. G. Knight, pp. 39–54. Cambridge University Press, Cambridge.
Woese, C. R. & Fox, G. E. (1977a). Phylogenetic structure of the prokaryotic domain: the primary kingdoms. *Proceedings of the National Academy of Sciences, USA,* **74,** 5088–90.
Woese, C. R. & Fox, G. E. (1977b). The concept of cellular evolution. *Journal of Molecular Evolution,* **10,** 1–6.
Woese, C. R., Gibson, J. & Fox, G. E. (1980). Do genealogical patterns in purple photosynthetic bacteria reflect interspecific gene transfer? *Nature,* **283,** 212–14.
Woese, C. R., Luehrsen, K. R., Pribula, C. D. & Fox, G. E. (1976a). Sequence characterization of 5S ribosomal RNA from eight Gram positive procaryotes. *Journal of Molecular Evolution,* **8,** 143–53.
Woese, C. R., Magrum, L. J. & Fox, G. E. (1978). Archaebacteria. *Journal of Molecular Evolution,* **11,** 245–52.
Woese, C. R., Magrum, L. J., Gupta, R., Siegel, R. B., Stahl, D., Kop, J., Crawford, N., Brosius, J., Gutell, R., Hogan, J. J. & Noller, H. F. (1980b). Secondary structure model for bacterial 16S ribosomal RNA: phylogenetic, enzymatic and chemical evidence. *Nucleic Acid Research,* **8,** 2275–94.
Woese, C. R., Maniloff, J. & Zablen, L. B. (1980a). Phylogenetic analysis of the mycoplasmas. *Proceedings of the National Academy of Sciences, USA,* **77,** 494–8.
Woese, C. R., Sogin, M., Stahl, D., Lewis, D. J. & Bonen, L. (1976b). A comparison of the 16S ribosomal RNAs from mesophilic and thermophilic bacilli: some modifications in the Sanger Method for RNA sequencing. *Journal of Molecular Evolution,* **7,** 197–213.
Zablen, L. B., Bonen, L., Meyer, R. & Woese, C. R. (1975a). The phylogenetic status of *Pasteurella pestis. Journal of Molecular Evolution,* **4,** 347–58.
Zablen, L. B., Kissil, M. S., Woese, C. R. & Buetow, D. E. (1975b). Phylogenetic origin of chloroplasts and prokaryotic nature of its ribosomal RNA. *Proceedings of the National Academy of Sciences, USA,* **72,** 2418–22.
Zillig, W., Stetter, K. O. & Janekovic, D. (1979). DNA dependent RNA polymerase from the Archaebacterium *Sulfolobus acidocaldarius. European Journal of Biochemistry,* **96,** 597–604.
Zillig, W., Stetter, K. O., Wunderl, S., Schulz, W., Priess, H. & Scholz, I. (1980). The *Sulfolobus*-'Caldariella' group: Taxonomy on the basis of the structure of DNA-dependent RNA polymerases. *Archives of Microbiology,* **125,** 259–69.
Zuckerkandl, E. & Pauling, L. (1965a). Evolutionary divergence and convergence in proteins, pp. 97–166. In *Evolving genes and proteins,* Ed. H. Bryson & H. J. Vogel, Academic Press, New York.
Zuckerkandl, E. & Pauling, L. (1965b). Molecules as documents of evolutionary history. *Journal of Theoretical Biology,* **8,** 357–66.

THE ORIGIN AND EARLY EVOLUTION OF THE EUKARYOTIC CELL

T. CAVALIER-SMITH

Department of Biophysics, University of London King's College, 26–29 Drury Lane, London WC2B 5RL

INTRODUCTION

The most momentous event in the history of life since the origin of photosynthesis must surely have been the appearance of the first eukaryotic cell. Two decades have elapsed since the fundamental nature of the many differences in cellular organization between prokaryotes (bacteria and blue-green algae) and eukaryotes (all other cellular organisms) first became indisputable (Picken, 1960; Stanier & Van Niel, 1962). Yet we are still far from understanding the causes of this great evolutionary step.

Because the fossil record is so unhelpful we must build speculations concerning the origin of eukaryotes on the widest biological base, taking account of well-established principles of evolutionary and ecological theory as well as genetics, cellular and molecular biology. An adequate theory must have three elements: (i) a unified phylogeny for prokaryotes and eukaryotes; (ii) a detailed description of the structural, genetic, physiological and developmental changes that must have occurred during the prokaryote-eukaryote transition; and (iii) explicit consideration of the selective forces responsible for all the postulated changes. Much discussion of eukaryote origins has been piecemeal, dealing in isolation with the origin of mitochondria (Raff & Mahler, 1972; Mahler & Raff, 1975; Mahler, 1980, 1981), plastids (Lee, 1972; Bogorad, 1980; Doolittle & Bonen, 1981), repetitive DNA (Reanney, 1974), split genes (Doolittle, 1978; Reanney, 1979) and meiosis (Maynard Smith, 1978).

The great drawback of this approach is that it altogether neglects interaction between the various cellular components, which may be of profound significance for cellular evolution. If, as I believe, (Cavalier-Smith, 1975, 1977, 1980a) such interaction offers the key to understanding the origin and biological significance of the characteristic eukaryotic properties, we should begin by summarizing the basic features whose origin we have to explain.

THE FUNDAMENTAL DIFFERENCES BETWEEN PROKARYOTES AND EUKARYOTES

Many of these differences are well known from earlier discussion (Stanier, 1970; Carlile, 1980), some such as split genes (Crick, 1979) have been only recently discovered, while others, notably the basic difference in DNA replication pattern, have been long known but only recently discussed in an evolutionary context (Cavalier-Smith, 1980a). It is important to distinguish universal qualitative differences from those that are merely widespread or quantitative. In this respect it is worth emphasizing that neither plastids nor mitochondria are universally found in eukaryotes: too much discussion has centred on their origin(s), and too little on the truly universal eukaryote characters (Table 1) which may conveniently be grouped under five headings:

The nuclear envelope and the uncoupling of transcription and translation

The defining characteristic of eukaryotes (Dougherty, 1957) is their possession of a nuclear envelope (Fig. 1a), made up of two phospholipid membranes which are both penetrated by a proteinaceous pore complex (Maul, 1977) which has a distinctive eightfold symmetry. The inner and outer envelope membranes are in topological continuity with each other at the periphery of each pore complex, which allows small molecules to pass in both directions across the pore (Feldherr, 1972; Franke, 1974; Paine, Moore & Horowitz, 1975; Siebert, 1978). In interphase, DNA is attached to the inner face of the inner membrane, and ribosomes are usually found on the outer face of the outer membrane which often shows continuity with, and probably does not differ fundamentally from, the rough endoplasmic reticulum of the cytoplasm. DNA and RNA synthesis, and RNA processing, occur exclusively within the nucleus, whereas protein synthesis occurs exclusively in the cytoplasm. Thus RNA has to be transported out of the nucleus, probably through the nuclear pores by an energy requiring process (Clawson, Koplitz, Moody & Smuckler, 1980). Though ribosomal subunits are largely assembled in the nucleolus they are somehow prevented from attaching to nascent messenger when in the nucleus.

In prokaryotes on the other hand (Fig. 1b) the nucleoplasm is not separated from the cytoplasm by an envelope, so the DNA and

Table 1. *Characters universally present in eukaryotes and universally absent from prokaryotes*

1. Nuclear envelope and nuclear pore complex
2. Uncoupling of transcription and translation
3. Exocytosis
4. Rough endoplasmic reticulum (RER): ribosomes absent from plasma membrane
5. Golgi cisternae
6. Lysosomes
7. Peroxisomes and other 'microbodies'
8. Actin microfilaments
9. Tubulin microtubules
10. Directed intracytoplasmic transport of vesicles
11. Separate RNA polymerase for mRNA, rRNA and tRNA
12. Absence of operons with polygenic messenger RNAs
13. Co-ordinate control of transcription during cell cycle
14. Nucleolus for rRNA processing and ribosome subunit assembly
15. 5'-methyl-guanosine capping of messenger RNA
16. Split genes and RNA splicing
17. Nucleocytoplasmic RNA transport
18. 80S ribosomes with larger rRNAs, extra proteins, and lacking Shine–Dalgarno sites for base pairing between mRNA and rRNA.
19. Many replicons per chromosome ('plural replicons')
20. Cell cycle with discrete S-phase and absence of overlap between successive rounds of replication
21. Chromosome segregation by mitosis: directed movement involving microtubules attached to kinetochores
22. Linear chromosomes instead of a single circular one

ribosomes can mingle freely in the same compartment. This means that ribosomes are free to attach to and translate messengers that are still being synthesized and still attached to the DNA by RNA polymerase molecules. This means that most ribosomes in a prokaryote cell are physically attached to the DNA. Those ribosomes that make proteins that have to be cotranslationally inserted into or across membranes (i.e. integral membrane proteins or secretory proteins (Blobel, 1980; Emr, Hall & Silhavy, 1980)) are also physically attached to the plasma or other membranes. An important consequence of this is that the prokaryote chromosome is attached to the plasma membrane not only directly at its replicon origin, replication forks and perhaps replicon terminus (Yoshikawa *et al.*, 1979) but also indirectly at a number of sites by membrane-bound messengers. In eukaryotes by contrast there is complete uncoupling between transcription and translation and neither the DNA nor the nucleus are necessarily attached to the plasma membrane.

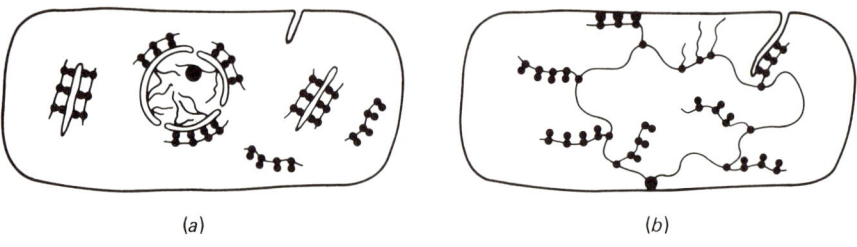

Fig. 1. The fundamental difference between prokaryotes and eukaryotes. (a) In eukaryotes the nuclear envelope separates the nucleoplasm, where DNA replication, transcription and RNA processing occur, from the cytoplasm where protein synthesis occurs. However, mRNA and nascent ribosomal subunits have to be transported out of the nucleus through the nuclear pore complexes before they are able to interact to make proteins. (b) Prokaryotes have no nuclear envelope, so replication, transcription and translation all occur in the cytoplasm. Messenger RNA needs no processing and ribosomes can immediately start translating nascent messenger while it is being made and is still attached to the DNA by RNA polymerase. Because ribosomes making plasma membrane integral proteins are attached to the membrane, their mRNA physically links the DNA to the plasma membrane.

Exocytosis, secretion and cytoplasmic compartmentation

In prokaryotes secretory proteins are cotranslationally secreted directly across the plasma membrane to the exterior of the cell by ribosomes attached to the inner face of the plasma membrane (Blobel, 1980; Emr *et al.*, 1980). Eukaryote ribosomes are never attached to the plasma membrane: secretory proteins are instead cotranslationally translocated into the cisternae of the rough endoplasmic reticulum (RER) or the perinuclear cisterna. These cisternae with one clearly established exception (Carothers, 1972) are not in direct topological continuity with the cell exterior. Instead of direct continuity between the RER and plasma membrane, smooth vesicles bud off from the RER, migrate through the cytoplasm, and fuse with the plasma membrane so as to liberate their contents to the exterior. Fusion of secretory vesicles with the plasma membrane, called exocytosis by Dupraw (1968), is also used for the secretion of molecules other than proteins, such as synaptic transmitters in nerves (Marchbanks, 1979), and is unknown in prokaryotes but universally present in eukaryotes.

It is probable that all eukaryotes not only show exocytosis, but also have specialized smooth membrane-bound cisternae that lack ribosomes on their surface – the Golgi cisternae – that often play an intermediary role in secretion between internal RER membranes and the plasma membrane. In most groups these are stacked into polarized arrays known as dictyosomes, but the non-flagellated fungi (the Eufungi, Cavalier-Smith, 1981) lack such stacking and have only single Golgi cisternae (Beckett, Heath & McLaughlin, 1974). All eukaryotes also appear to possess two kinds of cytoplasmic organelles in which enzymes are segregated from the cytosol inside membrane-bounded bags: i.e., lysosomes and peroxisomes, both of which probably arise by budding from the RER in a similar way to secretory vesicles.

Microfilaments, microtubules and cytoplasmic motility

All eukaryotes possess 5 nm microfilaments consisting of actin (Pollard & Weihing, 1974), and 25 nm microtubules consisting of tubulin (Roberts & Hyams, 1979). Molecules superficially resembling actin (Searcy, Stein & Green, 1978) and tubules of various kinds have been reported in a variety of prokaryotes but in no case has it been demonstrated that either are truly homologous in structure or function with the eukaryote organelles. Since actin microfilaments and microtubules have both been implicated in eukaryote intracytoplasmic motility, which is unknown in any prokaryote, the best working hypothesis is that actin, tubulin and their associated ATPases (myosin and dynein) are restricted to eukaryotes. All eukaryotes are able to move secretory or other membrane-bounded vesicles in a directed fashion through the cytoplasm, and all are able similarly to move chromosomes during mitosis. Microtubules and microfilaments are both involved in some way in directed vesicle transport, and since both are found in mitotic spindles the same may be true of directed chromosome transport.

Most eukaryote cells have a gel-like cytoplasm in which Brownian motion is not seen: organelles are held in place by a cytoskeleton of which microfilaments and microtubules form an important part. Even non-membrane-bound ribosomes do not diffuse freely, but are attached to a matrix (Lenk, Ransom, Kaufmann & Penman, 1977). Some eukaryote cells have a more fluid cytoplasm with active cytoplasmic streaming. Though cytoplasmic streaming and amoeboid movement both depend on actin, neither, contrary to what is often stated or implied (Margulis, 1974), are universal eukaryote properties, whereas actin, tubulin, and directed intracellular vesicle and chromosome transport appear to be universal properties of eukaryotes that are altogether absent in prokaryotes.

Transcription, RNA processing, RNA transport, and ribosomes

Eukaryote transcription differs from that of prokaryotes in that there are separate RNA polymerases for mRNA, rRNA and tRNA instead of a single polymerase for all three (Chambon, 1975). The control of mRNA synthesis is fundamentally different. Operons containing clustered genes coding for a single polygenic messenger are absent. Instead genes for related enzymes are widely scattered and there is coordinate control of all mRNA synthesis during the cell cyle (Frazer & Nurse, 1979). Genes fall into three classes, those transcribed at a low, at a medium, and at a high rate (Hereford & Rosbash, 1977), and the overall level of transcription is constant for most of the cell cycle, but abruptly doubles once every cycle at a time that is independent of the time of DNA replication.

In prokaryotes the stable RNA molecules (tRNA, rRNA) are processed by specific nucleases which cleave or shorten the primary transcript to produce the mature functioning molecule, but the short-lived mRNA is immediately functional and can be translated even while it is being transcribed. In eukaryotes the relatively stable mRNA is also extensively processed before transport to the cytoplasm and a methyl-guanosine cap is added in reversed orientation to the 5' end of the messenger. Moreover the precursors of some eukaryote mRNA, tRNA and rRNA molecules must have internal segments of the molecule deleted by RNA splicing during processing, because these genes have extra sequences, or 'introns', inserted at one or more points. Such split genes and the process of RNA splicing are unknown in prokaryotes but are also present in mitochondria and chloroplasts (Abelson, 1979). Processing of rRNA and assembly of ribosomal subunits occurs in a specific organelle, the nucleolus, that is absent in prokaryotes. Eukaryote ribosomes contain four rather than three RNA molecules, many more different proteins, and have different antibiotic sensitivities and a different initiation mechanism (Nakashima, Darzynkiewicz & Shatkin, 1980) from prokaryote ribosomes. Moreover the 18S RNA lacks the 3' OH dodecanucleotide present in prokaryote 16S RNA which forms complementary base pairs with the messenger to ensure efficient initiation (Shine & Dalgarno, 1975).

DNA replication, mitosis and the cell cycle

In prokaryotes, chromosome segregation, which probably occurs by

Table 2. *Widespread but non-universal characters of eukaryotes that are universally absent from prokaryotes*

1. DNA associated with 4–5 species of histones to form bead-like nucleosomes
2. Mitochondria containing unique ribosomes, having a modified genetic code, and a DNA plasmid coding for mitochondrial rRNA and about 10 inner membrane polypeptides
3. Sex, meiosis (chromosome pairing by a synaptinemal complex) and syngamy
4. 9 + 2 cilia and 9-triplet centrioles
5. Endocytosis (phagocytosis and pinocytosis)
6. Capacity to harbour cellular endosymbionts
7. Chloroplasts or other plastids
8. Amoeboid motion and cytoplasmic streaming
9. Circadian rhythms

membrane and wall growth (Davern, 1979), and replication overlap in time during the cell cycle, but in eukaryotes they are temporally dissociated in the cell cycle and involve different mechanisms. Eukaryote chromosome segregation is by a mitotic spindle that consists of two kinds of microtubules: kinetochore microtubules and centrosomal microtubules. Replication is confined to interphase, as a distinct S phase, and successive rounds of replication never overlap temporally as they do in rapidly growing bacteria (Davern, 1979). Moreover there are several separate chromosomes, not just one, and these are linear rather than circular and have numerous distinct units of replication (replicons) instead of a single one.

Other eukaryote properties

Characters that are widespread in eukaryotes but not universal are listed in Table 2. It is clearly important to decide which of these were present in the first eukaryote but have since been lost in the ancestors of the species that now lack them, and which of them were absent in the first eukaryote and have only evolved subsequently. I suggest that the first three characters (nucleosomes, mitochondria and sex) were primitively present, but that nucleosomes and histones were lost in free-living dinoflagellates, and that mitochondria were lost in several lines of anaerobic protozoa and in the one anaerobic fungus (Emerson & Held, 1969), and that sex was lost in numerous different lines. I shall argue that none of the remaining characters are primitive and all evolved well after the origin of the first eukaryotic cell.

As well as the positive characters shown in Tables 1 and 2, eukaryotes universally lack the non-universal prokaryotic characters

Table 3. *Some non-universal prokaryote characters universally absent from eukaryotes*

1. Gas vacuoles
2. Flagellin-based extracellular flagella with a proton driven rotary motor in the plasma membrane
3. Peptidoglycan cell walls
4. Nitrogen fixation
5. Denitrification
6. Anoxygenic photosynthesis
7. Anaerobic respiration
8. Chemoautotrophy

Table 4. *Eukaryote characters that differ only quantitatively from prokaryotic ones*

1. Many more copies of rDNA and tDNA
2. Large genome size with much DNA that probably does not code for proteins
3. Larger cell volumes
4. More repeated DNA (interspersed or tandemly arranged direct repeats, and inverted repeats)
5. More frequent multicellularity and cell differentiation
6. Sterols abundant and widespread
7. Polyunsaturated fatty acids abundant and widespread

shown in Table 3. They also differ quantitatively from prokaryotes in the respects shown in Table 4. Since there is overlap between the cell volumes and genome sizes of prokaryotes and eukaryotes (Cavalier-Smith, 1978a), and some prokaryotes do have a few repeated sequences (Calos & Miller, 1980) or sterols and polyunsaturated fatty acids (Ragan & Chapman, 1978), characters 2–7 should not, as they often have been, be regarded as distinguishing features of prokaryotes and eukaryotes.

THEORIES OF EUKARYOTE ORIGIN

The occurrence of all the 22 features listed in Table 1 in all eukaryotes in which they have been sought shows that the several million species of eukaryote must have had a common ancestor possessing all of these characters. This common ancestor, the first eukaryotic cell, could in principle have originated in any one of the four ways shown in Fig. 2.

The many features common to both prokaryotes and eukaryotes,

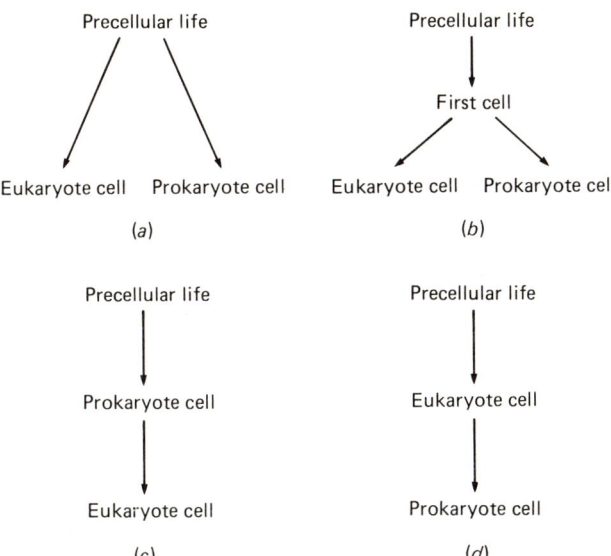

Fig. 2. The possible phylogenetic relationships between prokaryotes, eukaryotes, and precellular living systems.

such as the genetic code, the phospholipid bilayer structure of cell membranes, the mechanisms of movement of DNA replication forks, active transport, energy transduction by means of ATP and NAD, and a great deal of intermediary metabolism, indicate that the divergence between prokaryotes and eukaryotes occurred only after the origin of a cell with a modern genetic code, a DNA containing chromosome or chromosomes, plasma membrane and active transport, and efficient energy transduction and intermediary metabolism. This point is profoundly important not only because it eliminates the first possibility (Fig. 2a) but because it means that the transition between prokaryote and eukaryote is a problem, not of precellular evolution of which we have no direct knowledge, but of cellular evolution involving structures whose basic properties can actually be studied in living species.

The second possibility supposes that there was no direct transition between eukaryotes and prokaryotes but that both evolved from a common ancestor that possessed a mixture of prokaryotic and eukaryotic characters, or else was so different from both prokaryotes and eukaryotes that it would deserve neither name. On this hypothesis the fact that every extant organism has either all or none of the 22 universal eukaryotic characters of Table 1 could be explained only in three ways. One is that the very first evolutionary divergence after the formation of the first cell produced two separate lines, one with entirely eukaryotic and the other with entirely prokaryotic properties. The second is that the descendants of the first cell underwent extensive diversification but that every single line that had a mixture of prokaryotic and eukaryotic properties (or that retained the purely hypothetical ancestral properties that were neither prokaryotic or eukaryotic) became extinct leaving only the prokaryotes and eukaryotes. Both hypotheses seem highly improbable. The third, and most plausible explanation, is that all 22 characters were either simultaneously gained (Fig. 2c) or simultaneously lost (Fig. 2d). Darnell (1978) regards the differences between prokaryote and eukaryote as so great that one cannot envisage a transition from prokaryote to eukaryote (Fig. 2c), and so, like Doolittle (1978) and Reanney (1974), he implies that eukaryotes evolved by the simplification of prokaryotes (Fig. 2d). I do not see the logic of this; nothing is gained merely by reversing the direction of the transition if the problem is simply the magnitude of the differences, unless there is reason to think that the transition is easier in one direction than the other. The theory of a eukaryote origin for prokaryotes would have to explain not only how all 22 eukaryote characters were simultaneously lost, but also why this occurred: no one has even attempted to do this. Since, as discussed below, one can see advantages in the gain of all those characters but none in their loss I prefer the conventional view that eukaryotes evolved from prokaryotes (Fig. 2c). A second argument against the alternative view is that it supposes that the first cells were eukaryotic. This seems to me exceedingly improbable since almost all the 22 eukaryotic characters involve complex additions to the basic molecular machinery that must have been present in the common ancestor of prokaryotes and eukaryotes: to evolve these all directly from precellular systems would seem immensely more difficult than evolving first only the common features mentioned earlier.

In fairness to Doolittle, Darnell & Reanney one must point out that none of them explicitly suggests that the ancestor of eukaryotes had all the 22 characters in Table 1: they do not even mention, let alone discuss the origin of the majority of these characters. Their suggestions therefore do not constitute a theory of the origin of eukaryote cells but only of certain features of transcription, RNA processing and translation (i.e. only about five of the 22 characters in Table 1), and since these features never occur apart from the 17 other eukaryotic characters it is misleading to discuss them in isolation. The view that differences in the 'reactions that result in mRNA formation is the chief evolutionary basis that sets eukaryotes apart from prokaryotes' (Darnell, 1978) is entirely unjustified as it ignores the vast majority of the fundamental differences set out in Table 1.

Absence of a symbiotic theory for the origin of the eukaryote cell

The controversies in the 1970s concerning the symbiotic or non-symbiotic origins of mitochondria and plastids have diverted attention away from the much more fundamental issue of how the universal and more basic properties of eukaryotes (Table 1) evolved, which has been discussed by only a few authors, such as Allsopp (1969), Cavalier-Smith (1975, 1977, 1978b, 1980a) and Taylor (1976). One reason for the popularity of symbiotic theories is that the profound differences between prokaryotes and eukaryotes and the absence of any intermediates or organisms with a mixture of prokaryotic and eukaryotic characters seemed inexplicable in terms of a slow and gradual evolutionary process which has been the dominant view of evolution since the nineteenth century. Symbiosis seemed to offer a way of transforming a cell rather suddenly from one type to another. Though I think it probable, but not certain, that plastids did originate endosymbiotically as Schimper (1883) and Mereschkowsky (1905) first argued, it seems to me improbable that mitochondria did so (Cavalier-Smith, 1980a). But even if both organelles originated endosymbiotically this would do nothing whatever to explain the origin of the eukaryotic cell, for it would leave totally unexplained the origin of every single one of the 22 universal eukaryote properties of Table 1. There are hundreds of species of eukaryotes lacking both mitochondria and plastids. Even the theory that cilia evolved from ectosymbiotic spirochaetes and that spindle and other microtubules evolved from cilia (Sagan, 1967; Margulis, 1970) which – as I have discussed elsewhere (Cavalier-Smith, 1978b, 1981), is extremely implausible – would explain the origin only of two of the characters (microtubules and mitosis), leaving the origin of the remaining 20 a mystery.

Hence there is no symbiotic theory for the origin of the eukaryotic cell; the numerous statements to the contrary have prevented people from even discussing its origin. Proponents of symbiotic theories for the origin of one or more of the three non-universal eukaryotic cytoplasmic organelles implicitly or explicitly accept the *autogenous* origin of at least 20 of the basic eukaryote characters, and that cell compartmentation by membranes played a key role in this process. Where they differ from each other, and from those who have argued for the autogenous origin of these three organelles, is in the nature of the organism in which cell compartmentation to form nucleus, Golgi, ER, lysosomes and peroxisomes occurred, and the timing of the change. As the earlier ideas have been extensively reviewed (e.g. Taylor, 1974, 1980; Dodson, 1979) this article aims instead to develop some newer ones.

THE NATURE OF THE FIRST EUKARYOTE

Because the selective forces acting on an organism are most fundamentally determined by its mode of nutrition, it is important to try to decide how the first eukaryotic cell got its food. There are four basic possibilities: it could have been (1) a simple phototroph like a red alga, (2) a mixed phagotroph and phototroph like many chrysophyte or cryptophyte algae, (3) a simple phagotroph like an amoeba, or (4) a non-phagotrophic, non-photosynthetic osmotroph like a yeast. The first two theories in their simplest form invoke a single prokaryote ancestor capable of oxygenic photosynthesis and aerobic respiration, and derive both mitochondria and plastids autogenously. The second two theories postulate a non-photosynthetic prokaryote ancestor, and necessitate a symbiotic origin for plastids after the evolution of the basic eukaryote properties. A symbiotic or autogenous origin for mitochondria is equally compatible with all of these theories; although for theories (1) and (4) symbiosis could occur only if the cell first evolved phagocytosis, making eight possibilities in all as shown in Table 5. Of these eight possibilities the first two may reasonably be rejected on the ground that no organisms with plastids but lacking mitochondria exist, and that there is no obvious reason why if any ever did, all of them should have become extinct. Though examples of each of the remaining six exist, I shall argue that the first eukaryote was not photosynthetic, not anaerobic and not phagotrophic; and so must have been a fungus possessing mitochondria (Cavalier-Smith, 1980*a*).

There is now considerable evidence for a close similarity between the respiratory mechanisms and cytochromes of eukaryotes and of

Table 5. *The eight basic possibilities for the trophic and bioenergetic properties of the first eukaryote*

	Extant examples	Photo-synthesis	Phago-trophy	Mitochondria
1. Anaerobic alga	none	O^a	A^b	S^c
2. Anaerobic algal phagotroph	none	O	O	S
3. Aerobic alga^d	common	O	A	O
4. Aerobic algal phagotroph^d	frequent	O	O	O
5. Anaerobic fungus	rare	S	A	S
6. Anaerobic phagotroph^d	frequent	S	O	S
7. Aerobic fungus^d	common	S	A	O
8. Aerobic phagotroph	common	S	O	O

^{a} O represents presence in the first eukaryote.
^{b} A represents later autogenous evolution.
^{c} S represents later acquisition by endosymbiosis.
^{d} The four possibilities that have been seriously advocated, namely, (3) the classical autogenous view (e.g. Allsopp, 1969; Chadefaud, 1976), (6) the serial endosymbiosis theory (Margulis, 1970, Raven, 1970), (4) my 'autogenous' proposal (Cavalier-Smith, 1975, 1977), (7) the fungal theory presented here and elsewhere (Cavalier-Smith, 1980a, 1981).

the purple non-sulphur photosynthetic bacteria (e.g. *Rhodopseudomonas spheroides*) and their non-photosynthetic relatives such as *Paracoccus* (Whatley, John & Whatley, 1979). Likewise there is strong evidence for close homologies between the photosynthetic apparatus of the eukaryotic red algae and of the prokaryotic cyanophytes and between that of the eukaryotic green algae and of the prokaryotic prochlorophytes (Doolittle & Bonen, 1981). The homologies are so great that there are only two possible explanations. One is that either eukaryotes were derived autogenously from a common ancestor of the purple non-sulphur bacteria and the cyanophyte/prochlorophyte line that had both *Rhodopseudomonas*-like respiration and oxygenic photosynthesis (Cavalier-Smith, 1975, 1977; Uzzell & Spolsky, 1981), or else eukaryotic algae acquired either plastids (chloroplasts) or mitochondria or both by symbiosis. I now (Cavalier-Smith, 1980a) favour the symbiotic hypothesis for plastid origin for four reasons. (i) there is no evidence that the above mentioned common ancestor ever existed. (ii) The presence of a peptidoglycan wall and a plastid-sized genome in the cyanelles of *Cyanophora* (Herdman & Stanier, 1977) show that eukaryote cells (unlike prokaryotes) are able to acquire plastid-like organelles by endosymbiosis. (iii) The differences between prochlorophytes and cyanophytes, coupled with their respective resemblance to

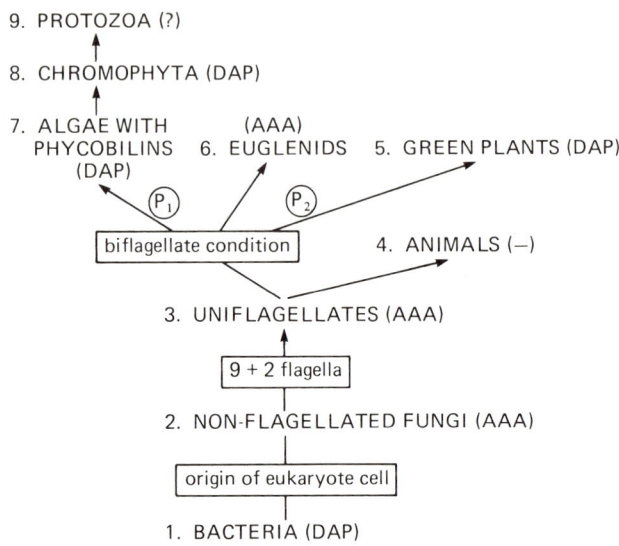

Fig. 3. Simplified phylogeny of the nine kingdoms of organisms as proposed by Cavalier-Smith (1981), showing how a eufungal origin for eukaryotes rationalizes the distribution of lysine metabolic pathways if the following assumptions are made. The diaminopimelic acid (DAP) pathway present in bacteria and all other prokaryotes was lost when DAP itself was lost from the cell wall when the wall was converted from peptidoglycan to chitin during the formation of the first eukaryote eufungus (Cavalier-Smith, 1980a). This ancestor evolved the new amino-adipic acid (AAA) pathway by reversal of the catabolic pathway for lysine. This AAA pathway is found in all the Eufungi and in their descendants the uniflagellate chytridiomycete fungi and the Euglenophyta, but has been lost in animals presumably because phagotrophy give them sufficient lysine in their food. When phagotrophic biflagellate descendants of the chytrids acquired plastids by endosymbiosis the DAP pathway was re-acquired from their photosynthetic endosymbionts and the AAA pathway was lost: endosymbiosis of cyanophytes (P_1) produced the phycobilin-containing eukaryotes, and endosymbiosis of prochlorophytes (P_2) produced the green algae. When the Euglenophyta secondarily acquired a eukaryotic green algal plastid by endosymbiosis they did not retain its DAP coding genes and so did not lose the AAA pathway. Their probable relatives the trypanosomes (here grouped with them in the Euglenida) appear to have retained the DAP pathway instead. The lysine metabolism pathway of other protozoa is uncertain (Ragan & Chapman, 1978). The Oomycete and Hyphochytrid fungi which I classify in the Chromophyta with the chlorophyll-c-containing algae on account of their flagella and mitochondrial characters (Cavalier-Smith, 1981) have the DAP pathway, which implies that they, unlike the eufungi and chytrids are derived from algae by the loss of plastids: I therefore predict that the chromophyte algae, whose lysine metabolism has not been studied, will have the DAP pathway.

green algal and red algal plastids, strongly suggests that these two plastids originated independently from prokaryotic ancestors. Since the basic eukaryotic properties of green and red algae show that the rest of their cells must have a common non-prokaryotic ancestor this implies that at least one of them acquired plastids endosymbiotically. (iv) Acceptance of the chloroplast endosymbiosis theory of

Schimper (1883) and Mereschkowsky (1905) makes it possible to explain the puzzling distribution of lysine biosynthetic pathways in eukaryotes and to construct a unified phylogeny for prokaryotes and eukaryotes which is highly plausible (Fig. 3).

Since there are reasons for thinking that plastids originated symbiotically (well reviewed by Phillips & Carr, 1977) but no positive reasons for thinking that mitochondria did so, the simplest hypothesis to explain the phylogenetic data is that mitochondria evolved autogenously and plastids were later acquired by the independent endosymbiosis of a cyanophyte and a prochlorophyte. Not only is there no reason to postulate a symbiotic origin for mitochondria, but the extensive evidence for the basically aerobic nature of the rest of the eukaryotic cell casts strong doubt (Raff & Mahler, 1975) on the idea that the first eukaryote was anaerobic. The similarities between mitochondria and plastids may reasonably be attributed to convergence (Gillham & Boynton, 1981) rather than a common mode of origin.

If these arguments are accepted we are only left with two possibilities: that the first eukaryote was either an aerobic fungus feeding osmotrophically or else an aerobic phagotroph. The fact that phagocytosis is unknown in prokaryotes and is not even a universal property of eukaryotes casts some doubt on the traditional idea that it is a primitive character present in the first eukaryote. As I emphasized earlier (Cavalier-Smith, 1975) extensive phagocytosis could not have evolved until after the origin of exocytosis which is needed to replace the plasma membrane internalized in a phagosome. I therefore argued that exocytosis evolved before phagocytosis and was the pivotal step in the origin of eukaryotes that made all subsequent changes possible. The origin of exocytosis still seems to me the pivotal step, but I now think it more probable that secretion of wall materials rather than digestive enzymes may have been its more important original function and that the first eukaryote was a fungus with a cell wall rather than a naked phagotroph. Phagotrophy and the amoeboid or ciliary motility that normally accompanies it seem to me to be complex and advanced rather than primitive eukaryotic characters. An indication of the complexity of amoeboid motion is that the slime mould *Dictyostelium* has 17 actin genes (Ng & Abelson, 1980), whereas the Hemiascomycete yeast *Saccharomyces*, which I think may be very similar to the ancestral eukaryote, has only one actin gene.

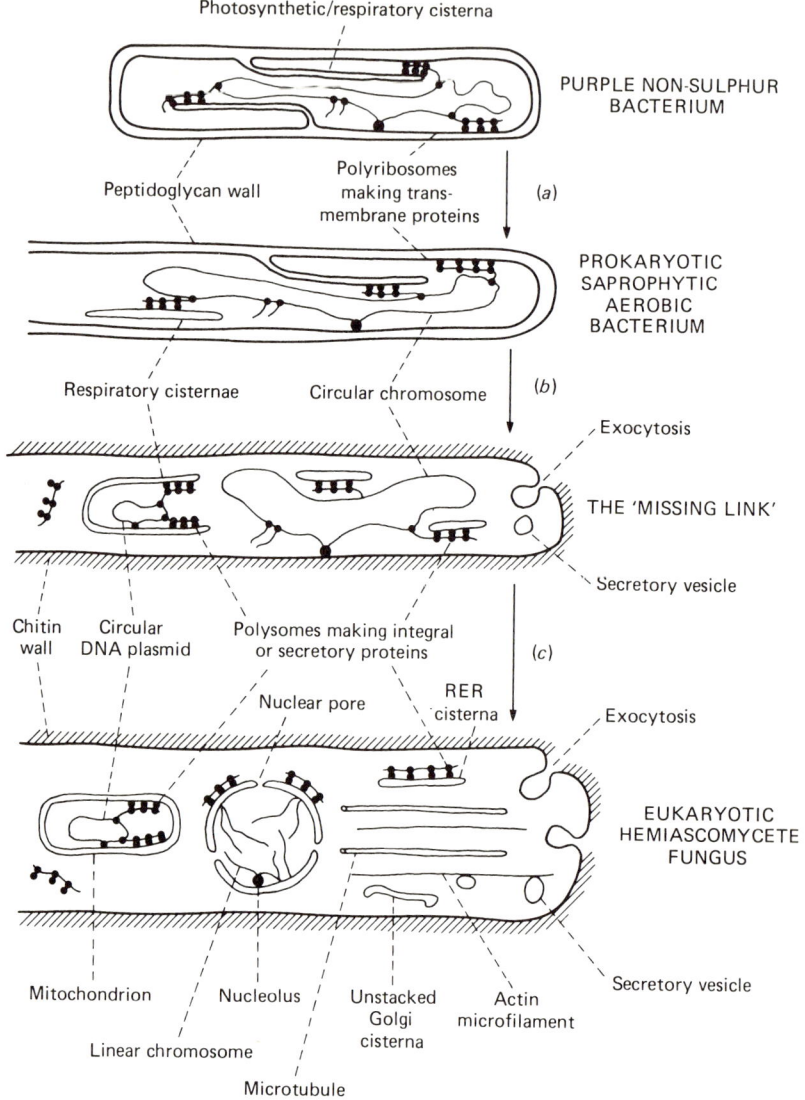

Fig. 4. Outline of the fungal theory of the origin of eukaryotes. A eukaryote fungus like the filamentous Hemiascomycetes *Eremascus* or *Dipodascopsis* can be derived in three main stages from a purple non-sulphur bacterial ancestor (e.g. *Rhodopseudomonas spheroides*) possessing plate-like cisternae capable of both anaerobic photosynthesis and mitochondrion-like aerobic respiration. (*a*) Photosynthesis is lost to produce a saprophytic bacterium. Integral membrane and secretory proteins are made by ribosomes attached both to the plasma membrane and to the respiratory cisternae: this intermediate stage is still fully prokaryotic and differs from *Paracoccus* only by having distinct flattened respiratory cisternae and a filamentous mode of growth. (*b*) Mutational loss of the enzyme making muramic acid from N-acetyl glucosamine led to the conversion of the bacterial peptidoglycan cell wall to a fungal wall of chitin (a polymer of *N*-acetyl glucosamine), and to the origin of exocytosis to allow the more rapid secretion of chitin at the growing hyphal tip. The origin of exocytosis was coupled

The fungal theory of eukaryote origin

If we therefore reject not only what Margulis (1970) calls the Botanical Myth (that the first eukaryotes were photosynthetic algae), but also the Zoological Myth that they were phagotrophic amoebae we are left with the fungal theory of eukaryote origins (Cavalier-Smith, 1980a), which allows a very much simpler transition from prokaryote to eukaryote. Fungi, like most non-photosynthetic prokaryotes, have cell walls and feed on dissolved organic materials, so no major change in way of life would be needed during the transition from an aerobic saprophytic bacterium to an aerobic saprophytic fungus. The only changes that would be needed are a change in wall chemistry and the acquisition of the basic eukaryote properties of Table 1, plus cell compartmentation to form mitochondria, and the acquisition of histones by nuclear DNA, as summarized in Fig. 4.

The fungi of the groups Zygomycetes, Hemiascomycetes, Euascomycetes and Basidiomycetes which I collectively call the Eufungi (Cavalier-Smith, 1981) have many features that suggest they are primitive eukaryotes. Like prokaryotes they lack 9 + 2 cilia, centrioles, plastids, endocytosis, and the capacity to harbour cellular endosymbionts. They have small cell volumes and small genome sizes that overlap with those of prokaryotes and they lack appreciable repetitive DNA and the nucleosomes of the Ascomycetes are smaller than in any other eukaryotes. Unlike almost all other eukaryotes the Golgi cisternae of Eufungi are not stacked into dictyosomes. Ascomycetes and Basidiomycetes synthesize a far narrower range of polyunsaturated fatty acids than do other eukaryotes; like prokaryotes they are unable to make the more complex ones (Ragan & Chapman, 1978). The Zygomycetes and

Caption to Figure 4 (*cont.*)

with the specialization of membrane-bound cisternae in three different directions: (*i*) as respiratory cisternae, (*ii*) as rough endoplasmic reticulum, and (*iii*) as smooth secretory vesicles or chitosomes; at this stage the plasma membrane could diverge from the cisternae by losing its capacity to bind ribosomes. (*c*) The origin of microtubules and microfilaments to move the secretory vesicles to the hyphal tip led to compartmentation of the cell to protect the DNA and mRNA from damage by the resulting cytoplasmic streaming; rough endoplasmic reticulum cisternae fused together to generate the nuclear envelope and perinuclear cisterna, whilst respiratory cisternae folded back on themselves and fused to generate a double-membraned mitochondrial envelope inside which were trapped ribosomes and plasmids coding for rRNA, tRNA and a handful of respiratory membrane integral proteins. The outer mitochondrial membrane lost its respiratory functions because of its separation from the ribosomes inside the mitochondrion, making the integral membrane respiratory proteins. The basic theory is described in more detail elsewhere (Cavalier-Smith, 1980a).

Hemiascomycetes have relatively simple methods of sexual reproduction and primitive mitotic mechanisms in which the nuclear envelope remains intact, the nucleolus usually persists, and a chromatin condensation cycle is often absent. Overall the Hemiascomycetes such as *Saccharomyces* have the largest number of apparently primitive characters, including in some species a mitotic spindle that persists during interphase. Their common ancestor could have been the first eukaryote.

COADAPTATION OF CHROMOSOME REPLICATION AND SEGREGATION

It is often said that mitosis evolved because of the need to segregate the larger genomes and the several chromosomes of eukaryotic cells for which the prokaryotic segregation mechanism was not sufficient. In my view, however, (Cavalier-Smith, 1975, 1980a), larger genomes and a multiplicity of chromosomes per genome could not have evolved before the origin respectively of plural replicons and mitotic segregation and therefore could not have provided a selective force for the establishment of mitosis. It has often been assumed but never explained why the prokaryote mechanism might be unable to segregate efficiently large genomes or chromosomes. The chromosome of *E. coli* is about five times and those of some blue-green algae about twenty times the size of those of *Saccharomyces*. There seems to me nothing about the prokaryotic *segregation* mechanism by membrane and wall growth that would in principle prevent the segregation of genomes 10, 100 or even 1 000 000 times larger than the largest found in prokaryotes. On the other hand one can readily see why the prokaryote *replication* mechanism prevents the evolution of large genomes, as indicated below.

The limitations of the prokaryote replication pattern

The prokaryote genome is organized as a single replicon with a single origin of replication. This means that the time taken for the two replication forks to move from origin to the terminus (Fig. 5) limits the rate of reproduction and is therefore of key evolutionary importance. Increases in genome size will increase the minimum length of the cell cycle T_c in direct proportion. This will similarly

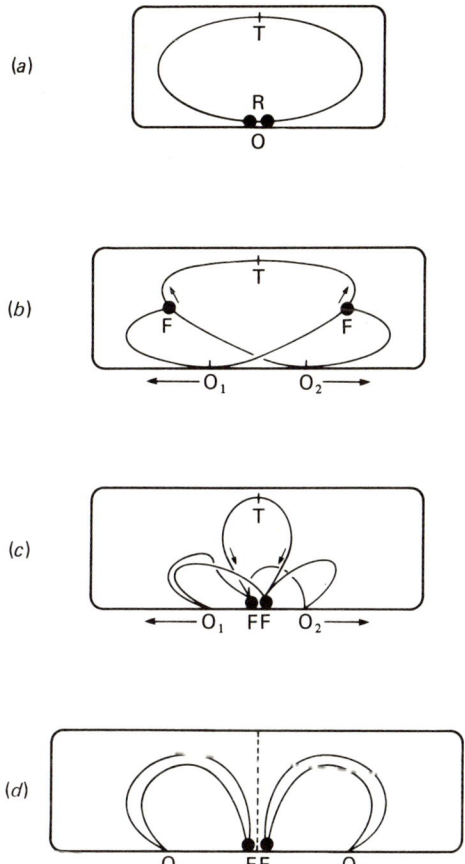

Fig. 5. DNA replication and segregation in the bacterial cell cycle. (*a*) An unreplicated circular chromosome is attached by its origin (O) to the cell membrane. The replication machinery (R) or 'replisome' (Bleecken, 1971) binds to the origin and bidirectional replication begins. (*b*) The two replication forks (F) move away from each other (small arrows) towards the terminus (T), while localized membrane growth (large arrows) separates the two daughter origins (O_1 and O_2). For clarity the replication forks are shown free in the cytoplasm: in reality they are probably attached to the plasma membrane by the replication machinery. (*c*) The replication forks 'move' because the unreplicated DNA is pulled (small arrows) through the stationary replisome. (*d*) The terminus meets the replisome and its replication is completed. It is then in the correct position to give a direct signal for division by septum formation (dotted line) in the plane separating the two daughter termini. In life the DNA molecule is very much longer in relation to cell size than shown. Segregation without chopping up the DNA therefore depends also on the folding of the DNA.

decrease the specific reproductive rate, r, which is inversely proportional to cell cycle length as indicated by the relationship $r = \ln lb/T_c$ where l is mean cell viability and b is the number of daughter cells produced per division (Cavalier-Smith, 1980*b*). Since selection acts to maximize effective reproductive rates it will strictly limit

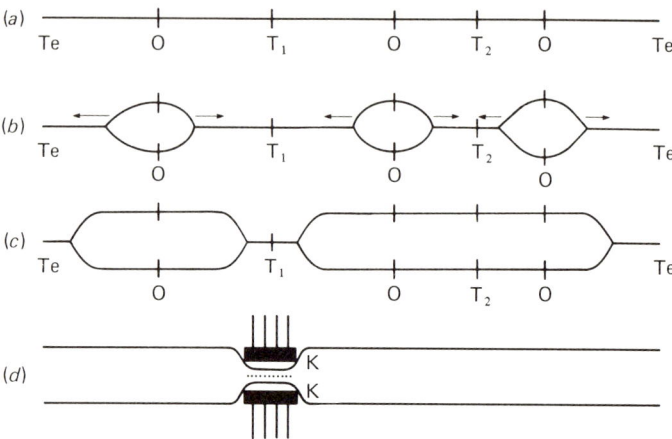

Fig. 6. DNA replication in a eukaryote chromosome consisting of three replicons. (a) Unreplicated chromosome with three origins (O) and four termini (T, Te). (b) Simultaneous initiation of replication at the three origins creates six replication forks which move bidirectionally (arrows) towards the termini. (c) Replication terminates sooner at T_2 than at the other termini because the forks have less far to travel. (d) Special mechanisms (Cavalier-Smith, 1974) are needed for termination at the telomeres (Te) and to ensure that the DNA to which the kinetochore (K) attaches remains unreplicated, or that the two daughters remain firmly associated throughout the G_2 phase of the cell cycle and until mitosis occurs.

evolutionary increases in prokaryote genome size beyond that needed to maintain viability at a reasonably high level. In eukaryotes, where there are many replicons per genome (Fig. 6), their number can be indefinitely increased as the genome size increases, so this limitation on genome size is absent. This explains why the genome size of some eukaryotes can increase so readily in evolution, in some cases to a level 50000 times that of the largest prokaryote genomes (Cavalier-Smith, 1978a). In *E. coli* the minimum travel time of replication forks, T_f, is 40 min. If the genome were increased to the maximum level known in eukaryotes T_f would become 10^4 h or a year, a prohibitively long time. Thus large genomes could not evolve until after the origin of plural replicons.

The strength of selection for minimum replication times in prokaryotes is shown by the way rapidly growing bacteria reduce their cell cycle time T_c below the fork travel time T_f by initiating a series of rounds of replication at the origin of replication before the earlier replication forks have reached the terminus. It might be thought that this mechanism could allow considerable increase in genome size without decreasing the growth rate, but this is not so, for as shown in Fig. 7 it generates highly branched chromosomal DNA in which genes near the origin are present in more copies per cell than those near the terminus: the origin/terminus copy ratio (O/T) will increase with growth rate according to the formula $O/T = 2^{Tf/Tc}$, so in an *E. coli* with a generation time of 20 min this ratio is 4. If however a 20-min generation time were to be sustained in the face of a mere tenfold increase in genome size one would require an origin/terminus ratio of 2^{20} (i.e. over 4 000 000) which would be quite impossible, for the average chromosome mass increases exponentially with the degree of branching and would have become about a million times the genome size – 10^4 times the mass of an actual *E. coli* cell! This *reductio ad absurdum* makes it clear that multiple initiation of replication at a single origin can only slightly offset the basic tendency in prokaryotes for cell cycle length to increase in direct proportion to genome size, which explains why even the chromosome of the prokaryote with the largest genome is only four times the length of that of *E. coli*: even a modest fourfold increase in genome size would require an O/T ratio of 256 for a doubling time of 20 min. Not surprisingly, prokaryotes with large genomes never reproduce this fast – their T_c is of the order of 1 day.

THE ORIGIN OF MITOSIS

Though the prokaryote segregation mechanism may not itself prevent an increase in genome size, it may be indirectly responsible for this limitation because it prevents the evolution of plural origins of replication. Transposition of DNA sequences is such a common event in prokaryotes (Calos & Miller, 1980) that in most and perhaps all species origins of replication must from time to time occur at more than one site on the chromosome. The fact that organisms with plural origins are not observed reveals strong stabilizing selection against them. I argue that this arises because the prokaryote segregation and division control mechanisms depend on the existence of single origin and a single terminus of replication as indicated in Fig. 7. Mutants with plural origins and termini would be unable to segregate chromosomes accurately and would suffer chromosome fragmentation if replication of one terminus led to division before other termini had replicated. It is hard to see how the branched chromosomes of rapidly growing bacteria could be accurately segregated if origins and termini existed at more than one chromosomal locus.

The same considerations may preclude the evolution of more than

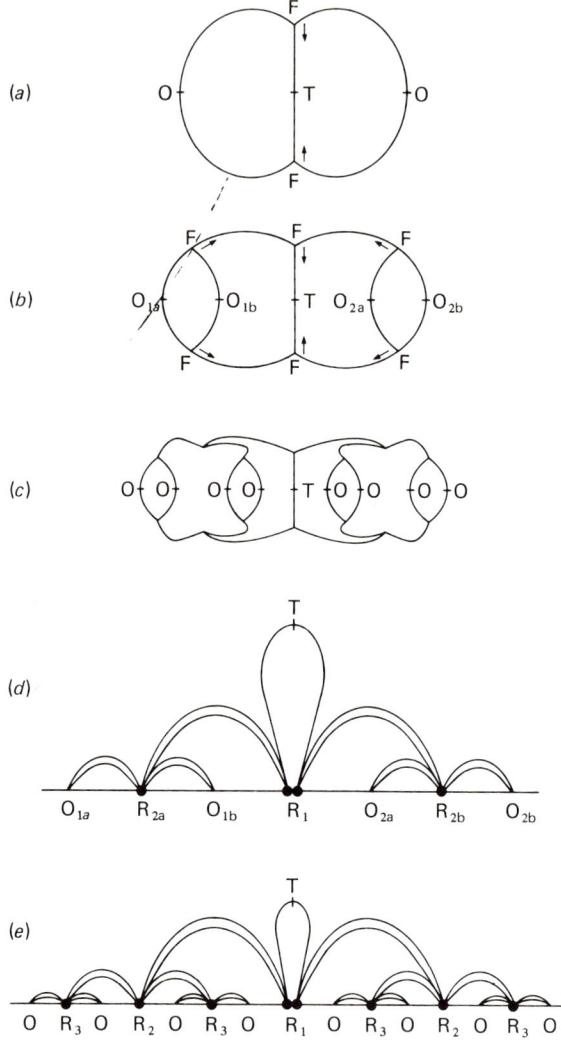

Fig. 7. Diagram showing the consequences of the decreased time interval between successive initiations of DNA replication that occurs in faster growing bacterial cells. (a) In slowly growing cells (those where $T_c \geq T_f$) there is normally only a pair of replication forks (F), which move as shown by arrows, and a pair of origins (O), as in Fig. 5. (b) In faster growing cells (where $T_c < T_f$) a second round of replication is initiated before the first is terminated so there are six forks and four origins. (c) In still faster growth a third round of replication can be initiated giving 14 forks and 8 origins (O), but still only one terminus (T). The total amount of DNA per replicating chromosome therefore increases markedly with growth rate (Maaløe & Kjeldgaard, 1966); so does the ratio between the number of copies of a gene near the origin and those near the terminus. Where there are several origins to be segregated it is essential that the correct ones go into each daughter cell; for example in Fig. 7(b) O_{1a} and O_{1b} must go into the same daughter during the cell division following the next termination, or the chromosome will break. (d) Such segregation can be achieved (Dingman, 1974) if the origins are attached in a row in the correct order on the rigid cell membrane. R_1, R_{2a} and R_{2b} will all become sites of septum formation. Division at R_1 will automatically include O_{1a} and O_{1b} in

one chromosome per genome in organisms with the prokaryote segregation and division mechanisms.

Mitotic segregation, in marked contrast, provides a way of segregating chromosomes irrespective of their numbers and of the numbers of replicons they contain. The origin of mitosis, therefore, represents a key step in the origin of the eukaryote genome, because thereafter the number of chromosomes and the number of replicons per chromsosome would be free to increase to an optimum level. I suggest that the basic mitotic mechanism, using microtubules and perhaps also microfilaments for chromosome segregation, originated as shown in Fig. 8 when the early protofungus still had a single circular chromosome and lacked a nuclear envelope. There may initially (Fig. 8b) have been no centrosomes or centrosomal microtubules, but only a single kinetochore microtubule attached to each daughter chromosome that was able to slide relative to actin filaments attached to the plasma membrane. I suggest that kinetochores evolved from the prokaryotic chromosome terminus and its attachment mechanism to the plasma membrane. This would allow chromosome attachment to the membrane to persist during the transition from segregation by membrane growth to that by microtubule movement, and make the transition smoother and less error prone. Moreover it would automatically ensure that chromosome segregation could not begin until the terminus, and therefore the whole chromosome, was replicated.

This hypothesis is more attractive than earlier ones postulating that centrosomal microtubules evolved first (Pickett-Heaps, 1974; Cavalier-Smith, 1975, 1978b), because with centrosomal microtubules only, and no specific attachment sites for chromosomes on either the plasma membrane or the primitive spindle, segregation would surely often be unsuccessful. If, as the fungal theory proposes, microtubules and microfilaments both evolved efficient sliding mechanisms for the transport of secretory vesicles before the origin of mitosis, then one can invoke them at the earliest stages of mitotic evolution and not have to rely on the rather vague pushing forces resulting from microtubule assembly that earlier theories postulated.

Caption to Figure 7 (*cont.*)

one daughter and O_{2a} and O_{2b} in the other, as is required, but division at R_2 would break the chromosome. Therefore the signal to divide must be strictly local: when T meets R_1 it allows division to occur there only. Division at R_{2a} and R_{2b} can only occur after the two daughter termini have contacted them. (*e*) The same mechanism can ensure the correct segregation of still more origins: after the first round of replication is complete, division occurs at R_1; after termination of the second round at R_2; and after the third round at R_3.

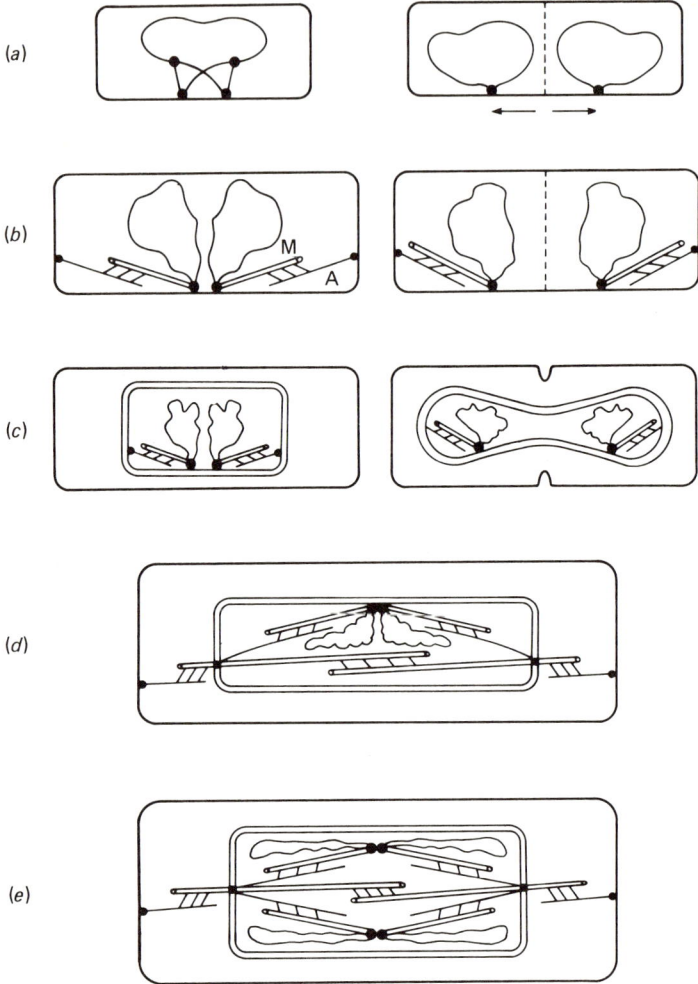

Fig. 8. A theory for the origin of mitosis. (a) Bacterial chromosome segregation by wall and membrane growth (b) Segregation by a kinetochore microtubule (M) sliding relative to filamentous actin (A) attached to the rigid fungal cell surface. If microtubule and microfilament sliding mechanisms had been perfected earlier for secretory vesicle transport, the key step in the origin of mitosis would be the evolution of kinetochores, either by mutation of the membrane attachment site of the bacterial replicon terminus to form a microtubule nucleating centre (Cavalier-Smith, 1978b) or the binding of this site to a separate pre-existing membrane-associated microtubule nucleating centre: the chromosome would stay attached to the plasma membrane throughout the transitional stage. (c) When the nuclear envelope evolved by invagination and budding off of the regions of the plasma membrane to which the chromosome was attached and the fusion of the resulting cisterna with rough endoplasmic reticulum cisternae (Cavalier-Smith, 1975, 1980a), the two chromosomes would be segregated to the opposite ends of the nucleus by an unchanged mechanism. (d) New mechanisms must have evolved to regulate nuclear growth during the cell cycle, while nuclear division was made more efficient by the origin of centrosomes and a framework of centrosomal microtubules, perhaps by the duplication of kinetochores to form an extra microtubule nucleating centre still embedded in the nuclear envelope but no longer attached to the DNA. Perhaps because of its

I now suggest that centrosomes and centrosomal microtubules evolved only when the nuclear envelope formed and the chromosome lost its attachment to the plasma membrane. When this first happened, in order to protect the DNA from shearing damage (Cavalier-Smith, 1980a), the kinetochores could be segregated by the sliding of their microtubules relative to actin attached to the inner surface of the nuclear envelope and daughter chromosomes would be enclosed in separate nuclei by the division of the nuclear envelope (Fig. 8c). Though this would achieve chromosome segregation it would not in itself provide an efficient means for nuclear division and the segregation of daughter nuclei into separate daughter cells: I suggest that this was the original function of centrosomal microtubules. Centrosomes could evolve if the primitive kinetochore underwent an evolutionary doubling such that both copies remained attached to the nuclear envelope but one lost its attachment to the chromosome, and evolved into a centrosomal plaque like that of *Saccharomyces* (Fig. 8d) by becoming a bipolar structure with a microtubule attached on each side. Such a structure could segregate the daughter nuclei into daughter cells by essentially the same mechanism used by its kinetochore ancestor to segregate daughter chromosomes into daughter nuclei. On this view the two-component eukaryote mitotic spindle is seen as having a two-stage origin; first the origin of kinetochore microtubules to segregate chromosomes, and then of centrosomal microtubules to segregate nuclei. The origin of centrosomes with several microtubules or microfilaments or both would allow chromosome number to increase by centromere transposition. The conventional idea that chromosome number increased before the origin of mitosis (Allsopp, 1969; Stanier, 1970) on the other hand creates immense difficulties for the transitional stages.

The above scheme suggests how the actual mechanisms of mitosis could have evolved with no major discontinuity. However this alone would not make it possible for the eukaryote condition of plural replicons and chromosomes to evolve: there also had to be changes

Caption to Figure 8 (*cont.*)

lack of attachment to DNA the new centrosome would be bipolar, able to nucleate microtubule growth into both nucleus and cytoplasm: the cytoplasmic centrosomal tubules could interact with cytoplasmic actin and pull the two halves of the nucleus apart, while the intranuclear centrosomal tubules could push them apart. (*e*) Finally, increase in the number of centrosomal microfilaments and microtubules would allow segregation of more than one chromosome, so the number of chromosomes could increase by fragmentation and kinetochore transposition; thus producing a mitotic apparatus like that seen today in yeast (Byers & Goetsch, 1974).

in the mechanisms co-ordinating the initiation and termination of DNA replication with respect to cell growth and division. Even though the mechanisms of co-ordination are hardly beginning to be understood, one can see that there are distinct requirements in prokaryotes and eukaryotes. In bacteria the replication of the chromosome terminus releases an earlier block to division (Donachie, 1974), thus allowing the cell to divide: in eukaryotes the existence of numerous replicon termini complicates matters for there must be some mechanism to ensure that chromosome division (and nuclear and cell division) does not occur until after every replicon has been terminated. It is also possible that some change might be needed in the control of replicon initiation to achieve simultaneous initiation of many separate replicons at the beginning of S-phase. As soon as such mechanisms evolved, a cell with mitosis would be free to evolve plural replicons, by the transposition of origins by transposable elements. This would be strongly favoured by selection because it would shorten DNA replication times.

THE ORIGIN OF SEX: SYNGAMY AND MEIOSIS

The phylogenetic distribution of the sexual process suggests that it evolved very early in eukaryote evolution and that present-day asexual eukaryotes have been derived from sexual ancestors (Cavalier-Smith, 1975; Heywood, 1975). Sex consists of alternating syngamy and meiosis. The origin of syngamy presents no problem, for somatic cell fusion is common in ascomycete fungi. Mycelial organisms in which hyphae of different genetic origins readily intermingle during normal vegetative growth seem therefore to be more favourable for the origin of sex than are unicellular organisms: no new mechanism of cell fusion is needed. The selective advantage for fusion may have been twofold: (a) doubling in cell volume produced by diploidy would be advantageous during starvation (Cavalier-Smith, 1980*b*), which may explain why syngamy in protists including fungi normally follows nutrient depletion, and (b) complementation between different strains could allow survival under conditions of partial starvation where neither alone could survive.

What is less obvious (Maynard Smith, 1978) is how and why meiosis evolved. Now that one realizes the great evolutionary importance of cell size and cell cycle length (Cavalier-Smith, 1978*a*, 1980*b*), it is plausible that the original selective advantage of a reduction division was the halving of cell volume that haploidy allows compared with diploidy. If when the diploid resting spore undergoes germination in a nutrient-rich environment it reverts to a haploid cell volume it can produce twice as many daughters at the next division as if it remained diploid, so it will immediately have a twofold selective advantage over a competitor that does not undergo reduction. It will have an additional advantage because smaller cells can normally reproduce faster than larger ones when resources are not limiting. When nutrients run out there will again be an advantage in large size and in fusing in pairs. So the best tactics for an early eukaryote fungus subject to fluctuations in nutrient level would be to alternate between the large-celled diploid and small-celled haploid state. In the early evolution of sex, far from there being a twofold cost of meiosis, there would have been at least a twofold cost of not undergoing meiosis.

The origin of the meiotic mechanism is easiest to explain if it evolved in an early uninucleate cell that still had only a single circular chromosome, but where mitosis had evolved at least to the stage of Fig. 8b with kinetochore microtubules. It has often seemed a puzzle (Maynard Smith, 1978) that meiotic reduction requires two rather than just one cell division. The reason for this is readily understood if meiosis arose by modification of the normal division cycle. In the normal mitotic cycle the occurrence of mitosis must initiate a block to division that can only be relieved by initiating and completing a round of DNA replication. If syngamy originally occurred as is usual in protists (Goodenough, 1980) in G1 of the cell cycle following this block, then no cell division could occur until after further DNA replication had produced four chromatids. In principle one can imagine two ways to modify this mechanism to achieve reduction. One is the evolution of a new mechanism to switch on division independently of replication, the second is to delay the imposition of the block to cell division until after two divisions, rather than the normal one, have occurred. I would argue that delaying an existing process is evolutionarily simpler than evolving an entirely new one and that this is why meiosis has two divisions.

The second essential innovation to achieve reduction is chromosome pairing. I suggest that pairing and the delay in the block to division both arose by modification of the behaviour of the centromere/kinetochore region of the chromosome. I postulate that in normal cell division the anaphase separation of centromeres is what provides the signal to block further division until after a new round of replication. A mutation preventing anaphase separation during the first meiotic division would therefore prevent the onset of this block and so enable the second meiotic division to occur without an intervening DNA replication. If in addition the unseparated centromeres acquired the capacity to pair prior to the growth of kinetochore microtubules then one would have a functional meiosis: clearly there must also be a coupling between centromere pairing and the delay of centromere separation. My assumption that it is the centromere, rather than the telomeres or some other part of the chromosome, that plays the key role in meiotic segregation is supported by the finding that a 1.6 kilobase yeast centromeric DNA sequence, when inserted into the circular yeast plasmid, enables it to be meiotically segregated (Clarke & Carbon, 1980).

In view of the considerable uncertainty over the selective advantages or disadvantages of recombination (Maynard Smith, 1978) it is perhaps as well that one can explain the origin of syngamy and meiosis without having to invoke recombination or crossing over at all. However, prokaryotic chromosomes have such a high propensity for recombination that as soon as meiotic kinetochore pairing originated, homologous crossing over between the two paired chromosomes would inevitably occur. Since however odd numbers of crossovers would produce circular dimers that would be fragmented at segregation as discussed by Cavalier-Smith (1975) and shown in Fig. 9a, there would be strong selection against crossing over. During this period any mutation that converted the circular chromosomes into linear molecules would be strongly selected, for it would restore viability to cells with odd numbers of crossovers: a possible one-step mechanism for linearization is shown in Fig. 9.

THE ORIGIN OF HISTONES AND NUCLEOSOMES

The tight winding of the DNA around the histone core of the nucleosome (Klug *et al.*, 1980) will stabilize it in that conformation. One consequence of this is that double strand breaks in DNA, especially if they were slightly staggered, could be more easily repaired by DNA ligase than in prokaryotes where the broken ends of the DNA molecule could separate and probably never be rejoined, so killing the cells. The prokaryote DNA gyrase, which actively introduces negative supercoils into DNA, works by intro-

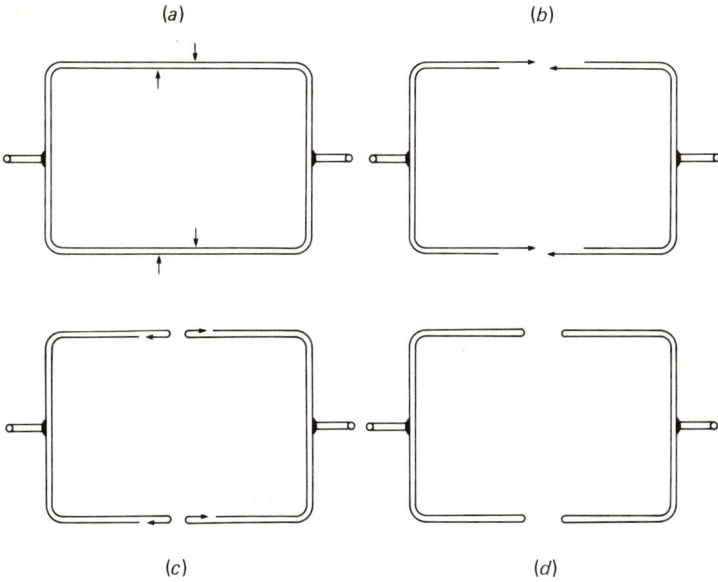

Fig. 9. The origin of linear chromosomes. (a) An odd number of sister chromatid exchanges during mitotic or meiotic segregation, or of meiotic crossovers between any two circular chromatids, would produce a circular dimer with two kinetochores which would be fragmented at anaphase, probably killing one or both daughters. Both daughters would be immediately rescued if an endonuclease made staggered cuts (indicated by arrows) in two homologous inverted repeats on opposite sides of the dimer. (b) Segregation would then pull apart the two chromosomes without breakage and the single-stranded ends could fold back on themselves. (c) Base pairing could form terminal hairpins. (d) These could undergo ligation by DNA ligase to form linear chromosomes with covalently cross-linked ends as found in *Saccharomyces* (Forte & Fangman, 1979) and postulated to be a universal property of eukaryote telomeres (Cavalier-Smith, 1974; Bateman 1975). Both the nicking and the ligation could have been achieved by a mutant DNA transposase, like those that carry out DNA splicing at the inverted repeat termini of transposons (Calos & Miller, 1980); such a mutant transposase could thereafter function in telomere replication by combining the functions of the ligase and sequence-specific nuclease postulated by the telomere replication theory (Cavalier-Smith, 1974; Bateman, 1975).

ducing temporary double strand breaks into the DNA (Mizuuchi, Fisher, O'Dea & Gellert, 1980). It seems probable that when microfilament- and microtubule-based cytoplasmic transport evolved in the ancestral eukaryote the resulting cytoplasmic shearing motions would tend to pull apart transiently broken ends with lethal consequences. Duplications of the gene for the prokaryote DNA binding protein HU (Rouvière-Yaniv & Gros, 1975; Haselkorn & Rouvière-Yaniv, 1976), leading to overproduction of the protein – as is characteristic for duplications (Hartley, 1979) – would be selected for, as it would allow it to bind all along the DNA rather than just at intervals, which would more effectively prevent the separation of broken ends. The duplicated genes could rapidly diverge to produce the five different nucleosomal histones, under selection to allow continued transcription and replication despite the greater protein/DNA ratio of the chromosome. The clustering of the histone genes (Kedes, 1979) is strongly indicative of such an origin by gene duplication.

Once nucleosomes had evolved the DNA would be stabilized in negative supercoils by bound histones, so there would be no further need for the active negative supercoiling by DNA gyrase which could be dispensed with, or perhaps evolve into a eukaryotic swivel DNA topoisomerase (Baldi, Benedetti, Mattoccia & Tocchini-Valentini, 1980). In mitochondria where microtubules and microfilaments are absent there would be no selection for histones, so both a bacterial kind of DNA binding protein (Caron, Jacques & Rouvière-Yaniv, 1979) and gyrase (Castora & Simpson, 1979) are retained. During the origin of nucleosomes there must have been rapid coevolution of the transcription system, leading to a divergence between prokaryotic and eukaryotic RNA polymerases, and the origin of separate polymerases for mRNA, tRNA and rRNA, and also of the DNA replication system leading to a shortening of the Okazaki fragments to the length of DNA contained in a single nucleosome (Tseng et al., 1979). Since then stabilizing selection will have strongly conserved the basic features of nucleosomes, and transcription and replication, for the same basic reason as the genetic code is conserved, namely the great difficulty of simultaneously changing numerous co-adapted molecules. The reason why the mitochondrial code was able to diverge from the normal one may be related to the very small number of molecules coded by mitochondrial DNA (Cavalier-Smith, 1980a).

A second advantage of histones is that they would also protect against the harmful effects of single strand breaks. In the absence of histones a single strand break leads to the unfolding of that whole supercoiled domain. Since in bacteria each supercoiled domain is about 2% of the whole chromosome (Worcel & Burgi, 1972), this would produce a loop of about 20 μm. In prokaryotes this need not be serious but in eukaryotes with cytoplasmic streaming or vesicle transport it could easily lead to DNA breakage and death. Single-strand breaks could also increase the recombination frequency in zygotes by providing a substrate for recombination enzymes (Radding, 1978) which as discussed above would be disadvantageous until linear chromosomes had evolved. A further advantage of the evolution of histones may therefore have been the reduction of recombination. The fact that histones are bound to DNA gives considerable protection against nucleases (Seale, 1976) and must markedly cut down the frequency of both single and double strand breaks.

It seems significant that the only eukaryotes that lack histones, the free-living dinoflagellates (the Dinophyceae: Loeblich, 1976), protect their chromosomes from any direct contact with microtubules and microfilaments. They share with the parasitic dinoflagellates (the Syndiniophyceae; Loeblich, 1976) and the parabasalian flagellates, both of which have histones, a unique mitotic mechanism in which the spindle microtubules are entirely extra-nuclear and the chromosomes are attached to the microtubules by kinetochores embedded in the nuclear envelope (Heath, 1980). This protects them from shearing during both mitosis and cytokinesis. I suggest that the common ancestor of the free-living dinoflagellates suffered a chromosomal deletion eliminating the clustered histone genes, but was able to survive because of its unusual mechanism of mitosis which may be regarded as a preadaption for the successful loss of histones. Loss of histones in other eukaryotes where the mitotic spindle is in the same compartment as the DNA would probably be lethal. The alternative idea that the histone-less dinoflagellates are a primitive missing link between prokaryotes and eukaryotes (Dodge, 1966; Sleigh, 1979) does not seem plausible. Their cytoplasmic structure and that of the Parabasalia is at least as complex as that of any other eukaryote, and they have normal meiosis (Beam & Himes, 1980). Their DNA is fully eukaryotic in its sequence organization, having large amounts of repetitive DNA (Allen, Roberts, Loeblich & Klotz, 1975) and differs from that of prokaryotes and other eukaryotes in containing large amounts of hydroxy methyl uracil (Rae & Steele, 1978). It is clearly important to determine how histoneless dinoflagellates negatively supercoil their DNA. Do they have a nuclear DNA gyrase? If so, does it show signs of having evolved from the mitochondrial gyrase (presumably coded by the nucleus), from the prokaryote gyrase, or from some other eukaryotic topoisomerase? Phylogenetic and *a priori* arguments both make it unlikely that histones originated to achieve chromosome condensation during mitosis. Mitotic chromosome condensation is absent in most Hemiascomycetes (Heath, 1980), which as discussed earlier show many primitive characters yet have proper nucleosomal histones (Morris, 1980). *A priori* one would expect that it is simplest initially to evolve a system that is constant during the cell cycle, and only later to evolve extra controls such as histone phosphorylation (Bradbury, Inglis & Matthews, 1974) so as to modulate it at different stages. The fact that most Hemiascomycetes not only have a constant degree of chromatin condensation during the cell cycle, but also normally have a persistent nucleolus and nuclear envelope, and presumably also nuclear matrix, during mitosis, and that some such as *Saccharomyces* have a persistent spindle during interphase (Heath, 1980), strongly suggests that they have retained many primitive eukaryote characters. Anyone wishing to argue that these are derived characters would have to explain why many eufungi and also Euglenoids lost the more widespread cycle-modulated properties, and why and how the first eukaryote would have evolved a more elaborate mechanism requiring several additional controls.

I also doubt the validity of my earlier suggestion that histones originated to facilitate meiotic pairing and segregation (Cavalier-Smith, 1975). They would probably have helped meiotic segregation, by preventing single strand breaks from leading to unfolding and the breakage of DNA during cytokinesis just as in mitosis; but it is not obvious how they might help pairing. Moreover since free-living dinoflagellates have effective meiotic sexual mechanisms (Heywood, 1975; Beam & Himes, 1980) but no histones, it is implausible to invoke sex in explaining the origin of histones.

Nucleosome size and structure may prove to be of great value for understanding eukaryote phylogeny. I suggest that the reason why ascomycete fungi have the smallest nucleosomes of any eukaryote (Kornberg, 1977) is that they have retained the primitive condition of the first hemiascomycete eukaryotes, and that diversification of their descendants led to increases in nucleosome linker length as more complex functions, and especially more complex folding patterns, for chromatin evolved.

THE ORIGIN OF THE NUCLEAR ENVELOPE AND PORE COMPLEXES

The chief problem in the origin of the nuclear envelope is not the fusion of cytoplasmic cisternae to enclose the chromosomes, whether to protect them from damage by cytoplasmic shearing (Cavalier-Smith, 1980*a*) or to maintain higher concentration of nuclear or cytoplasmic macromolecules or complexes unable to traverse the pores (Cavalier-Smith, 1975). It is rather how to explain

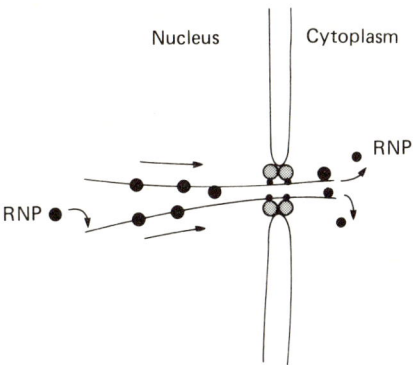

Fig. 10. Sliding filament or conveyor belt model for the active transport of ribonucleoprotein (RNP) particles through the nuclear pore complex. It is suggested that RNP is bound in the nucleus to actin microfilaments which are continually sliding from the nucleus to the cytoplasm and which may correspond with the 3–6 nm axial filaments seen in the electron microscope. (Franke, 1974; Maul, 1977). The filaments could slide relative to the eight protein subunits of the pore complex annulus, which I suggest are myosin- or dynein-like ATPases that actively push the actin filaments just as do muscle cross-bridges. The pore complexes are known to have ATPase activity (Maul, 1977). The actin microfilaments disassemble in the cytoplasm and release their RNP; the G-actin diffuses back into the nucleus where it is reassembled at the other end of the filament. ATP would be required for this treadmilling (Kirschner, 1980), as well as for the active sliding.

the origin of the nuclear pore complex (Maul, 1977) and their associated mechanism for the active transport of RNA (Clawson et al., 1980). The main handicap is our ignorance of the mechanism of transport and its structural organization — perhaps the major gap in our understanding of the fundamental mechanisms of the eukaryotic cell. In order to stimulate thought and experiment I hazard two highly speculative suggestions: a sliding-filament model for trans-pore RNA transport (Fig. 10), and that the pore complex originated by the aggregation of membrane-bound ATPase molecules that had evolved earlier for the cytoplasmic transport of secretory vesicles by means of microfilaments, microtubules or both.

THE ORIGIN OF SPLIT GENES AND RNA SPLICING

Following the origin of plural replicons, there would have been a weakening of the stabilizing selection that prevents the increase in prokaryote genome size, for increased genome size would no longer necessarily lengthen overall replication times. Duplication and transpositions would therefore be likely to increase the amount of non-coding or 'selfish' DNA or both (Cavalier-Smith, 1978a, 1980a, *c*; Doolittle & Sapienza, 1980; Orgel & Crick, 1980) which would suddenly provide an ideal environment (Cavalier-Smith, 1978*a*) for the multiplication of transposable DNA sequences (Cavalier-Smith, 1980*a*). This makes plausible the suggestion that RNA splicing to remove introns first evolved as a defence mechanism against the accidental insertion of transposable elements into coding sequences (Cavalier-Smith, 1978*a*; Crick, 1979) and that once evolved such a mechanism could never be lost because of the impossibility of simultaneous deletions of introns from numerous genes. It is easy to imagine that RNA splicing evolved by the mutation of a DNA transposase protein (Calos & Miller, 1980) enabling it to splice out RNA instead of DNA. One possibility is that mutation occurred in the ancestor of all eukaryotes after a lethal insertion had occurred in a coding gene and that by 'suppressing' the lethality by making a functional messenger it rescued the cell. Alternatively it might have arisen initially by intragenomic selection (Cavalier-Smith, 1980*c*) in a transposable virus-like element for which it may have served a useful function, e.g. in making several messengers for a small genome, as in adenovirus (Ziff, 1980). This second alternative is perhaps the more plausible since it does not depend on the chance rescue of a lethal intermediate. If splicing first evolved in an early eukaryote virus, inefficient intermediate stages would not have threatened the life of the cell, and could have been rapidly perfected by intragenomic selection. Once the virus perfected the splicing mechanism it could greatly expand its habitat by invading coding as well as non-coding DNA, without necessarily killing the cell, which may be how non-viral split genes originated. Since extensive transposition into and out of coding DNA would tend to have a variety of deleterious effects there would also be selection at the cellular level to restrict transposition to non-coding sequences. If such controls evolved early in eukaryote evolution, new insertions into coding DNA would be rare during subsequent evolution but established gene inserts would be stably inherited, especially if they subsequently acquired a function essential for maintaining the fitness of the cell, although there is no need to invoke such a function for introns in general. Once splicing evolved, nuclear RNAses would be needed to destroy the excised intron RNA. Capping of mRNA might then have evolved to protect mRNA from such degradation.

Though on this theory most gene inserts are 'fossils' remaining from this early phase of eukaryote evolution it would be incorrect to regard them now as selfish DNA (Doolittle & Sapienza, 1980) if, as

I have argued, the overall amount of DNA plays a key role in determining nuclear volume (Cavalier-Smith, 1978a, 1980b, c). According to the nucleoskeletal DNA theory (Cavalier-Smith, 1978a), one would expect the total length of introns in the genome of multicellular organisms to vary in proportion to overall genome size, but that the sequence (apart from the terminal regions that specify the splicing sites) and the length of individual introns would be free to evolve by random genetic drift, which seems often to be the case (Perler et al., 1980).

THE LINKAGE REVOLUTION AND TRANSCRIPTIONAL CONTROL

In prokaryotes relative gene dosage is a function not only of growth rate, but also of position on the chromosome, since genes near the origin will be present at higher doses than those near the terminus. It is reasonable to suggest that selection for maximum efficiency and rapid growth will strongly select against gene transpositions because they will alter relative gene dosage; in fact bacterial linkage maps do show strong conservatism (Riley & Anilionis, 1978) even though transposition is known to be frequent (Calos & Miller, 1980). But as soon as plural replicons evolved, relative dosage would no longer depend on gene position on the genetic map, nor would relative gene dosage vary much with growth rate, so stabilizing selection against transposition would cease. Since transposable elements seem to be universal inhabitants of cells, there would inevitably be evolutionary rearrangement of the genome. Initially the different genes within an operon would tend to stay together because of the need for their co-ordinate control. But co-ordinate control of transcription is possible for widely dispersed genes as shown by studies with fungi (Fincham, Day & Radford, 1979), and this could come about simply by the transposition of operator and promoter sequences to the front of each gene in an operon. With active transposable elements, and no selection against it, extensive transposition would inevitably occur and lead to the dispersed arrangement of related genes characteristic of eukaryotes. I regard the presence of multigenic operons in prokaryotes not as a device to allow co-ordinate control *per se,* but as one that automatically maintains the relative concentration of the different gene products irrespective of variations of growth rate. This could not be achieved in prokaryotes if their genes were scattered randomly around the chromosome, but it can in eukaryotes where relative gene dosage does not depend on their chromosomal location.

The eukaryote system of co-ordinate control of the rate of synthesis of all RNA (Frazer & Nurse, 1979), and the post-replication doubling of this rate, may also be a further device to ensure that the relative dosage of different messengers is constant during the cell cycle irrespective of variations in growth rate. I suggest it originated at the same time as transcription was undergoing rapid evolutionary change following the origin of histones. It will be interesting to see if this mechanism is present in dinoflagellates: if it is, this would show that histones themselves are not essential for it.

Although there was probably a drastic revolution in the genetic map at that time, linkage relations appear to be relatively stable in eukaryotes (De Grouchy, Turleau & Sinaz, 1978) as in prokaryotes. This is probably because meiotic pairing imposes strong stabilizing selection against transpositions and translocations. Therefore the linkage revolution discussed above was probably restricted to the brief phase of early eukaryote evolution, perhaps even in a single species, intervening between the origin of plural replicons and the origin of meiosis.

RIBOSOME DIVERSIFICATION

The fungal theory of the origin of the eukaryote cell (Cavalier-Smith, 1980a) supposes that eukaryote ribosomes have a dual origin. Mitochondrial ribosomes (mitoribosomes) and cytoplasmic ribosomes (cytoribosomes) presumably diverged from a common ancestor when the bacterial cytoplasm became compartmented into mitosol and cytosol by the folding back on themselves of respiratory cisternae during the transition from aerobic bacterium to aerobic fungus that created the first eukaryote. Plastid ribosomes on the other hand would have been acquired much later by the conversion of endosymbiotic prokaryotic algae into plastids. Such a dual origin is supported by the close resemblance of the 70S plastid ribosomes to the 70S ribosomes of prokaryotes, and the strong divergence of mitoribosomes and 80S cytoribosomes both from each other and from prokaryote ribosomes. Plastid and prokaryote ribosomes can exchange subunits and still show some function (Gillham, 1978).

They also show very strong ribosomal RNA homologies (Doolittle & Bonen, 1981) and similar drug sensitivities: but mitoribosomal and cytoribosomal subunits are not functionally interchangeable with those of prokaryotes.

Though the drug sensitivities of mitoribosomes have many similarities with those of prokaryotes, their size and ribosomal RNA are strongly divergent. In fungi, green plants, and protozoa mitoribosomes are 80S-like cytoribosomes: in animals they sediment at 55–60S because their rRNA molecules are much smaller and the amount of protein much greater than in other eukaryotes (Gillham, 1978). The size of the RNA molecules is much more variable in mitoribosomes and cytoribosomes than in prokaryote or plastid ribosomes (Table 6). In many eukaryotes, especially in animals, mito rRNA is markedly smaller than cyto rRNA. The fact that in Ascomycetes alone they are almost exactly the same size (1.3 M daltons) is readily explained by the fungal theory. Ascomycetes, which I have argued on other grounds include the most primitive eukaryotes, have retained the ancestral condition found in the very first eukaryote, whereas as other eukaryote groups evolved divergent selection pressures (perhaps aided by the founder effect: Dobzhansky, Ayala, Stebbins & Valentine, 1977) often caused a decrease in the size of mitochondrial rRNA and an increase in the size of cytoplasmic rRNA. I suggest that this divergence occurred rapidly during the origin of certain of the eukaryote kingdoms, rather than gradually, since Table 6 suggests that rRNA size may be rather uniform within each of the major kingdoms, though many more data are needed to justify such a generalization.

Though the mitoribosomes and cytoribosomes and their RNA are similar in size in Ascomycetes they differ in many other properties (Gillham, 1978; Mahler, 1980, 1981). One reason for the greater divergence of cytoribosomes than of mitoribosomes from their common prokaryotic ancestor may be that, unlike mitochondrial or prokaryote ribosomes, they have to be actively transported from the nucleus through the nuclear pores and had to acquire new binding sites to allow this (Cavalier-Smith, 1975, 1980a). A second fundamental reason may be that a mechanism is required to prevent newly made ribosomal subunits from translating primary transcripts of mRNA within the nucleus since, once split genes evolved, translation within the nucleus would produce many faulty proteins. The rapid export of the large subunit to the cytoplasm may help to prevent translation within the nucleus, but I suggest that the

Table 6. *Variation in the size of the RNA of the large ribosomal subunit. Some values are means, others ranges or single determinations. The units are Mdaltons*

		Cytoribosomes	Mitoribosomes
Prokaryotes	1.07–1.12[a]		
Plastids	1.07–1.11[a]		
	1.05–1.34[b]		
Eukaryotes			
Eufungi			
Ascomycetes (2spp)		1.305[c]	
Saccharomyces		1.3[b]	1.3[b,d]
Neurospora			1.28[b,d]
Basidiomycetes (3spp)		1.315[c]	
Zygomycetes (2spp)		1.335[c]	
Uniflagellata			
Chytridiomycetes (4spp)		1.34[c]	
Animalia			
rat/human		1.7[b], 1.75[a]	0.53[d], 0.54[b]
cattle			0.54[b]
Xenopus		1.5[a]	0.55[d], 0.58[b]
Drosophila		1.4[a]	0.48[d]
Biliphyta			
Rhodophyta		1.21[e]	
Verdiplantae (green plants)			
Chlamydomonas		1.3[b]	
pea		1.3[b]	
maize			1.25[b]
'higher plants'		1.27–1.31[a]	1.1–1.25[d]
Chromophyta			
Oomycetes (2spp)		1.415[c]	
Hyphochytrids		1.36[c]	
Euglenida			
Euglena		1.5[a]	0.93[b]
Protozoa			
Acrasiales		1.42[c]	
Myxomycete		1.45[c]	
Tetrahymena		1.3[a], 1.4[b]	0.9[d]

[a] Loening (1970).
[b] Gillham (1978).
[c] Léjohn (1974).
[d] Mahler (1980).
[e] Howland & Ramus (1971).

changeover from the prokaryotic protein initiation mechanism involving base-pairing between the 5' end of the mRNA and 16S rRNA (Shine & Dalgarno, 1975) to the basically different eukaryote system using a mRNA with a 5' guanosine cap (Nakashima, Darzynkiewicz & Shatkin, 1980) was associated with the need to prevent the characteristically prokaryotic protein synthesis on nascent mRNA.

MITOCHONDRIAL PLASMID EVOLUTION

The term plasmid was coined (Lederberg, 1952) for the then hypothetical hereditary determinants of mitochondria and plastids and other 'plasmagenes', so it is entirely appropriate to use it for mitochondrial and plastid DNA which have a similar size and organization to bacterial plasmids (Meyer, 1973; Gillham, 1978). In the case of plastids this similarity in size is probably the result of the loss of DNA from the chromosome of the cellular endosymbiont from which they are derived in a similar way to the cyanelle of *Cyanophora* (Herdman & Stanier, 1977). But there is no evidence for such reduction in the case of mitochondria: all the evidence fits the hypothesis that mitochondrial DNA evolved from a plasmid in the first eukaryote (Meyer, 1973; Raff & Mahler, 1972, 1975; Mahler, 1980, 1981; Cavalier-Smith, 1975, 1980a). Mitochondrial DNA codes for only six or seven polypeptides (Mahler, 1981), namely three subunits of cytochrome oxidase, one or two of the ATPase, apocytochrome *b*, and one ribosomal protein in addition to mitochondrial rRNA and tRNA. The unique mitochondrial genetic code (Mahler, 1980, 1981; Anon, 1980), which uses many fewer tRNA molecules than the normal code used by prokaryotes and by eukaryote cytoribosomes, and presumably plastids, is best explained as a 'frozen accident' (Crick, 1968) dating from the time when the ancestral plasmid containing these genes was first separated from the mitosol by the folding and fusing of respiratory cisternae (Cavalier-Smith, 1980a; Mahler, 1980a, b).

If when the plasmid was first trapped inside the mitochondrial envelope it contained genes for fewer than the 32 tRNA molecules required for the normal code, a simplification and modification of the code would be forced on it if it were to survive. I would argue that such changes were only possible because it coded for so few polypeptides. Mahler (1980, 1981) suggested that the mitochondrial code may be a primitive relic rather than derived, and that its origin predated the divergence between prokaryotes and eukaryotes. This seems highly implausible as it would require a changeover from the mitochondrial code to the more general code in a cell coding for hundreds or thousands of proteins. Moreover unless one postulates that mitochondria and the rest of the eukaryote cell came together by symbiosis, one would have to postulate that prokaryotes evolved from eukaryotes: the mitochondria would have to de-evolve and transfer respiratory enzymes to the plasma membrane and the

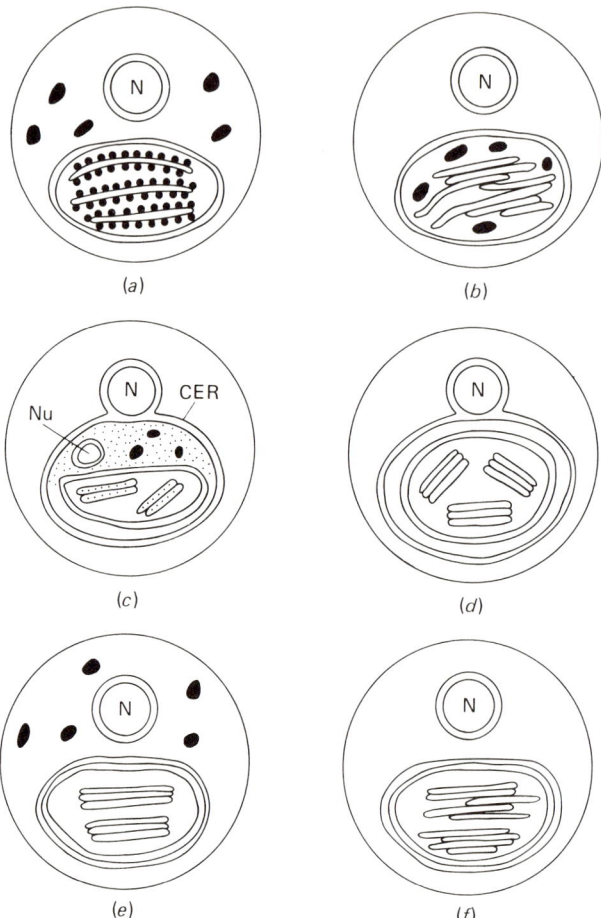

Fig. 11. The diverse origins of plastids. (a) Those of the Rhodophyta and Glaucophyta have unstacked thylakoids bearing phycobilisomes on their outer surface as in cyanophytes from which they probably arose. (b) The chloroplasts of the green plants (Verdiplantae) have stacked thylakoids with chlorophyll-b and no phycobilins; unlike all other plastids they contain starch (shown in black) which is probably directly inherited from their endosymbiotic ancestors – the Prochlorophyta. The so-called 'starch' of the Rhodophyta, Glaucophyta, Cryptophyta and Dinophyta more closely resembles glycogen than it does the true starch of the Verdiplantae and prokaryote algae and is located outside the plastids. It is probably therefore of host origin (Whatley, 1981), and inherited like the glycogen of animals and protozoa from the ancestral Eufungi which characteristically have glycogen carbohydrate reserves. The Chromophyta and Euglenophyta lack starch or other α-1,4 glucans and instead store β-1,3 glucans outside the plastid. This polymer may also ultimately be of fungal ancestry since Eufungal cell walls characteristically contain β-1,3 glucans as well as chitin. (c) Cryptophyte plastids contain thylakoids stacked in pairs with phycobilins inside the intra thykaloid spaces, as well as chlorophyll-c, and are separated from the chloroplast endoplasmic reticulum (CER) by a space containing starch, granules resembling ribosomes, and a structure resembling a minute nucleus (the nucleomorph (Nu) – Greenwood, 1974). These complex structures are most simply explained as the residues of a eukaryote endosymbiont: this hypothetical endosymbiont may have been a red alga or glaucophyte that evolved chlorophyll-c during or after the endosymbiotic event (Cavalier-Smith, 1981). (d) The chromo-

conversion from the mitochondrial to the general code would have to occur in two stages, the first involving all the genes but six to make a eukaryote and the second the six mitochondrial genes to make a prokaryote! The siting of the gene for the mitochondrial ATPase proteolipid in yeast mitochondrial DNA, but in the nuclei of *Neurospora* (Mahler, 1980) and *Aspergillus* (Turner, Imam & Küntzel, 1979), may reflect a fundamental divergence between Hemiascomycetes and Euascomycetes very early in eukaryote evolution; determining its site in other eukaryotes could greatly clarify phylogeny. An early divergence between Hemiascomycetes and Euascomycetes is also supported by the amino-acid sequences of the proteolipid (Dayhoff & Schwartz, 1981) and of cytochrome *c* (Fitch & Margoliash, 1970); those of yeast and *Neurospora* differ as much as either does from the homologous mitochondrial proteins of animals or green plants.

The existence of a separate mitochondrial code does not disprove Wallin's (1922, 1927) symbiotic theory of mitochondrial origin but it places severe constraints on it; the hypothetical mass transfer of respiratory genes to the nucleus (Margulis, 1970) must have occurred before the change in the code. However the discovery of the profound differences between prokaryotes and mitochondria, which are unexplained by the symbiotic theory, but have a plausible explanation in the autogenous fungal theory, deprives the symbiotic theory of its earlier attractions. The reduction in the size of mitochondrial DNA of animals compared with the yeast *Saccharomyces* and *Neurospora* was formerly taken as support for the idea of gradual gene transfer to the nucleus, but it is now clear that the extra DNA in these fungi does not code for proteins. It is dubious that it is 'selfish DNA' (Doolittle & Sapienza, 1980; Orgel & Crick, 1980) as has been suggested (Reid, 1980) because *Schizosaccharomyces pombe* which is larger and slower growing than *Saccharomyces* has much less mitochondrial DNA. The reasons for the approximately sixfold variation in mitochondrial DNA size (Gillham, 1978) are still obscure: why does *Chlamydomonas* have the smallest known mitochondrial DNA and flowering plants the largest? Is the great diversity in eufungal mitochondrial DNA sizes a

Caption to Figure 11 (*cont.*)

phyte plastid with its chlorophyll-*c*-containing thylakoids stacked in threes could have evolved from that of cryptophytes simply by loss of phycobilins and of the nucleomorphs, starch and ribosomes from the space inside the CER. (*e*) The chlorophyll-*c*-containing dinoflagellate chromoplast. (*f*) The chlorophyll-*b*-containing Euglenophyte chloroplast. These could have evolved by the endosymbiosis of chromophyte and chlorophyte algae respectively (Gibbs, 1981).

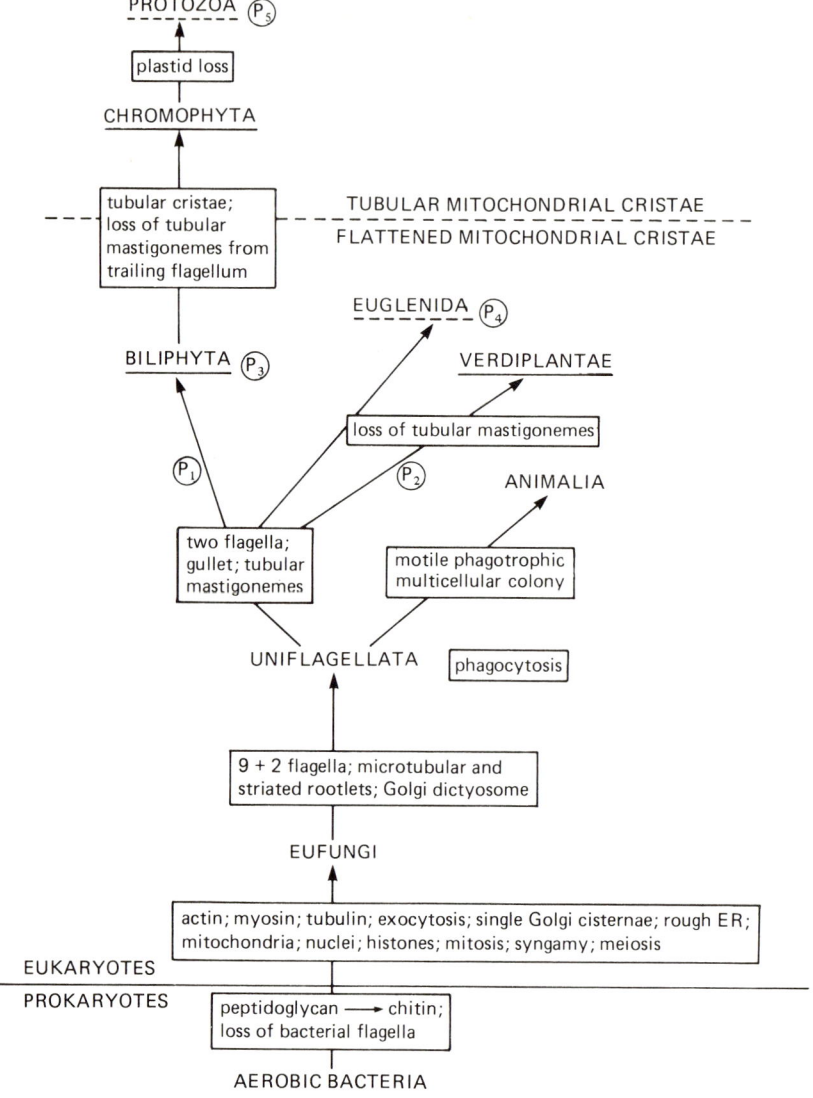

Fig. 12. A simplified phylogeny showing the relationship between the eight major groups, or kingdoms, of eukaryotes and their bacterial ancestors. The major evolutionary changes responsible for the origin and diversification of eukaryote cells are indicated: changes that probably occurred autogenously are shown in rectangular boxes and those that probably resulted from endosymbiosis, namely five independent chloroplast symbioses ($P_1 - P_5$) which include the origin of Dinoflagellate plastids within the Protozoa, are shown in the circles. Three eukaryote kingdoms (continuous underlining) are predominantly photosynthetic. Two (dashed underlining) contain only one photosynthetic phylum each. These phyla are the Dinophyta within the Protozoa and the Euglenophyta within the Euglenida. Three kingdoms (not underlined) are entirely non-photosynthetic. This eight-kingdom classification of eukaryotes (Cavalier-Smith, 1981, and in preparation) aims to reflect their cellular diversity more accurately than the traditional two kingdoms of 'plant and animals', or the more recent four

sign of ancient divergences in these primitive eukaryotes or of evolutionary lability?

THE EVOLUTION OF CILIA, PHAGOCYTOSIS AND PLASTIDS

There is not space here properly to discuss the evolution of these non-universal eukaryotic properties. Cilia and phagocytosis probably both evolved in an ancestor of the Chytridiomycete fungi, as discussed in detail elsewhere (Cavalier-Smith, 1980a, 1981), whereas plastids probably had multiple symbiotic origins as summarized in Fig. 11. The ultimate test of theories of eukaryote origins is whether they can enable us to devise a plausible unified phylogeny for both prokaryotes and eukaryotes that contravenes no established facts or biological principles. An attempt to construct such a phylogeny is shown in simplified form in Fig. 12. According to this phylogeny plastids were first acquired symbiotically by fully eukaryotic aerobic ciliated eukaryotes with well-developed mitochondria, as Margulis (1970) argued. The idea that plastids were acquired by anaerobic eukaryotes before mitochondria (Stanier, 1970) is unappealing because no eukaryotes have plastids but lack mitochondria. A second objection is that both cyanophytes and prochlorophytes are aerobes with well-developed respiratory chains which have been lost in plastids – if their host was anaerobic it would have had a selective advantage in retaining the respiratory system in its newly acquired plastids, and little or none in converting aerobic bacterial symbionts into mitochondria.

GRADUALISM VERSUS QUANTUM EVOLUTION

In the nineteenth century, even before Darwin, evolution was seen as a gradual inexorable process driven forward by some inner momentum, or the mere passage of time. Today many biochemists (e.g. Darnell, 1978; Doolittle, 1980; Fox et al., 1980) and mathematicians (e.g. Schwartz & Dayhoff, 1978) appear to have similar

Caption to Figure 12 (cont.)
kingdoms of Whittaker & Margulis (1978), which are cytologically markedly heterogenous and probably polyphyletic. Though originally based mainly on cell ultrastructure, photosynthetic pigments, storage carbohydrates, cell wall chemistry, and lysine metabolism, the present phylogeny is consistent with other biochemical characters (Ragan & Chapman, 1978) and is also remarkably congruent with the tree for cytochrome c (Dayhoff & Schwartz, 1981).

gradualist views, where the amount of divergence between two molecules or organisms is held directly to reflect the passage of time, and so suggest that the tremendous differences between prokaryotes and eukaryotes must reflect a divergence very long ago – perhaps 3.8×10^9 years ago (Doolittle, 1980). But on such a view one would expect large eukaryotic cells and multicellular organisms to have appeared in the fossil record much earlier than they do (Cavalier-Smith, 1980*a*). The alternative view presented here of a sudden revolution in cell structure is radically different from the gradualist view, and is more in keeping with the paleontological evidence that evolution goes in bursts of innovation and adaptive radiation interspersed with very long periods with little substantial change (Simpson, 1953; Eldredge & Gould, 1972; Stanley, 1979), and with present-day understanding of the mechanism of evolution (Williams, 1966; Cavalier-Smith, 1980*a*). Studies of evolution in progress, for example of industrial melanism or drug resistance, also show that evolutionary rates are not uniform, and that conditions can suddenly switch from strong conservatism maintained by stabilizing selection to rapid change in a defined direction when selective pressures alter (Dobzhansky *et al.*, 1977). In the cases mentioned it was environmental change that altered the selective pressures. But genetic changes also can alter selection pressures. According to the fungal theory (Cavalier-Smith, 1980*a*) it was a coincidence of mutations, rather than changed external circumstances, that triggered off the events that led to the first eukaryote, perhaps (Schopf, 1978) about 1 500 million years ago (Cavalier-Smith, 1980*a*). I have tried here to outline how, once exocytosis and cytoplasmic motility evolved, this would have inevitably led to a drastic transformation of the entire genetic mechanism and cell structure so as to produce all 22 of the universal eukaryote properties of Table 1 as well as mitochondria, nucleosomes, and meiosis. Once these structures and mechanisms were perfected, which could have occurred rapidly and in a single ancestral species, the basic properties of eukaryote cells would have been maintained by stabilizing selection during their subsequent diversification.

I suggest that the transition was so sudden that all intermediates between prokaryote and eukaryote were rapidly extinguished by intraspecific competition, and that the origin of the eukaryote cell is the most profound and far-reaching case of quantum evolution (Simpson, 1953; Stanley, 1979) known to us. On the gradualist view one might expect a great variety of intermediates with varying numbers of the 22 properties to have arisen and survived. In my view quantum evolution, rather than nineteenth-century gradualism (Doolittle, 1980), is the revolutionary concept in evolutionary cell biology, that has yet to be assimilated by those biochemists who accept it only if it happened near the dawn of life (Doolittle, 1980).

There are probably many defects in this theory, that will be revealed by future comparative study and criticism: I hope nonetheless that it may stimulate a more holistic treatment of cell evolution and show the value of seeking to understand the actual selective forces involved.

CONCLUSIONS

The eukaryote cell may have originated relatively suddenly, around 1 500 million years ago, when an aerobic bacterium was transformed into a hemiascomycete fungus following the changes in cell-wall chemistry from peptidoglycan to chitin. The drastic transformation of cell organelles that this entailed, including the compartmentation of the cell into nucleus, cytoplasm and mitochondria, may have been triggered by the origin of exocytosis, and of the cytoplasmic transport of secretory vesicles by microfilaments and microtubules. Mitosis, nucleosomes, and meiosis probably evolved early, but cilia later. Plastids were probably acquired by several independent endosymbiotic events after the evolution of phagocytosis in ciliated cells.

I thank M. Dayhoff, W. F. Doolittle, S. Gibbs, H. R. Mahler and J. Whatley for valuable discussion and, together with other colleagues too numerous to mention, for sending me preprints.

REFERENCES

ABELSON, J. (1979). RNA processing and the intervening sequence problem. *Annual Review of Biochemistry*, **48**, 1035–69.

ALLEN, J. R., ROBERTS, T. M., LOEBLICH, A. R. III. & KLOTZ, L. C. (1975). Characterisation of the DNA from the dinoflagellate, *Crypthecodinium cohnii* and implications for nuclear organisation. *Cell*, **6**, 161–9.

ALLSOPP, A. (1969). Phylogenetic relationships of the Procaryota and the origin of the eucaryotic cell. *New Phytologist*, **68**, 591–612.

ANON. (1980). Maverick mitochondria. *Nature, London*, **287**, 9–10.

BALDI, M. I., BENEDETTI, P., MATTOCCIA, E. & TOCCHINI-VALENTINI, G. P. (1980). In vitro catenation and decatenation of DNA and a novel eukaryotic ATP-dependent topoisomerase. *Cell,* **20,** 461–7.
BATEMAN, A. J. (1975). Simplification of palindromic telomere theory. *Nature,* **253,** 379–80.
BEAM, C. A. & HIMES, M. (1980). Sexuality and meiosis in dinoflagellates. In *Biochemistry and Physiology of Protozoa.* 2nd edn, Vol 3. ed. M. Levandowsky & S. H. Hutner, pp. 171–206. New York: Academic Press.
BECKETT, A., HEATH, I. B. & McLAUGHLIN, D. J. (1974). *An Atlas of Fungal Ultrastructure.* London: Longmans.
BLEECKEN, S. J. (1971). 'Replisome' – controlled initiation of DNA replication. *Journal of Theoretical Biology,* **32,** 81–92.
BLOBEL, G. (1980). Intracellular protein topogenesis. *Proceedings of the National Academy of Sciences, USA,* **77,** 1496–500.
BOGORAD, L. (1980). Regulation of intracellular gene flow in the evolution of eukaryotic genomes. In *Origins of Chloroplasts,* ed. J. A. Schiff & H. Lyman. Amsterdam: Elsevier North Holland.
BRADBURY, E. M., INGLIS, R. J. & MATTHEWS, H. R. (1974). Control of cell division by very lysine rich histone (FI) phosphorylation. *Nature,* **247,** 257–61.
BYERS, B. & GOETSCH, L. (1974). Duplication of spindle plaques and integration of the yeast cell cycle. *Cold Spring Harbor Symposia on Quantitative Biology,* **38,** 123–31.
CALOS, M. P. & MILLER, J. H. (1980). Transposable elements. *Cell,* **20,** 579–95.
CARLILE, M. J. (1980). From prokaryote to eukaryote: gains and losses. In *The Eukaryotic Microbial Cell. Symposium of the Society for General Microbiology,* **30,** 1–40, ed. G. W. Gooday, D. Lloyd & A. P. J. Trinci. Cambridge University Press.
CARON, F., JACQUES, C. & ROUVIÈRE-YANIV, J. (1979). Characterisation of a histone-like protein extracted from yeast mitochondria. *Proceedings of the National Academy of Sciences, USA,* **76,** 4265–9.
CAROTHERS, Z. (1972). Studies of spermatogenesis in the hepaticae: III Continuity between plasma membrane and nuclear envelope in androgonial cells of *Blasia. Journal of Cell Biology,* **52,** 273–82.
CASTORA, S. J. & SIMPSON, M. V. (1979). Search for DNA gyrase in mammalian mitochondria. *Journal of Biological Chemistry,* **254,** 1193–5.
CAVALIER-SMITH, T. (1974). Palindromic base sequences and replication of eukaryote chromosome ends. *Nature, London,* **250,** 467–70.
CAVALIER-SMITH, T. (1975). The origin of nuclei and of eukaryotic cells. *Nature, London,* **256,** 463–8.
CAVALIER-SMITH, T. (1977). Mitocondri e cloroplasti: un problema evolutivo. In *Scienza e Technica,* **77,** 305–18. Milan: Mondadori.
CAVALIER-SMITH, T. (1978*a*). Nuclear volume control by nucleoskeletal DNA, selection for cell volume and cell growth rate, and the solution of the DNA C-value paradox. *Journal of Cell Science,* **34,** 247–68.
CAVALIER-SMITH, T. (1978*b*). The evolutionary origin and phylogeny of microtubules, mitotic spindles and eukaryote flagella. *BioSystems,* **10,** 93–114.
CAVALIER-SMITH, T. (1980*a*). Cell compartmentation and the origin of eukaryote membranes organelles. In *Endocytobiology; Endosymbiosis and Cell Biology,* ed. W. Schwemmler & H. E. A. Schenk, pp. 831–916. Berlin: de Gruyter.
CAVALIER-SMITH, T. (1980*b*). *r*- and *K*-tactics in the evolution of protist developmental systems: cell and genome size, phenotype diversifying selection, and cell cycle patterns. *BioSystems,* **12,** 43–59.

CAVALIER-SMITH, T. (1980c). How selfish is DNA? *Nature*, **285**, 617–18.
CAVALIER-SMITH, T. (1981). The evolutionary origin and phylogeny of eukaryote flagella. In *Eukaryotic and Prokaryotic Flagella, Symposium of the Society for Experimental Biology*, ed. W. B. Amos & J. G. Duckett, **35**, in press. Cambridge University Press.
CHADEFAUD, M. M. (1976). Sur l'origine des plastes, les plastes 'cyanelloïdes' et la 'classe?' des Glaucophycées. *Comptes Rendus. Academie des Sciences, Paris,* **283**, Série D, 1029–32.
CHAMBON, P. (1975). Eukaryotic nuclear RNA polymerases. *Annual Review of Biochemistry,* **44**, 613–38.
CLARKE, L. & CARBON, J. (1980). Isolation of a yeast centromere and construction of functional small circular chromosomes. *Nature,* **287**, 504–9.
CLAWSON, G. A., KOPLITZ, M., MOODY, D. E. & SMUCKLER, E. A. (1980). Effects of thioacetamide treatment on nuclear envelope nucleotide triphosphatase activity and transport of RNA from rat liver nuclei. *Cancer Research,* **40**, 75–9.
CRICK, F. H. C. (1968). The origin of the genetic code. *Journal of Molecular Biology,* **38**, 367–79.
CRICK, F. (1979). Split genes and RNA splicing. *Science,* **204**, 264–71.
DARNELL, J. E. (1978). Implications of RNA–RNA splicing in evolution of eukaryotic cells. *Science,* **202**, 1257–60.
DAVERN, C. I. (1979). Replication of the prokaryotic chromosome with emphasis on the bacterial chromosome replication in relation to the cell cycle. In *Cell Biology: A Comprehensive Treatise.* Vol 2, ed. D. M. Prescott & L. Goldstein, 131–69. New York: Academic Press.
DAYHOFF, M. O. & SCHWARTZ, R. M. (1981). Evidence on the origin of eukaryotic mitochondria from protein and nucleic acid sequences. *Annals of the New York Academy of Science,* (in the press).
DE GROUCHY, J., TURLEAU, C. & SINAZ, C. (1978). Chromosomal phylogeny of primates. *Annual Review of Genetics,* **12**, 289–328.
DINGMAN, C. W. (1974). Bidirectional replication: some topological considerations. *Journal of Theoretical Biology,* **43**, 187–95.
DOBZHANSKY, T., AYALA, F. J., STEBBINS, G. L. & VALENTINE, J. W. (1977). *Evolution.* San Francisco: Freeman.
DODGE, J. D. (1966). The Dinophyceae. In *The Chromosomes of the Algae,* ed. M. B. E. Godward, pp. 96–115. London: Arnold.
DODSON, E. O. (1979). Crossing the eukaryote-prokaryote border: endosymbiosis or continuous development? *Canadian Journal of Microbiology,* **25**, 651–74.
DONACHIE, W. (1974). Cell Division in Bacteria. In *Mechanism and Regulation of DNA Replication,* ed. A. R. Kolber & M. Kohiyama, pp. 431–45. New York: Plenum.
DOOLITTLE, W. F. (1978). Genes in pieces: were they ever together? *Nature,* **272**, 581–2.
DOOLITTLE, W. F. (1980). Revolutionary concepts in evolutionary cell biology. *Trends in Biochemical Sciences,* **5**, 146–9.
DOOLITTLE, W. F. & BONEN, L. (1981). Molecular sequence data indicating an endosymbiotic origin for plastids. *Annals of New York Academy of Sciences. In press.*
DOOLITTLE, W. F. & SAPIENZA, C. (1980). Selfish genes, the phenotype paradigm and genome evolution. *Nature,* **284**, 601–3.
DOUGHERTY, E. C. (1957). Neologisms needed for structures of primitive organisms. *Journal of Protozoology,* **4**, (Suppl): 14.
DUPRAW, E. J. (1968). *Cell and Molecular Biology.* London and New York: Academic Press.

ELDREDGE, N. & GOULD, S. J. (1972). Punctuated equilibria: an alternative to phyletic gradualism. In *Models in Paleobiology.* ed. T. J. M. Schopf, pp. 130–45. San Francisco: Freeman.

EMERSON, R. & HELD, A. A. (1969). *Aqualinderella fermentans* gen. et sp. nov., a phycomycete adapted to stagnant waters II. Isolation, culture and gas relationships. *American Journal of Botany,* **56,** 1103–20.

EMR, S. D., HALL, M. N. & SILHARY, T. J. (1980). A mechanism of protein localization: the signal hypothesis and bacteria. *Journal of Cell Biology,* **86,** 701–11.

FELDHERR, C. M. (1972). Structure and function of the nuclear envelope. In *Advances in Cell and Molecular Biology,* ed. E. J. Dupraw, pp. 273–307. New York: Academic Press.

FINCHAM, J. R. S., DAY, P. R. & RADFORD, A. (1979). *Fungal genetics,* 4th edn. Oxford: Blackwell.

FITCH, W. M. & MARGOLIASH, E. (1970). The usefulness of amino acid and nucleotide sequences in evolutionary studies. *Evolutionary Biology,* **4,** 67–109.

FORTE, M. A. & FANGMAN, W. L. (1979). Yeast chromosomal DNA molecules have strands which are cross-linked at their termini. *Chromosoma,* **72,** 131–50.

FOX, G. E., STACKEBRANDT, E., HESPELL, R. B., GIBSON, J., MANILOFF, J., DYER, T. A., WOLFE, R. S., BALCH, W. E., TANNER, R. S., MAGRUM, L. J., ZABLEN, L. B., BLAKEMORE, R., GUPTA, R., BONEN, L., LEWIS, B. J., STAHL, LUEHRSEN, K. R., CHEN, K. N. & WOESE, C. R. (1980). The phylogeny of prokaryotes. *Science,* **209,** 457–63.

FRANKE, W. W. (1974). Structure, biochemistry, and functions of the nuclear envelope. *International Review of Cytology.* Suppl 4, 71–236.

FRAZER, R. S. S. & NURSE, P. (1979). Altered patterns of ribonucleic acid synthesis during the cell cycle: a mechanism compensating for variation in gene concentration. *Journal of Cell Science,* **35,** 25–40.

GIBBS, S. P. (1981). The chloroplasts of some algal groups may have evolved from endosymbiotic algae. *Annals of the New York Academy of Sciences, In press.*

GILLHAM, N. W. (1978). *Organelle Heredity.* New York: Raven.

GILLHAM, N. W. & BOYNTON, J. E. (1981). Evolution of organelle genomes and protein synthesising systems. *Annals of the New York Academy of Sciences, In press.*

GOODENOUGH, U. W. (1980). Sexual microbiology: mating reactions of *Chlamydomonas reinhardii, Tetrahymena thermophila* and *Saccharomyces cerevisiae.* In *The Eukaryotic Microbial Cell. Society for General Microbiology Symposium,* **30,** ed. G. W. Gooday, D. Lloyd & A. J. P. Trinci, pp. 301–328. Cambridge University Press.

GREENWOOD, A. D. (1974). The Cryptophyta in relation to phylogeny and photosynthesis. *8th International Congress of Electron Microscopy, Canberra,* pp. 566–7.

HARTLEY, B. S. (1979). Evolution of enzyme structure. *Proceedings of the Royal Society B,* **205,** 443–52.

HASELKORN, R. & ROUVIÈRE-YANIV, J. (1976). Cyanobacterial DNA-binding protein related to *Escherichia coli* HU. *Proceedings of the National Academy of Sciences, USA,* **73,** 1917–20.

HEATH, I. B. (1980). Variant mitoses in lower eukaryotes: indicators of the evolution of mitosis. *International Review of Cytology,* **64,** 1–80.

HERDMAN, M. & STANIER, R. Y. (1977). The cyanelle: chloroplast or endosymbiotic prokaryote? *Federation of European Microbiological Societies Letters,* **1,** 7–12.

HEREFORD, L. M. & ROSBASH, M. (1977). Number and distribution of polyadenylated RNA Sequences in yeast. *Cell,* **10,** 453–62.

HEYWOOD, P. (1975). Algal sexuality. *Nature, London,* **259,** 425.

HOWLAND, G. P. & RAMUS, J. (1971). Analysis of blue-green and red algal ribosomal-RNAs by gel electrophoresis. *Archiv für Mikrobiologie*, **76**, 292–8.

KEDES, L. H. (1979). Histone genes and histone messengers. *Annual Review of Biochemistry*, **48**, 837–70.

KIRSCHNER, M. W. (1980). Implications of treadmilling for the stability and polarity of actin and tubulin polymers in vivo. *Journal of Cell Biology*, **86**, 330–4.

KLUG, A., RHODES, D., SMITH, J., FINCH, J. T. & THOMAS, J. O. (1980). A low resolution structure for the histone core of the nucleosome. *Nature*, **287**, 509–16.

KORNBERG, R. D. (1977). Structure of Chromatin. *Annual Review of Biochemistry*, **46**, 931–54.

LEDERBERG, J. (1952). Cell genetics and hereditary symbiosis. *Physiological Reviews*, **32**, 403–30.

LEE, R. E. (1972). Origin of plastids and the phylogeny of algae. *Nature, London*, **237**, 44–6.

LÉJOHN, H. B. (1974). Biochemical parameters of fungal phylogenetics. *Evolutionary Biology*, **7**, 79–125.

LENK, R., RANSOM, L., KAUFMANN, V. & PENMAN, S. (1977). A cytoskeletal structure associated with polyribosomes obtained from Hela cells. *Cell*, **10**, 67–78.

LOEBLICH, A. R. III. (1976). Dinoflagellate evolution: speculation and evidence. *Journal of Protozoology*, **23**, 13–28.

LOENING, V. (1970). The mechanism of synthesis of ribosomal RNA. In *Organisation and Control in Prokaryotic and Eukaryotic cells. Symposium of the Society for General Microbology*, 20, ed. H.P. Charles & B. C. J. G. Knight, pp. 77–106. Cambridge University Press.

MAALØE, O. & KJELDGAARD, N. O. (1966). *Control of Macromolecular Synthesis*. New York: Benjamin.

MAHLER, H. (1980). Nonsymbiotic hypotheses of mitochondrial origin and their relevance to cell research. In *Endocytobiology: Endosymbiosis and Cell Biology*, ed. W. Schwemmler & H. A. E. Schenk, 869–92. Berlin: de Gruyter.

MAHLER, H. R. (1981). Mitochondrial evolution: organisation and regulation of mitochondrial genes. *Annals of the New York Academy of Sciences*. (In press.)

MAHLER, H. R. & RAFF, R. A. (1975). The evolutionary origin of the mitochondrion: a non-symbiotic model. *International Review of Cytology*, **43**, 2–124.

MARCHBANKS, R. M. (1979). Role of storage vesicles in synaptic transmission. In *Secretory Mechanisms. Symposia of the Society for Experimental Biology*, **33**, ed. C. R. Hopkins & C. J. Duncan, pp. 251–276. Cambridge University Press.

MARGULIS, L. (1970). *Origin of Eukaryotic Cells*. New Haven: Yale University Press.

MARGULIS, L. (1974). Five kingdom classification and the origin and evolution of cells. *Evolutionary Biology*, **7**, 45–78.

MAUL, G. G. (1977). The nuclear and cytoplasmic pore complex: structure, dynamics, distribution, and evolution. *International Review of Cytology*, Suppl. **6**, 75–186.

MAYNARD SMITH, J. M. (1978). *The Evolution of Sex*. Cambridge University Press.

MERESCHKOWSKY, C. (1905). Über Natur und Ursprung der Chromatophoren im Pflanzenreiche. *Biologisches Zentralblatt*, **25**, 593–604.

MEYER, R. R. (1973). On the evolutionary origin of mitochondrial DNA. *Journal of Theoretical Biology*, **38**, 647–63.

MIZUUCHI, K., FISHER, L. M., O'DEA, M. H. & GELLERT, M. (1980). DNA gyrase action involves the introduction of transient double-strand breaks into DNA. *Proceedings of the National Academy of Sciences, USA*, **77**, 1847–51.

Morris, N. R. (1980). Chromosome structure and the molecular biology of mitosis in eukaryotic microorganisms. In *The Eukaryotic Microbial Cell. Society for General Microbiology Symposium,* **30**, ed. G. W. Gooday, D. Lloyd & A. P. J. Trinci, pp. 41–76. Cambridge University Press.

Nakashima, K., Darzynkiewicz, E. & Shatkin, A. J. (1980). Proximity of mRNA 5'-region and 18S rRNA in eukaryotic initiation complexes. *Nature, London,* **286**, 226–30.

Orgel, L. & Crick, F. H. C. (1980). Selfish DNA: the ultimate parasite. *Nature, London,* **284**, 604–7.

Ng, R. & Abelson, J. (1980). Isolation and sequence of the gene for actin in *Saccharomyces cerevisiae*. *Proceedings of the National Academy of Sciences, USA,* **77**, 3912–6.

Paine, P. L., Moore, L. C. & Horowitz, S. B. (1975). Nuclear envelope permeability. *Nature, London,* **254**, 109–14.

Perler, F., Efstratiadis, A., Lomedico, P., Gilbert, W., Kolodner, R. & Dodgson, J. (1980). The evolution of genes: the chicken prepro-insulin gene. *Cell,* **29**, 555–66.

Phillips, D. O. & Carr, N. G. (1977). Nucleic acid analysis and the endosymbiotic hypothesis. *Taxon,* **26**, 3–42.

Picken, L. (1960). *The Organisation of Cells and Other Organisms*. Oxford: Clarendon Press.

Pickett-Heaps, J. (1974). The evolution of mitosis and the eukaryotic condition. *BioSystems,* **6**, 37–48.

Pollard, T. D. & Weihing, R. R. (1974). Actin and myosin and cell movement. *CRC Critical Reviews in Biochemistry,* **2**, 1–65.

Radding, C. M. (1978). Genetic recombination: strand transfer and mismatch repair. *Annual Review of Biochemistry,* **47**, 847–80.

Rae, P. M. M. & Steele, R. E. (1978). Modified bases in the DNAs of unicellular eukaryotes: an examination of distributions and possible roles, with emphasis on hydroxymethyluracil in dinoflagellates. *BioSystems,* **10**, 37–53.

Raff, R. A. & Mahler, H. R. (1972). The non-symbiotic origin of mitochondria. *Science,* **77**, 575–82.

Raff, R. A. & Mahler, H. R. (1975). The symbiont that never was: an inquiry into the evolutionary origin of the mitochondrion. In *Symbiosis. Society for Experimental Biology Symposium,* **29**, ed. D. H. Jennings & D. L. Lee, pp. 41–92. Cambridge University Press.

Ragan, M. A. & Chapman, D. J. (1978). *A Biochemical Phylogeny of Protists*. New York: Academic Press.

Raven, P. A. (1970). A multiple origin for plastids and mitochondria. *Science,* **169**, 641–46.

Reanney, D. C. (1974). On the origin of prokaryotes. *Journal of Theoretical Biology,* **48**, 243–51.

Reanney, D. C. (1979). RNA splicing and polynucleotide evolution. *Nature, London,* **277**, 598–600.

Reid, R. A. (1980). Selfish DNA in 'petite' mutants. *Nature, London,* **285**, 620.

Riley, M. & Anilionis, A. (1978). Evolution of the Bacterial Genome. *Annual Review of Microbiology,* **32**, 519–60.

Roberts, K. & Hyams, J. S., eds (1979). *Microtubules*. London: Academic Press.

Rouvière-Yaniv, J. & Gros, F. (1975). Characterisation of a novel, low-molecular-weight DNA-binding protein from *Escherichia coli*. *Proceedings of the National Academy of Sciences, USA,* **72**, 3428–32.

Sagan, L. (1967). On the origin of mitosing cells. *Journal of Theoretical Biology,* **14**, 225–74.

SCHIMPER, A. F. W. (1883). Uber die Entwickelung der Chlorophyllkörner und Farbkörper. *Botanische Zeitung*, **41**, 105–14.
SCHOPF, J. W. (1978). The evolution of the earliest cells. *Scientific American*, **239**, 84–102.
SCHWARTZ, R. M. & DAYHOFF, M. O. (1978). Origins of prokaryotes, eukaryotes, mitochondria, and chloroplasts. *Science*, **199**, 395–403.
SEALE, R. L. (1976). Studies on the mode of segregation of histone nu bodies during replication in Hela cells. *Cell*, **9**, 423–9.
SEARCY, D. G., STEIN, D. B. & GREEN, G. R. (1978). Phylogenetic affinities between eukaryotic cells and a thermophilic mycoplasma. *BioSystems*, **10**, 19–28.
SHINE, J. & DALGARNO, L. (1975). Determination of cistron specificity in bacterial ribosomes. *Nature, London*, **254**, 24–8.
SIEBERT, G. (1978). The limited contribution of the nuclear envelope to metabolic compartmentation. *Biochemical Society Transactions*, **6**, 5–9.
SIMPSON, G. G. (1953). *The Major Features of Evolution*. New York: Columbia University Press.
SLEIGH, M. A. (1979). Radiation of the Eukaryote Protista. In *The Origin of Major Invertebrate Groups*, ed. M. R. House, pp. 23–53. London, New York: Academic Press.
STANIER, R. Y. (1970). Some aspects of the biology of cells and their possible evolutionary significance. In *Organisation and Control in Prokaryotic and Eukaryotic Cells. Society for General Microbiology Symposium*, **20**, ed. H. P. Charles & B. C. J. G. Knight, pp. 1–38. Cambridge University Press.
STANIER, R. Y. & VAN NIEL, C. B. (1962). The concept of a bacterium. *Archiv für Mikrobiologie*, **42**, 17–35.
STANLEY, S. M. (1979). *Macroevolution: pattern and process*. San Francisco: Freeman.
TAYLOR, F. J. R. (1974). Implications and extensions of the serial endosymbiosis theory of the origin of the eukaryotes. *Taxon*, **23**, 229–58.
TAYLOR, F. J. R. (1976). Autogenous theories for the origin of eukaryotes. *Taxon*, **25**, 377–90.
TAYLOR, F. J. R. (1980). The stimulation of cell research by endosymbiotic hypotheses for the origin of eukaryotes. In *Endocytobiology: Symbiosis and Cell Research*, ed. W. Schwemmler & H. A. E. Schenk, pp. 917–47. Berlin: de Gruyter.
TSENG, B. Y., GRAFSTROM, R. H., REVIE, D., OERTEL, W. & GOULIAN, M. (1979). Studies on early intermediate in the synthesis of DNA in animal cells. *Cold Spring Harbor Symposia on Quantitative Biology*, **43**, 263–70.
TURNER, G., IMAM, G. & KÜNTZEL, H. (1979). Mitochondrial ATPase complex of *Aspergillus nidulans* and the dicyclohexylcarbodiimide-binding protein. *European Journal of Biochemistry*, **97**, 565–71.
UZZEL, T. & SPOLSKY, C. (1981). Two data sets: alternative explanations and interpretations. *Annals of the New York Academy of Sciences, In press*.
WALLIN, I. E. (1922). On the nature of mitochondria. III. The demonstration of mitochondria by bacteriological methods. IV. A comparative study of the morphogenesis of root-nodule bacteria and chloroplasts. *American Journal of Anatomy*, **30**, 451–71.
WALLIN, I. E. (1927). *Symbionticism and the Origin of Species*, London: Baillière, Tindall & Cox.
WHATLEY, J. M. (1981). Chloroplast evolution – ancient and modern. *Annals of the New York Academy of Sciences. In press*.
WHATLEY, J. M., JOHN, P. & WHATLEY, F. R. (1979). From extracellular to intracellular: the establishment of mitochondria and chloroplasts. *Proceedings of the Royal Society of London Series B*, **204**, 165–87.

WHITTAKER, R. H. & MARGULIS, L. (1978). Protist classification and the Kingdoms of Organisms. *BioSystems,* **10,** 3–18.

WILLIAMS, G. C. (1966). *Adaptation and Natural Selection.* Princeton University Press.

WORCEL, A. & BURGI, E. (1972). On the structure of the folded chromosome of *E. coli. Journal of Molecular Biology,* **71,** 127–38.

YOSHIKAWA, H., YAMAGUCHI, K., SEIKI, M., OGASAWARA, N. & TOYADA, H. (1979). Organisation of the replication-origin region of the *Bacillus subtilis* chromosome. *Cold Spring Harbor Symposia on Quantitative Biology,* **43,** 569–76.

ZIFF, E. B. (1980). Transcription and RNA processing by the DNA tumour viruses. *Nature, London,* **287,** 491–9.

SPORULATION IN EVOLUTION

IAN W. DAWES

Department of Microbiology, University of Edinburgh, Edinburgh, EH9 3JG, UK

INTRODUCTION

Microbial differentiation processes, especially those involving sporulation, are of considerable interest in cell biology and have often been discussed as possible models in which to study mechanisms of control that might apply during cell development in higher systems. Despite this interest, there has been surprisingly little direct comment either on how sporulation processes evolved, or on their possible rôle in the evolution of development and specialization of cells in higher organisms. This may be due in part to the lack of detailed knowledge at the molecular level of the underlying mechanisms in the more extensively-studied sporulation systems and also to the highly speculative nature of such an enterprise, given the absence of a detailed fossil record covering the period when this primitive form of cell specialization probably emerged. It may also be due to the feeling, by the unenlightened, that sporulation represents an evolutionary backwater of little consequence to the mainstream of evolution. The aim of this article, therefore, is not only to consider when and how sporulation processes evolved, but to show how sporulation in one form or another may have been central to the evolution of important aspects of eukaryotic development, particularly the evolution of meiosis and of stable diploid systems. Various evolutionary aspects of sporulation which have been reviewed or discussed include: evolution of differentiation and development (Bullough, 1967; Bonner, 1958a); metabolic control in differentiation (Wright, 1970; Mandelstam, 1971); bacterial endospore formation (Bisset, 1950; Foster, 1956; Lewis, 1969; Slepecky, 1972; Gould & Dring, 1979); the place of spores in nature (Sussman & Halvorson, 1966); evolution of genetic systems (Darlington, 1958; Raper & Flexer, 1970; Esser, 1974).

The subject is complicated by the existence of many types of spores. There are fewer prokaryotic spore forms than there are eukaryotic types, but still enough to prompt discussions of their diversity (Cross, 1970; Slepecky & Leadbetter, 1977; Hanson,

Table 1. *Several definitions of spore*

spore 1836 1. *Bot.* One of the minute reproductive bodies characteristic of flowerless plants. 2. *Zool.* and *Biol.* A very minute germ or organism 1876.
The Shorter Oxford English Dictionary

spore. A 1-celled, or sometimes several-celled, very small reproductive body produced in plants, bacteria and Protozoa.
Henderson's Dictionary of Biological Terms, 9th edn

spore. Single-celled or several-celled reproductive body that becomes detached from the parent and gives rise either directly or indirectly to a new individual. Usually microscopic, of many different types and produced in a variety of ways. Thin or thick-walled; often serving for very rapid increase in the population of the species, produced in enormous numbers and distributed far and wide by wind, water, animals; others are resting spores, the means of survival through an unfavourable period. Occurring in all groups of plants, particularly in Fungi; in Bacteria; and in Protozoa.
Penguin Dictionary of Biology

1979), whereas for fungi (Weber & Hess, 1975) and algae (Bold & Wynne, 1978; Erben, 1962), the range of spores and sporulation processes is rich and extremely varied. It should be noted that the term 'spore' has been rather generously applied to a wide range of structures, especially in lower eukaryotes, so much so that it is probably easiest to define spores by what they are not. Dictionary definitions can be delightfully misleading; that from the Shorter Oxford English Dictionary, while of little use, is not without irony (see Table 1). The Penguin Dictionary of Biology definition, by virtue of its extent, is probably most accurate and highlights the fact that spores can be assigned many functions; as agents of dispersal, of reproduction or of survival under unfavourable conditions. This definition misses one of the main features of spores, i.e. that they develop as morphologically distinct structures.

Some types of spore, and sporulation process, are outlined in Table 2 and Fig. 1. The table is not comprehensive, but it does illustrate some general features of spores; that there are a number of mechanisms for their formation and that, in eukaryotes, spores can be formed in association with sexual or asexual processes.

Fig. 1. Life cycles of some sporulating organisms, illustrating a number of different sporulation processes. In addition, endospore formation in bacteria, and ascospore formation in *Saccharomyces* are depicted in Figs. 2 and 4.

Eimeria schubergi

Azotobacter

Xenococcus (Pleurocapsalean Cyanobacterium)

Streptomyces

Chlorella

Phycomyces

Chlamydomonas

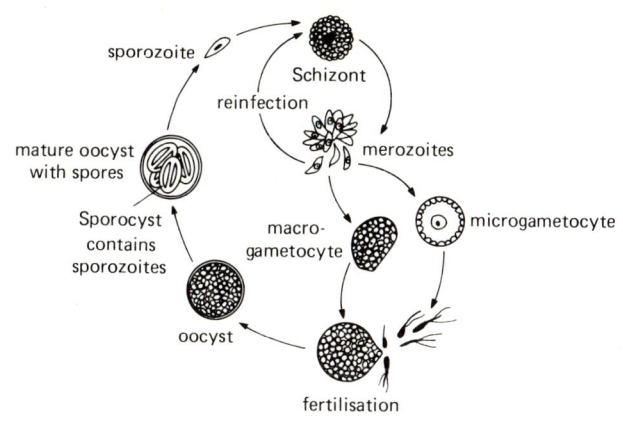

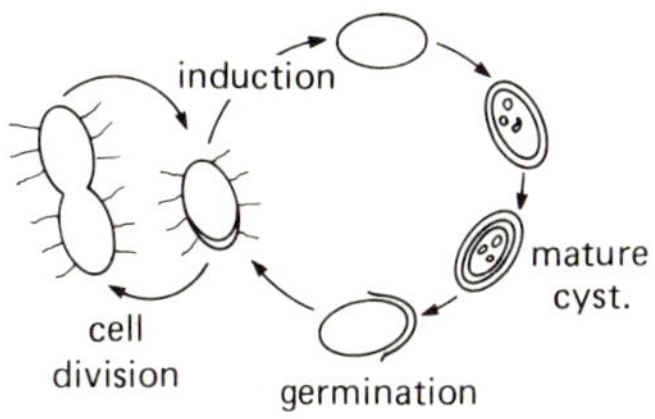

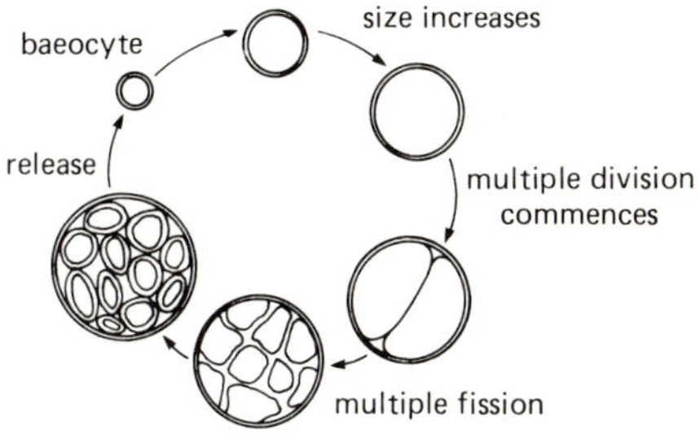

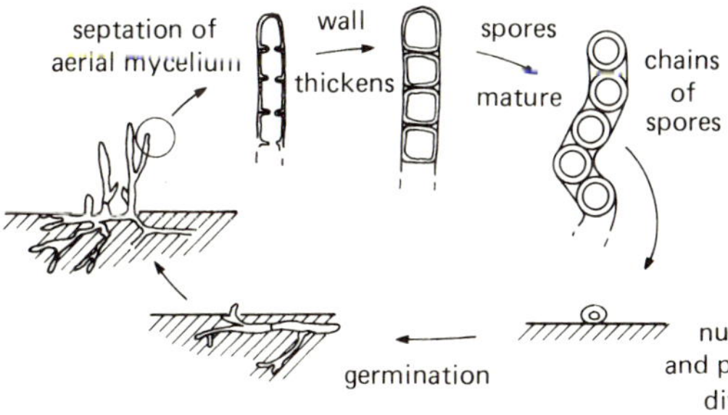

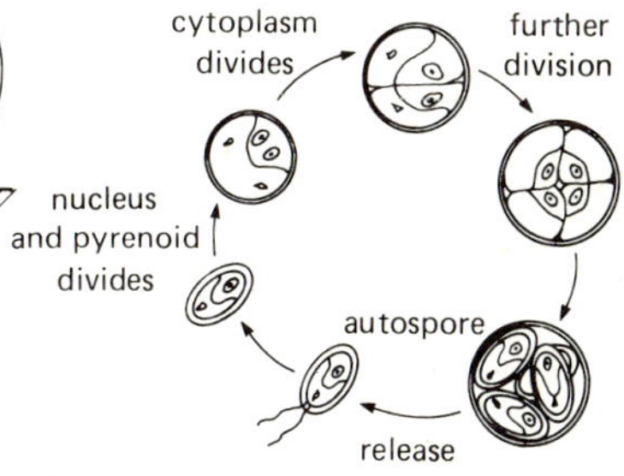

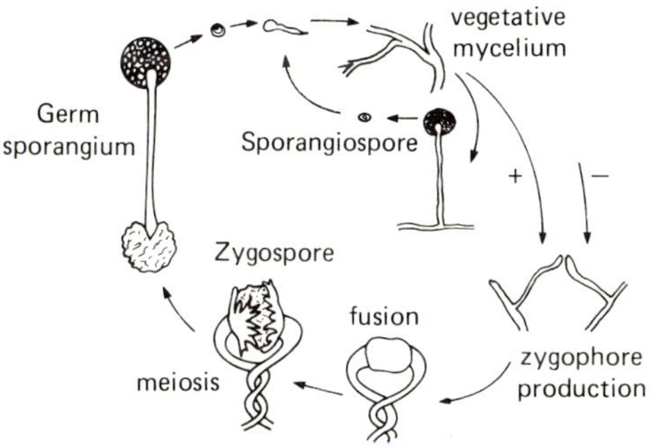

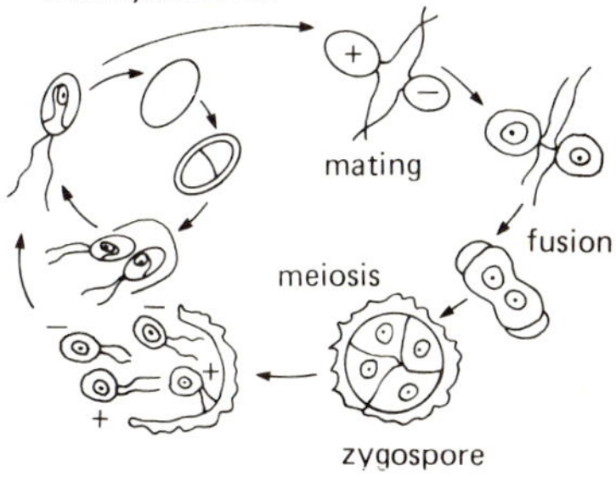

Table 2. *Representative spores and sporulation processes*

Spore type	Representative organisms	Mechanism of production	General references
1. PROKARYOTIC			
a. *Endogenously formed*			
(i) endospore	*Bacillus, Clostridium, Sporosarcina, Sporolactobacillus, Desulfotomaculum Thermoactinomyces*, etc.	Acentric invagination of cell membrane, followed by engulfment of prespore and development of wall layers. 1 spore per cell.	Slepecky (1972) Cross (1970)
(ii) holdfasts/endospore	*Arthromitus* (?)	Acentric invagination and engulfment. 2 holdfasts per cell formed by division of the included body. The two holdfasts sometimes are enclosed in a single endospore.	Chase & Erlandsen (1976)
(iii) baeocytes	Pleurocapsalean cyanobacteria	Reproductive units formed by multiple cell divisions. Many baeocytes per cell.	Waterbury (1979) Waterbury & Stanier (1978)
(iv) zoospores	Actinomycetes: *Planomonospora, Planobispora, Dactylosporangium*	Formed within parent hypha which constitutes an elongated sporangium.	Cross & Attwell (1975)
b. *Exogenously formed*			
(i) cysts heterocysts/akinetes	*Azotobacter, Beijerinckia, Methylosinus, Methylobacter,* Filamentous Cyanobacteria	Either by direct thickening of the vegetative cell wall, or by the deposition of additional wall layers.	Sadoff (1975) Carr & Bradley (1973) Whittenbury et al. (1970)
(ii) exospores, arthrospores	*Methylosinus, Rhodomicrobium Streptomyces, Nocardia* and other actinomycetes	Deposition of additional wall layers around cells formed by budding or hyphal septation; one exospore per cell, can be present in chains of spores.	Whittenbury & Dow (1977) Cross (1970) Kalakoutskii & Agre (1976)
(iii) zoospores, aplanospores	*Actinoplanes, Spirillospora Streptosporangium*, etc.	Formed by septation of hyphae within a surrounding envelope. One to many spores per vesicle.	Cross (1970)

2. FUNGAL
a. *Endogenously formed*

(i) ascospore	Ascomycetes	Asexual spores formed by enclosure of haploid nuclei after meiosis. Double-layered membrane from endoplasmic reticulum surrounds spores, wall synthesised between layers. 4, 8 or 16 per ascus.	Moens (1971)
(ii) zoospores, sporangiospores meiospores	lower fungi (Phycomycetes)	Nuclei in a multinucleate sporangium are enclosed by vesicle-derived membranes. Few to many spores per sporangium.	Lovett (1975)

b. *Exogenously formed*

(i) zygospores	Zygomycetes	Sexual spores formed by thickening of wall formed after fusion of two gametangia. New walls formed inside old.	Gooday (1973)
(ii) oöspores	Oomycetes	Sexual spores formed after fertilization *in situ* of oospheres produced endogenously in the oogonium.	Alexopoulos (1962)
(iii) basidiospores	Basidiomycetes	Sexual spores formed by a specialized budding of the basidium in which meiosis occurs.	Burnett (1968)
(iv) conidia	mainly Ascomycetes	Asexual spores produced singly at the tip or side of the hypha by budding.	Turian & Bianchi (1971)
(v) arthrospores, oidia, blastospores		Asexual spores produced by septation or fragmentation of hypha.	
(vi) chlamydospores		Asexual spores formed by the thickening of the wall of a hyphal cell.	Alexopoulos (1962)

Spore type	Representative organisms	Mechanism of Production	General references

3. ALGAL

In a recent treatment of phycology (Bold & Wynne, 1978) the term spore has been confined largely to asexually produced structures. Zygotes are often resting stages and the terms 'zygospore' and 'oospore' have been used in earlier literature (Erben, 1962). Some spores may be produced either exogenously or endogenously.

(i) sporangiospores, polyspores paraspores, seirospores, unispores	all groups	Spores produced within specialized cells or groups of cells.	Bold & Wynne (1978) Erben (1962)
(ii) tetraspores	Rhodophycophyta	Haploid spores produced in fours within a diploid cell following meiosis.	
(iii) akinete	Chlorophycophyta	Thickening of the vegetative cell wall.	
(iv) zoospores, hypnospores aplanospores, androspores	Chlorophycophyta, Chrysophycophyta, Phaeophycophyta	Flagellated (or ontogenetically potentially so) asexual reproductive spores usually formed by division of protoplast within parent cell wall.	
(v) auxospores	Bacillariophyceae (Diatoms)	Result from fusion of gametes; not a resting stage, increasing in volume soon after formation.	
(vi) monospores	Rhodophycophyta	Exospores formed by unequal division or direct transformation of the vegetative cell to a spore.	
(vii) autospores	Chlorophycophyta,	Reproductive cells that resemble the parent cell, and are formed within it by division of its protoplast.	
(viii) statospores	Chrysophyceae	A resting stage formed by delimitation of part of the vegetative cell protoplast within a membrane derived from the cell membrane.	

4. PROTOZOAL

The typical resting stage of most protozoa is a cyst, but for the phylum Sporozoa it is described as a spore

(i) spores	Sporozoa	spores formed within the zygote, each spore contains one to eight sporozoites which are released on germination.	

There are two general ways that spores are formed; *endospores* are generated within the cytoplasm of a 'mother cell', while *exospores* and *cysts* are formed by the deposition of coat layers or the thickening of a pre-existing cell wall external to the existing cell membrane. Within these categories there are significant differences in mechanism; for example, bacterial endospores can be of two types, 'true' endospores formed as one spore per cell by a specific process of membrane invagination, or *baeocytes*, formed in Pleurocapsalean cyanobacteria by multiple divisions of a larger cell (Waterbury & Stanier, 1978; Waterbury, 1979). Bacterial exospores include *cysts*, e.g. in *Azotobacter* (Sadoff, 1975) or *Methylobacter* (Whittenbury *et al.*, 1970) and the *myxospores* of the myxobacteria (Dworkin, 1972) formed by the thickening of the external wall structures, as well as more complicated forms resulting either from septation or fragmentation of hyphae within a surrounding envelope e.g. *Nocardia* or *Streptomyces* (Cross, 1970) or by budding off and thickening of the bud wall, e.g. in *Methylosinus* (Whittenbury *et al.*, 1970).

In addition to this diversity of morphology, there are also some differences in the environmental or physiological conditions that have been reported to influence or trigger the sporulation process. In most cases, however, sporulation is a response directly or indirectly to the depletion of nutrients, as was proposed as early as 1890 for *Bacillus* endospore formation (Büchner, 1890). Other examples include; sporulation in *Streptomyces* (Kalakoutskii & Agre, 1976), zoospore formation in the water mould, *Blastocladiella* (Lovett, 1975), ascospore formation in *Saccharomyces* (Miller, 1973a, b) and conidiation in *Neurospora* (Stine & Clark, 1967). For *Azotobacter* encystment and myxospore formation, development can be *induced* by specific compounds, and in the case of *Azotobacter* the inducer, β-hydroxybutyrate, is metabolized (Sadoff, 1975). This does not, however, necessarily imply that under normal circumstances encystment is not induced by a restriction of the nutrient supply. There are cases in which starvation is not apparently directly involved. For example, zygospore formation in fungi and algae and ascospore formation in some ascomycetes follow the fusion of haploid gametes. Even in these cases, however, starvation does play a part, since gamete formation and conjugation are often induced by starvation conditions, e.g. in *Schizosaccharomyces pombe*, in *Mucor* (Gooday, 1973) and in *Chlamydomonas* (Sager & Granick, 1954).

Within this rich tapestry of sporulation processes it is a daunting task to trace the threads of evolution of each type. In this article, the discussion is therefore concerned with questions such as when sporulation evolved in prokaryotes and eukaryotes, how it evolved in a few selected cases, what advantages accrue from the ability to sporulate and finally the more speculative area of what part sporulation may have played in evolution.

THE EMERGENCE OF PROKARYOTIC SPORULATION

The fairly widely accepted time scale for evolution, proposed mainly from a consideration of the palaeontological record, puts the emergence of living organisms during the Precambrian era, between 3 and 4×10^9 years ago (Barghoorn, 1971; Schopf, 1978). The earliest living forms were probably prokaryotes, anaerobic heterotrophs from which anaerobic phototrophs and subsequently bluegreen bacteria developed. The oxygen generated by the latter led to a dramatic shift in the microbial environment, heralding the eventual emergence around 1 to 1.5×10^9 years ago of eukaryotes. While the time of first appearance of eukaryotes is subject to uncertainty, possibly as much as 2×10^9 years ago (Kazmierczak, 1979) based on the rather tenous assignment of fossils as eukaryotes, it is fairly clear from the fossil record that prokaryotes predominated until about 1.5×10^9 years ago (Cloud *et al.*, 1969; Barghoorn, 1971; Schopf, 1978).

Apart from the notable exception of heterocyst formation in cyanobacteria (Licari & Cloud, 1968; Schopf, 1974), the fossil record does not help in determining when cellular differentiation processes arose in prokaryotes. This is probably due, not to the absence of other spore forms at the time the earliest fossils were preserved, but to the difficulty of recognizing them, because of their small size, in the materials so far examined. Nonetheless, the finding of structures resembling heterocysts in blue-green bacteria-like fossils of age 2.2×10^9 years sets a reasonable latest date for the emergence of at least one form of exospore structure. Since heterocysts are considered to be the site of nitrogen fixation (Fay *et al.*, 1968; Carr & Bradley, 1973), providing the nitrogenase complex with an essential reducing environment in an oxygen-containing atmosphere, the heterocyst may have developed from a previous 'cyst' structure extant before substantial amounts of oxygen appeared in the environment about 2×10^9 years ago.

For other forms of prokaryote sporulation the problem can, to some extent, be resolved by considering the phylogeny of the sporulating organisms now extant and inferring at what stage in the evolution of bacteria spores may have first appeared. This task is made easier by the recent revolution in bacterial taxonomy engendered by the sequence analysis of 16S ribosomal RNA (Fox et al., 1980; Stackebrandt & Woese, this volume). Endospore formation occurs in a relatively narrow range of bacteria, although it is not restricted solely to the Gram-positive rods of the genera *Bacillus* and *Clostridium* (see Cross, 1970; Slepecky & Leadbetter, 1977; Hanson, 1979). Heat-resistant, dipicolinic acid (DPA)-containing spores are formed in Gram-positive cocci of the genus *Sporosarcina* (Thompson & Leadbetter, 1963), in the filamentous *Thermoactinomyces* (Cross, Walker & Gould, 1968; Cross & Attwell, 1975) and putative endospores have been reported in a number of other genera including *Sporolactobacillus* (Kitahara & Lai, 1967), *Thiobacillus* (Egorova & Deryugina, 1963), *Fusoporus, Arthromitus, Coleomitus, Bacillospira, Sporospirillum, Oscillospira, Metabacterium* (Robinow, 1960; Slepecky, 1972; Slepecky & Leadbetter, 1977, citing Bergey's Manual of Determinative Bacteriology, 1974) and *Desulfotomaculum* (Campbell & Postgate, 1965; Nazina & Pivovarova, 1979). Hanson (1979) has pointed out that these represent a physiologically diverse group of organisms, including aerobes and anaerobes, nitrogen fixers, facultative chemolithotrophs and sulphate reducers. In many of the less well-known cases, however, the morphology and biochemistry of the sporulation process have not been characterized in adequate detail to determine whether or not they resemble those seen in *Bacillus*.

On the basis of the proposed major lines of prokaryotic descent, and the more detailed phylogenetic relationships for eubacteria set out by Fox et al. (1980), it would appear that this well-characterized process of endospore formation arose very early in the progenitor of the Gram-positive eubacteria represented by *Clostridium* and its relatives (Fig. 5 in Fox et al., 1980; see also Stackebrandt & Woese this volume). It is interesting that this approach to phylogeny separates *Thermoactinomyces vulgaris* from the actinomycetes and puts them very close to *Bacillus subtilis* and *Bacillus sphaericus*; *Sporolactobacillus* is close to *Bacillus coagulans*.

A number of authors have suggested, on the basis of the wide range of physiology seen in the endospore formers, that this type of sporulation may well be far more widespread in bacteria than is now perceived. While this is doubtless the case, the one main example used to illustrate the extent of diversity, the putative Gram-negative spore formers of the genus *Desulfotomaculum*, highlights the need to check by electron microscopy the nature of both the vegetative and sporulation phases. A recent detailed electron microscopic examination of sporulation in *D. nigrificans* has indicated that the main stages of sporulation resemble those described for *Bacillus* spp., and that, despite its classification as a Gram-negative organism, its cell wall is devoid of an outer lipoprotein membrane and is typically that of a Gram-positive organism (Nazina & Pivovarova, 1979).

There is, however, an interesting organism with a multi-layered wall resembling cytologically that seen in Gram-negatives. This undergoes a curiously-modified sporulation process with considerable affinities to that seen in *Bacillus* spp. The organism is a filamentous segmented bacterium found attached to the ileum of rats and mice (Chase & Erlandsen, 1976) and is discussed in more detail, together with the possibility of a relationship between the evolution of Gram-negative bacteria and endospore formation, in a later section.

Clearly, future research may help to resolve the present uncertainties. All organisms reported to form endospores need to be examined in detail by electron microscopy, and in their phylogenetic relationship to clostridia or other groups of bacteria, by all available criteria, e.g. 16S ribosomal RNA sequences, presence or absence of specific compounds such as DPA, modified peptidoglycan etc.

The origin of bacterial exospores is more difficult to date since they have arisen in diverse genera, there are a variety of different mechanisms (even within the actinomycetes there are several) and they have in many cases a simpler form of morphological specialization, i.e. the thickening of the vegetative cell-wall structure. Nonetheless, it is interesting that in the Gram-negatives, the cysts of *Azotobacter* and *Beijerinckia*, the exospores of the apparently unrelated myxobacteria, and the cysts of the methane oxidizer *Methylomonas* all contain $\alpha(1,3)$glucan (Sutherland & Mackenzie, 1977).

POSSIBLE COURSE OF ENDOSPORE EVOLUTION

Endospore formation in bacteria is an attractive system in which to consider how a process of cellular differentiation may have evolved.

The process appears to have arisen early in simple undifferentiated cells; it goes through morphological steps which are neither too many to be too complicated nor too few to be trivial. Moreover the process is by far the most extensively characterized example of cell differentiation.

Bacillus endospore formation has been reviewed in detail so recently (Hanson, 1979; Chambliss, 1979) that only the most salient features are summarized here.

The seven morphological stages of sporulation are outlined in Fig. 2. Sporulation is initiated under conditions in which a carbon source, a nitrogen source, or phosphate is limiting for growth. This is not an all-or-none response. The chance that a cell will sporulate rather than divide is a function of the nature and concentration of the limiting nutrient (Schaeffer *et al.*, 1965; Dawes & Thornley, 1970). Moreover, cells can be initiated to sporulate only at a particular stage in the cell division or DNA replication cycle (Dawes *et al.*, 1971; Mandelstam & Higgs, 1974). The early stages (I & II) resemble a modified cell division (Hitchins & Slepecky, 1969), except that septum formation is asymmetric, and cell wall synthesis between the septum layers is arrested. The smaller protoplast containing at least one complete chromosome is then engulfed by the larger (Fitz James, 1960) to form a 'forespore' enclosed within the mother cell (Stage III). This forespore is bounded by two membrane layers that have 'reversed polarity' due to the way invagination occurred. If peptidoglycan synthesis were to resume at this stage, it would be formed between the two forespore membranes. During Stage IV, this happens, the peptidoglycan of the cortex layer being synthesized. Stage V is characterized by the assembly, external to the outer forespore membrane, of a multi-layered spore coat structure (Aronson & Fitz-James, 1976); the mature spore form is Stage VI and lysis of the mother cell in Stage VII leads to the release of the spore.

This process in *Bacillus subtilis* involves fifty or more operons specific to sporulation (Hranueli *et al.*, 1974; Piggot & Coote, 1976) and the programmed expression of some genes (Coote & Mandelstam, 1973; Mandelstam, 1976). Some proteins are synthesized during sporulation that are not synthesized by vegetative cells, and the synthesis of some others is turned off during sporulation (Linn & Losick, 1976). How did such a multi-stage, highly-regulated process arise?

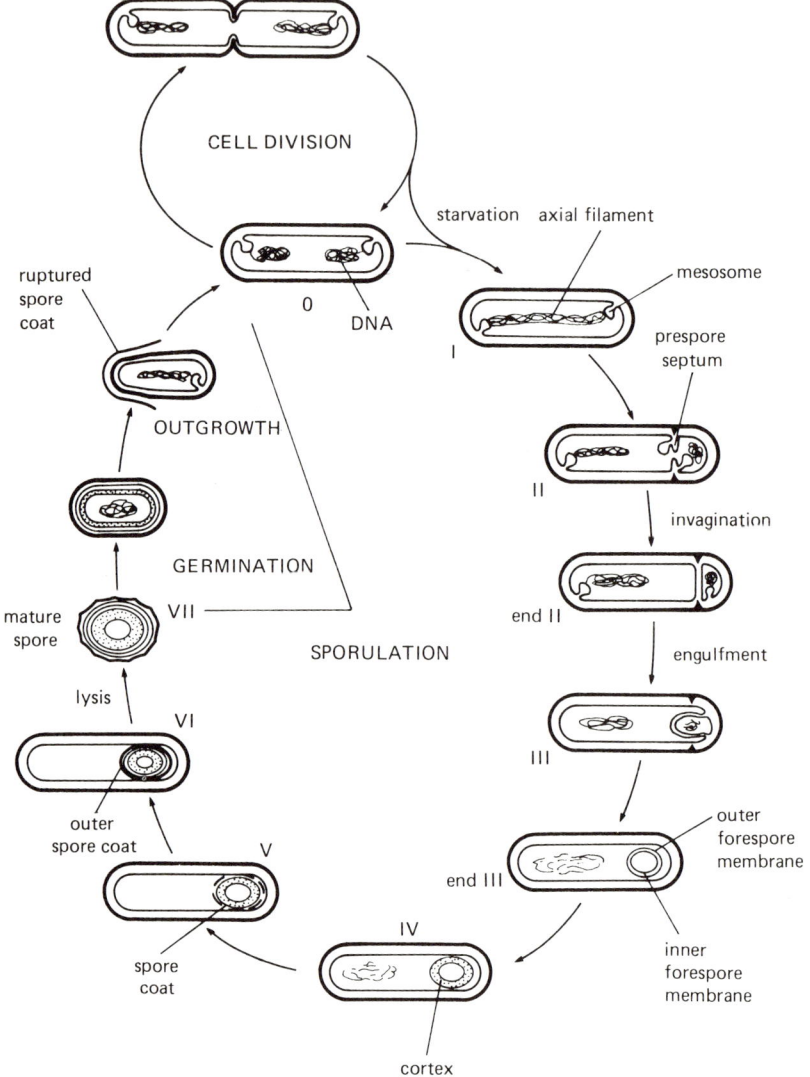

Fig. 2. Endospore formation, germination and outgrowth in *Bacillus* species. The seven stages of sporulation are indicated.

In trying to trace an evolutionary sequence of this nature, it is necessary to make the following postulates.

(i) The individual steps in the evolutionary pathway resulted from very few mutations, preferably only one.
(ii) The outcome of each step should be heritable, i.e. not interfere with normal reproduction of the micro-organism.

Once a step in the evolution of sporulation involved a process normally concerned with reproduction of the cell (i.e. septation, chromosome segregation) it had to be *conditional*, in the same way that a mutation affecting an essential gene function has to be, to allow survival of the mutant.

(iii) Most of the evolutionary steps should confer some additional selective advantage. In general terms, sporulation and vegetative cell division are mutually exclusive so that cells undergoing sporulation are unable to contribute directly to population growth. Sporulation is, in fact, a disadvantage under conditions in which growth is restricted by depletion of the nutrient supply since mutants blocked in the earliest stages of sporulation are selected (Aubert *et al.*, 1965; Michel *et al.*, 1968; Dawes & Mandelstam, 1970). Thus the acquisition of most of the successive steps in the evolution of sporulation (which is triggered by nutrient limitation) must have resulted in some very positive advantage(s), or back mutation and selection to asporogeny would have occurred rapidly.

The first stable intermediate

Since sporulation is triggered by starvation, it is likely that the first stable 'intermediate' to emerge provided the organism with an increased chance of surviving starvation. If, as is commonly thought, the earliest living forms were (anaerobic) heterotrophs, the selection pressure imposed by starvation would have arisen very early, especially in small enclosed pools subject to intermittent replenishment by the 'primaeval soup'.

Moreover, since it is most likely that the initial steps of this developmental sequence evolved first and that these appear to involve a modification of the normal process of cell division (see Hitchins & Slepecky, 1969; and a subsequent detailed review by Hitchins, 1978), the important differences between the two processes can be summarized as follows.

(i) Activation of one terminal septation site as opposed to a central one, in such a way that a fully-replicated chromosome is enclosed at one pole of the cell.

(ii) Inhibition of peptidoglycan synthesis in the developing spore septum, and possibly also the hydrolysis of some peptidoglycan that would normally form the new vegetative cell cross wall.

(iii) Engulfment of the spore protoplast by differential growth of the mother cell membrane (relative to cell wall and inner spore membrane growth); this may also involve a termination switch to shut off further membrane extension when engulfment is complete.

Mutations affecting some of the 'switches' controlling these processes have been identified, since the terminal phenotypes of some spoII and spoIII mutants display an obvious defect in the switching and/or positioning of wall and membrane synthesis e.g. some abortive Stage II mutants synthesize cell wall material in the developing septum, and a Stage III mutant over-produces the double-layered membrane during invagination (Waites et al., 1970).

It is proposed that the first stable step in the evolution of endospores, providing increased resistance to starvation, was the formation of an engulfed protoplast enclosed within the double membranes, i.e. the Stage III protoplast. Such compartmentation would confer considerable survival potential on the enclosed 'forespore' since metabolism in the forespore would be reduced due to the reversed polarity of the enclosing membranes, and contact with the external environment would no longer be direct.

At first sight, this appears to involve the simultaneous appearance of too many changes, and the initial emergence of a 'small protoplast', analogous to the Stage II prespore, would seem to be a more logical choice.

However, it is possible to explain how the ability to form a Stage III protoplast under starvation conditions could have been inherited by an early organism. The crucial step would have been the acquisition of a mutation causing the activation of a terminal cell septation site *under starvation conditions*. This fits with the observation that sporulation can only be initiated at a particular point in the cell division cycle (Mandelstam & Higgs, 1974) probably associated with DNA replication. It also implies that the initiation of sporulation should be of interest to those studying the cell division cycle; this has, in fact, been considered in more detail for sporulation in the eukaryote *Saccharomyces cerevisiae* (Shilo et al., 1978; Vezinhet et al., 1979). Also, mutations leading to asymmetric cell division producing either anucleate minicells, or cells of all ranges and sizes (Donachie, 1979) have been identified in *Escherichia coli* and in *B. subtilis* (Reeve et al., 1978) and mini-like cells occur in stationary phase cultures of *B. subtilis* (Neale & Chapman, 1970). Hitchins

(1978) has concluded that it is not yet possible to explain the asymmetric location of the sporulation septum in terms of a modified model for the control of cell division septation (because the process of symmetric septation is not yet understood). It should also be kept in mind that the controls operating on cell division in the progenitor of sporulating bacteria may have been less rigorous than they are at present. Even now it is quite common to see spore septa formed at both ends of the cell in *Bacillus subtilis*, only one of which is engulfed.

How then does one explain the inhibition of peptidoglycan synthesis and the process of engulfment appearing without many simultaneous mutations? One clue comes from the conditions of nutrient starvation that lead to triggering of sporulation, since not all do.

In *Bacillus* species it is limitation of either a main energy source, or a source of fixed nitrogen, but not of ions (with the possible exception of phosphate) or of auxotrophic requirements. Net peptidoglycan synthesis would not be possible if either sugars, or precursors to sugars, or fixed nitrogen was the substrate first to be exhausted. Fixed nitrogen may have indeed been in short supply in the early Precambrian since nitrate ions could not then have been formed (by the reaction of atmospheric nitrogen and oxygen) and ammonia would have been susceptible to u.v.-induced dissociation into nitrogen and hydrogen (Schopf, 1978) in the absence of a u.v.-absorbing ozone layer. Under these conditions of limitation, the cell membrane could have extended by the continued synthesis of lipid and this may have accounted for formation of the primitive Stage II septum lacking peptidoglycan by a process of imbalanced synthesis and buckling (Pitel & Gilvarg, 1971). It could also account for a primitive engulfment process since an asymmetric cell septation, giving rise to unequal protoplasts, would lead to unequal rates of membrane synthesis in the two, especially under nutrient-limited conditions. A greater rate of membrane lipid synthesis would occur in the larger one due to its possessing a larger surface area and a greater number of the components needed for lipid synthesis. In the absence of a balance between cell wall extension and lipid synthesis in the presence of an asymmetric septum, engulfment would be inevitable. A rigid mother cell wall is still necessary for engulfment since protoplasts of Stage II cells cannot carry out the process (Fitz-James, 1964).

In proposing this mechanism, there is no denial that these early stages of sporulation may *now* have come under considerable regulation; the initial process need only have been successful at low frequency to provide a distinct advantage towards survival of the primitive sporulating organism; one spore can generate a population. Subsequent changes would have brought the initiation of sporulation under regulation, probably making it respond to a metabolic state associated with guanine nucleotide deprivation (Elmerich & Aubert, 1975; Lopez, Marks & Freese, 1979).

Metabolic preconditions?

Wright (1970) has suggested that there may have been a requirement for a number of metabolic preconditions as first steps in the evolution of a simple differentiation process. Cells would be selected having the abilities: (i) to synthesize high levels of endogenous reserves; (ii) to utilize them efficiently by producing appropriate lytic enzymes, and (iii) to raise partial permeability barriers to protect the cell from the loss of small molecules resulting from the degradation of macromolecular reserves. There is no *a priori* reason for these to have been essential preconditions for the emergence of the 'prespore' outlined above, since there is no postulated requirement for large amounts of macromolecular synthesis. These attributes would even have led to an extension of the survival of vegetative cells and are not restricted to differentiating organisms under starvation conditions (Mandelstam, 1958). Endospore-forming bacteria synthesize a whole battery of *extracellular* lytic enzymes after they begin to sporulate (Schaeffer, 1969); these are presumably not to utilize their own reserves, but those of other organisms and, as such, would provide strong selective advantage once sporulation had evolved as a successful process. Nonetheless, the acquisition of reserve compounds and the ability to use them and other 'unnecessary' structures would have provided the metabolites for the synthesis of protective coats, and this brings us to the next step.

Subsequent steps

The next major step was probably the ability to switch on again the synthesis of peptidoglycan between the forespore membranes, to form a primordial cortex. This structure may initially have resembled cell wall in composition, evolving gradually to the present, more loosely cross-linked and highly charged structure (Warth & Strominger, 1969; Tipper & Gauthier, 1972). This cortex would have provided the 'prespore' with a vastly increased resistance to desiccation as well as starvation due to the osmotic properties of the peptidoglycan (Gould & Dring, 1975).

It is interesting that the spore peptidoglycan of *Bacillus sphaericus* resembles that of most other sporulating bacilli in containing *meso*-diaminopimelic acid in the peptide side-chain, whereas the vegetative cell wall polymer *B. sphaericus* contains lysine as the subterminal amino acid (Tipper & Gauthier, 1972). Since the spore form of the polymer is very similar throughout the *Bacillus* spp. tested, it seems likely that sporulation arose in an organism with peptidoglycan of Chemotype I (Ghuysen, 1967) and the cortical peptidoglycan in *Bacillus sphaericus* is more primitive than its vegetative counterpart. This may apply to other spore components. Still others may be vestiges of an earlier metabolism. For example, glucose dehydrogenase, which is lacking from vegetative cells of most *Bacillus* spp., is synthesized during sporulation and may function in spore germination (Prasad *et al.*, 1972); it may have had a more fundamental role in the primitive cell.

The spore coat

Lewis (1969) pointed out that the protein coat of the spore appears to be unique in the eubacteria insofar as other spore structures arise from homologous components of the vegetative cell, and that perhaps it came early on the evolutionary road to the spore as we know it. The spore coat protein, comprising 40 to 80% of the total protein (Aronson & Fitz-James, 1976), does not seem to have any counterpart in vegetative bacteria. In *Bacillus cereus*, the spore coat seems to be composed of a single major subunit polypeptide (of mol. wt. about 12 000) which is rich in cysteine residues and extensively cross-linked in the final structure. On the other hand, the thicker *B. subtilis* coats may be composed of more than one polypeptide. Species show more variation in the composition of their spore coats than in most other features. These differences are revealed not only in terms of chemical composition, but also in the arrangement of layers within the spore coats, and of subunits within the layers, as revealed by freeze-etch electron microscopy (Aronson & Fitz-James, 1976).

The spore coat could have appeared before or after the cortex in the evolutionary sequence. Cortex-less mutants, or phenotypic variants lacking the cortex in which the spore coats are deposited normally have been described (Balassa & Yamamoto, 1970; Coote, 1972). Similarly, mutants lacking normal coats in which cortex formation appears to be normal have been reported (Coote, 1972; Aronson & Fitz-James, 1975). Since the spore coat's main function now appears to involve the protection of the cortical structure (coatless spores are susceptible to germination by lysozyme, or by divalent cations) and since spores stripped of their coats retain most of their resistance to heat and desiccation and retain the metabolism characteristic of the dormant state, the spore coats may have evolved after the cortex. This would allow the spore to survive without requiring the enzyme-resisting properties of the mother cell wall. Lewis (1969) suggested that the acquisition of the coat protein allowed the evolution of a passive peptidoglycan of the cortical region into a functional pressure-generating, i.e. osmotically-active, cortex, as the coat took over the protective function of the peptidoglycan.

The coat protein displays a degree of regular assembly, probably much that is self-assembly, and a chemical resistance that is not found in other vegetative structures. Where may such a structure have originated? One highly speculative suggestion is based on the similarity between spore coat assembly and the assembly of the protein coats that form the capsids of bacteriophage. Maybe the spore coat originated from a lysogen carrying a defective phage genome. This is perhaps less fanciful than it seems at first sight. Many *Bacillus* and *Clostridium* species carry bacteriocins and some of these are of the defective-phage type (Tagg *et al.,* 1976). Moreover, a major feature of the regulation of phage gene expression is the alteration after infection of the transcription mechanisms of the host; for a number of phages, e.g. T4 and λ of *Escherichia coli* and phage SPO1 of *Bacillus subtilis* (Doi, 1977), this is achieved by modification of the host cell RNA polymerase. An analogous type of modification appears to occur during sporulation in *B. subtilis* to bring some genes under sporulation control (Segall & Losick, 1977; Sonenshein & Campbell, 1970). This may imply an external phage origin for some of the control of gene expression during *bacterial* sporulation. Obviously, the activation of such a defective phage must have come under some form of metabolic control at an early stage; such control may have already been in existence, since in phage λ one of the modes of repressor protein synthesis is sensitive to catabolite repression (Grodzicker, Arditti & Eisen, 1972).

Sporulation metabolism

With the development of each step, there was doubtless a number of regulatory and other changes which brought the process under tighter control. The metabolism of the sporulating cell and also of the germinating spore would have become directed towards the provision of energy and biosynthetic intermediates at the expense of existing vegetative polymers, so that intracellular proteases in particular may have increased in number and specificity (Reysset & Millet, 1972; Setlow, 1977).

The presence in the spore protoplast of high levels of DPA and calcium ions present as the CaDPA chelate, has been the subject of considerable speculation. It may be that DPA synthesis and concentration within the spore protoplast evolved as a mechanism for removing calcium ions from the modified cortical peptidoglycan, and the chelate, or at least the calcium-ion component, assumed a role in the stabilization of the protoplast membrane (Fitz-James, 1971; Ellar, 1978) or in the germination of the spore (Gould, 1977; Hanson, 1979), since spores of a heat-resistant DPA$^-$ mutant failed to germinate normally.

At many steps in the evolutionary pathway the 'mature spore' form would have become progressively more dormant. Concomitant with this increase in dormancy at the expense of some vegetative cell components would have been the need to accumulate reserve compounds which would be used very rapidly by the germinated spore. The low molecular weight compounds include the amino acids glutamate and, to a lesser extent, lysine and arginine, and also 3-phosphoglycerate. In some strains of *Bacillus subtilis,* sulpholactic acid also accumulates (Nelson & Kornberg, 1970; Wood, 1971). Why these compounds? One possibility is that, together with DPA, they are all related metabolically to the components of the 'normal' peptide side chain of the vegetative cell wall peptidoglycan. For example DPA and lysine are derived from the same pathway as diaminopimelic acid; 3-phosphoglycerate is a potential precursor of alanine; glutamate is a member of the side chain and arginine is one of the glutamate family of amino acids. Moreover, these compounds are taken up into the developing spore around Stage IV when cortex synthesis takes place (Singh, Setlow &

Setlow, 1977). As the cortical polymer became less cross-linked the pathways for synthesis of the peptide side chains would still have been active and able to provide metabolities for other functions even if the side chains were synthesized, but removed by peptidases.

The evolution of germination and outgrowth

The primordial spore could probably be stimulated to resume vegetative growth by a range of nutrients. Gradually, permeability barriers evolved, and with them a restriction of the number of molecular species that could trigger the first steps in the return to the vegetative state. The very condensed state of the eventual spore protoplast required the concurrent emergence of a process, i.e. germination, for returning the protoplast to a more hydrated swollen form. As a consequence, a variety of mechanisms for triggering germination evolved, each responding to a particular and very restricted set of molecules reflecting the various environments the species continued to occupy. Thus a number of *Bacillus* spp., including *B. subtilis, B. cereus* and some strains of *B. megaterium,* respond to L-alanine either alone, or in combination with ribonucleotides; other strains of *B. megaterium* respond to mixtures of D-glucose and nitrate ions, while *B. fastidiosus,* which can only grow on allantoin, glyoxylate or uric acid as energy source, will germinate only in uric acid (Den Dooren de Jong, 1929; Leadbetter & Holt, 1968; Slepecky & Leadbetter, 1977).

As the spore became a more specialized structure, with modified components for transcription, translation and other metabolic processes, successful return to the vegetative state would have required new functions specific to spore outgrowth. Mutations affecting the germination mechanisms and others solely affecting spore outgrowth have been isolated especially in *B. subtilis* (Galizzi *et al.,* 1973; Smith *et al.,* 1978).

EMERGENCE OF EUKARYOTIC SPORULATION

The very diverse nature of sporulation in microbial eukaryotes, and the uncertainty of when eukaryotes evolved (Stanier, 1970; Marguilis, 1970; Cavalier-Smith, 1975 and this volume) ensure that discussion of when and how differentiation of spore-like forms arose is highly speculative. This is especially the case when some structures described as spores, e.g. autospores in *Chlorella* (Griffiths &

Griffiths, 1969), are very similar to the mature vegetative cell and are either the result of a process of reproduction, or resemble cells from a reproductive phase of the life cycle. Other sporulation processes lead to more differentiated structures, as in many asexual resting spores of fungi, and sexual zygospores of fungi and algae. There is obviously a question of how one defines spores and sporulation; nonetheless, all forms currently described as spores, be they 'resting' or 'reproductive', may have relevance to the questions of how and when sporulation evolved. For example, some of the processes currently reproductive may have evolved following the loss of a division process during cell fission in an organism capable of forming multiple endospores. The converse may apply, that the fundamental process of mitosis arose in a 'sporulating' organism, and reproduction initially depended on the formation of multiple endospores.

The fossil record may eventually prove to be very rewarding in assessing when eukaryotes possessed structures that resembled spores, subject, of course, to the uncertainties in assigning eukaryotic status to fossil remains. Together with other evidence for the existence of eukaryotic fossils, Schopf (1978) lists several structures that may be relevant.

(i) Spheroidal microfossils exhibiting two-layered walls and having 'medial splits' on their surface which may represent an encystment stage of an eukaryotic alga; McMinn shale formation, Northern Australia, 1.46×10^9 years old.

(ii) A tetrahedral group of four small cells, resembling spores produced by mitotic cell division; Amelia dolomite of Northern Australia, ca. 1.5×10^9 years old.

(iii) A tetrahedral group of four spore-like cells which may have been the products of mitosis or possibly meiosis; Bitter Springs formation, Central Australia (Schopf & Blacic, 1971).

(iv) Spiny cells or algal cysts several hundred μm in diameter; Siberian shales, 0.95×10^9 years old.

This evidence is, of course, open to alternative interpretations, not only whether they are remains of spores or cysts, but even whether or not they record eukaryotic structures (Knoll & Barghoorn, 1975; see, however, Schopf & Ochler, 1976). The tetrahedral structures are very interesting since they resemble a number of eukaryotic forms. Initially, Schopf & Blacic (1971) suggested the exciting possibility that the fossil from the Bitter Springs formation,

Eotetrahedrion, represented a tetrad of algal spores derived by meiotic cleavage of a single parent cell, thereby implying an alternation of diploid and haploid generations and the existence of eukaryotic sexuality. Unfortunately, other interpretations are possible, as the authors were aware; this fossil may not even represent spores, since the tetrads seen in some green algae (notably *Tetracystis*) are the result of vegetative cell division in which some of the parental cell wall is passed on to the daughter cells. Since the individual members of each tetrad were enclosed within an outer wall, it seems likely that they arose by a form of sporulation. The above structures were found in the formations in which the earliest eukaryotes appear; earlier fossil records seem to contain only prokaryotic organisms.

If, as is widely accepted, eukaryotes originated from prokaryotes, it is possible that the eukaryotic progenitor was capable of sporulating. Gould & Dring (1979) pointed out that the engulfment process occurring during endospore formation in bacteria is strikingly similar to that suggested in one hypothesis for the origin of intracellular organelles in eukaryotes. In any event, most fungal and algal groups, and a number of protozoan ones, contain representatives that sporulate. Either the progenitor to eukaryotes was capable of sporulation, or sporulation processes arose a number of times in the evolution of different groups. The latter may seem more likely given the wide variation in asexual sporulation mechanisms in microbial eukaryotes, although in a later section the possibility that a sporulation process may have been required for the successful emergence of conjugation and meiosis is discussed.

Mechanisms in evolution of eukaryotic sporulation

For non-photosynthetic eukaryotes, there must have been considerable pressure for the selection of mechanisms to enable survival through periods of starvation. This could have been achieved by the evolution of a resistant phase in the life cycle, or the production of structures promoting dispersal. Both of these have evolved via sporulation processes.

It is not surprising, therefore, to find that nutrient deprivation in the eukaryotic, as in the bacterial system, triggers sporulation directly or is a major factor in its induction. Even in those instances in which more elaborate mechanisms have evolved, exemplified by the Mucorales in which the fungal 'hormone' trisporic acid initiates the process of zygophore formation, fusion and ultimate production of the zygospore (Gooday, 1973), the process takes place more readily under starvation conditions. Nutrient limitation induces conjugation in many other fungi in which meiosis directly follows nuclear fusion.

Sporulation versus vegetative reproduction

Apart from the simple exospore-forming process, there are not many eukaryotic sporulation processes that can be regarded as a modification of the 'normal' mode of cell division. Nonetheless, there are cases in which the relationship between cell division and sporulation can be investigated; the most extensively studied example is ascospore formation in *Saccharomyces cerevisiae*. The cytology of this process is outlined in Fig. 3. Since meiosis precedes sporulation in this organism, such studies include a comparison between mitosis and cell division on the one hand, and meiosis and sporulation on the other. In *S. cerevisiae*, a range of mutations affecting cell division (*cdc*) is available (Hartwell, 1974), and Shilo *et al.* (1978) have drawn attention to the similarities between the premeiotic and meiotic phases of yeast sporulation and the cell division cycle, by showing that a number of *cdc* genes, especially those concerned with the early stages of the cell cycle, are also involved in meiosis and spore formation. Moreover, initiation of sporulation can only occur during the G1 phase of the cell cycle, and mutations leading to a derepressed initiation of sporulation in diploids, cause an arrest in the G1 phase of the cell cycle in non-sporulating haploids (Vezinhet, Kinnaird & Dawes, 1979). The extent to which new gene functions are required during yeast sporulation is a question that has not yet been answered. While it has been estimated that about 50 genes are involved in yeast sporulation (Esposito, Frink, Bernstein & Esposito, 1972), pulse-labelling techniques and two-dimensional gel electrophoresis experiments have failed to reveal synthesis of proteins that are unique to a/α sporulating diploids (Trew, Friesen & Moens, 1979; Liske-Peterson, Kielland-Brandt & Nilsson-Tillgren, 1979; Kraig, Pearson & Haber, 1980). These results need, however, to be interpreted with caution since sporulating yeast cells erect permeability barriers to amino acids and nucleic-acid bases (Mills, 1972), and these increase during sporulation (Dawes, Wright, Vezinhet & Ajam, 1980). Changes in polypeptides specific to a/α cells have been

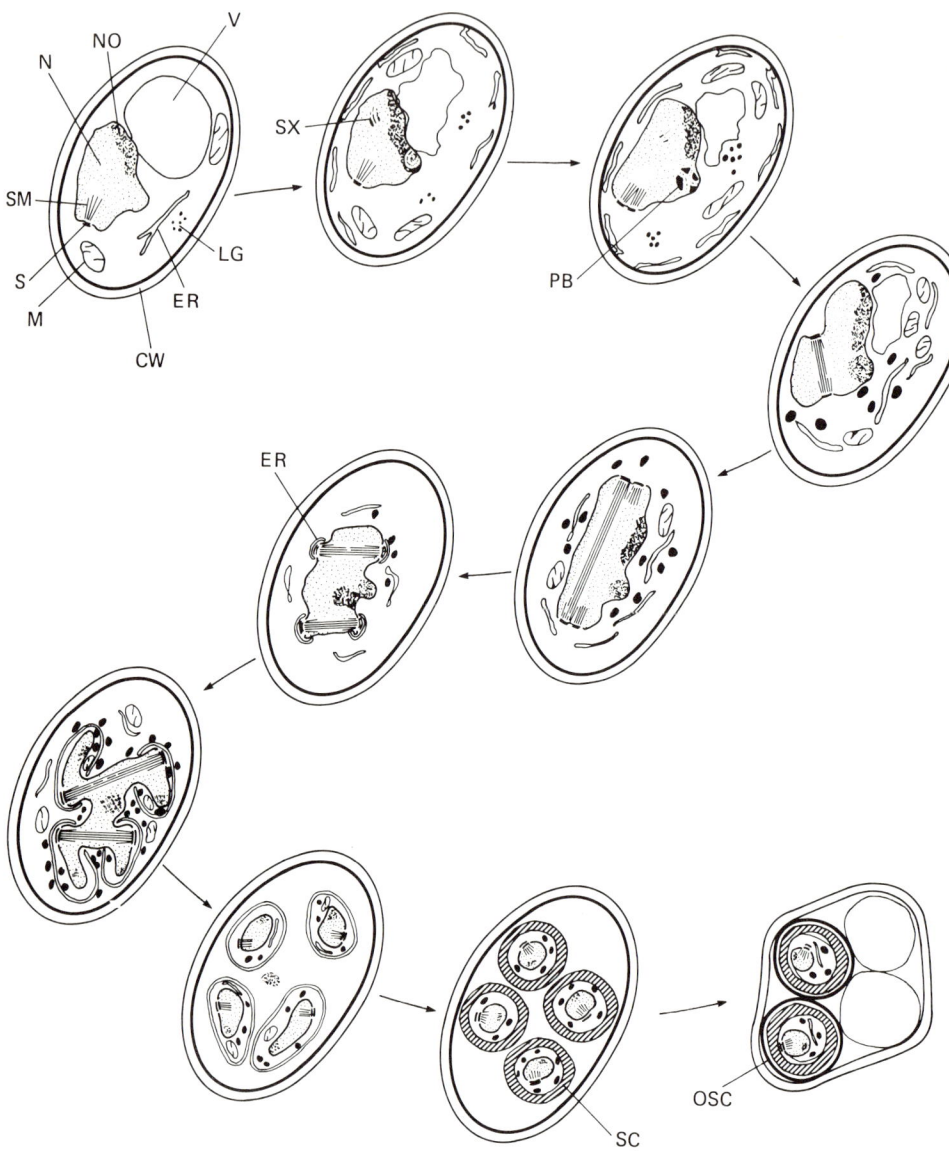

Fig. 3. The cytology of meiosis and sporulation in yeast. Redrawn after Esposito & Esposito (1974) and Fowell (1975). The structures represented are: cell wall (CW); endoplasmic reticulum (ER); nucleus (N); nucleolus (NO); mitochondrion (M); spindle plaque (SP) spindle microtubule (SM); lipid granule (LG); vacuole (V); synaptonemal complexes (SX); polycomplex body (PB); spore coat (SC); outer spore coat (OSC).

detected (Wright & Dawes, 1979), but under these conditions modifications to pre-existing proteins, as well as *de novo* synthesis, would have been detected. So far, only a few 'a/α specific' enzymes have been reported; these include an α-glucosidase for glycogen degradation (Clancy, Smith & Magee, 1980) and a β-glucanase (del Rey, Santos, Garcia-Acha & Nombela, 1980). Sporulation is, therefore, markedly dependent on vegetative cell functions. As in the case of endospore formation in bacteria, the developing spore becomes enveloped in a double-layered membrane (derived from the endoplasmic reticulum) and between the layers of this membrane is deposited a thick wall layer (Lynn & Magee, 1970) composed of glucan and mannan. These polymers are those found in the vegetative cell wall, but may be modified in the spore (I. W. Sutherland, personal communication). Moreover, a thin, electron-dense, outer spore-coat layer forms at a late stage in sporulation. This is probably composed largely of protein (Briley, Illingworth, Rose & Fisher, 1970) and is likely to be the determinant of the spore surface antigen (Snider & Miller, 1966). The relatively thick spore wall layers contain electronegative polymers (the phosphate of the mannan) and the yeast spore has increased resistance to solvents (Zakharov & Inge-Vechtomov, 1964; Dawes & Hardie, 1974) and to some extent, also to heat (Put & Sand, 1975), but is still very sensitive when compared with the bacterial endospore.

While there is some similarity between the formation of bacterial endospores and yeast ascospores, and yeast sporulation shows a marked dependence on vegetative cell division functions, it is not necessarily the case that ascospore formation arose directly as a modification of the present process of yeast cell division. In fact, it is possible that the converse may be true, that the unicellular yeast phase in Ascomycetes evolved after spore formation as a modification to the germination and outgrowth processes.

There are other, clearer, cases illustrating that once a diversity of life cycles and multicellular or hyphal forms had evolved, eukaryotic sporulation processes need not have arisen by the direct modification of a cell division process. An interesting case in the Mucorales is discussed by Bartnicki-Garcia (1970) and Gooday (1973). The asexual sporangiospores are formed within a sporangium on a hyphal sporangiophore, and they have a markedly different wall composition (mainly glucan) from that of the sporangiophore walls (chitin, chitosan and polyphosphates). The walls of the other differentiated structures, the zygospores and arthrospores, have similar components to the vegetative hyphae and the sporangiophores. Since the wall composition of fungi reflects clearly their phylogenetic relationships, Bartnicki-Garcia (1970) suggested that the sporangiospore wall represents an ancient trait, indicating a relationship with the Chytridomycete water moulds, and that the new vegetative cell wall formed inside the spore wall during germination is the more advanced type characteristic of the Zygomycetes. Thus sporangiospore formation in these organisms may have had an origin in a more ancient (cell division?) process.

The Mucorales also provide an example of one sporulation process (conidium formation) that may have evolved from another (sporangium formation) (Alexopoulos, 1962).

THE EVOLUTIONARY ADVANTAGES OF SPORULATION

The extreme resistance of bacterial spores has attracted considerable speculation on the advantages that sporulation confers, as well as argument as to whether spores can be ascribed a rôle in nature (Sussman & Halvorson, 1966). Some earlier authors were even unwilling to speculate on the functions of spores: Cook (1932) stated that 'the only interpretation that can be given as yet is that they form spores because they form spores'. Suffice to say that life would be more sterile without speculation – or sporulation.

One important point that is often overlooked, but yet is relevant to the argument of whether spores do have a rôle in nature, is that the ability to sporulate is not without disadvantages. Apart from the need to maintain an additional genetic burden during the vegetative phase of reproduction, and possibly also the requirement to maintain patterns of metabolism which limit growth rate (see Wright, 1970), there is the disadvantage that sporulation and cell division are usually mutually exclusive processes. Under conditions of nutrient limitation, or of periodic cycles of starvation and replenishment, asporogenous mutants are rapidly selected and such mutants are blocked at the start of sporulation (Michel, Cami & Schaeffer, 1968; Dawes, 1975). Thus, some of the conditions that trigger sporulation in a wide variety of organisms, prokaryotic and eukaryotic, are also those that favour selection of non-sporulating mutants. It seems likely that this is the reason why many non-sporulating variants have appeared e.g. Deuteromycotina, which has lost the sexual sporulation process: also the asporogenous strain

Bacillus megaterium KM, from which Fitz-James (1971) has recovered a mutant capable of producing slightly modified spores.

It can, therefore, be argued that sporulation in response to nutrient deprivation must provide some very positive advantages. When spores are formed in response to nutrient deprivation, the primary advantage would be their survival until nutrients are restored. Spores can extend the viability of an organism in time, and can also allow dispersal to a more favourable environment.

Survival in time

The unique resistance of bacterial endospores to extreme conditions, and the fact that only one spore is formed per cell seem to indicate that the advantage to the organism lies more in survival in time than, as suggested by Bisset (1950), in dispersal. The initial advantage conferred by the precursor to present spores may have been brief and have operated only in aqueous conditions; successive developments led to important extensions in both survival time and in the range of environments in which spores would remain viable. Probably the main advantage acquired was the ability to survive desiccation by osmotic control of the protoplast (Gould & Dring, 1975), since dehydration leads to a reduction in metabolism, and, therefore, to extension of the survival period and to an increase in the ability of vegetative cells to withstand heat (Keilin, 1958).

The extreme heat resistance of endospores has often been considered an enigma, or dismissed as a by-product of the acquisition of resistance to desiccation in an 'ametabolic state'; it may, however, have been an important advantage in the anoxic environment when sporulation developed. Temperatures at the earth's surface, especially dry surfaces, may have been much higher than at present due to the 'greenhouse' effect of carbon dioxide in the atmosphere. Similarly, the high radiation resistance of endospores may have been required to a greater extent than at present since in an anoxic environment ultraviolet radiation would have been more intense in the absence of an ozone layer.

Among the adaptations that prolong the dormant state of spores are those that delay germination. For example, those *Bacillus* spp. germinated by L-alanine produce during sporulation alanine racemase (Stewart & Halvorson, 1953), which catalyses the conversion of L-alanine to D-alanine, a potent inhibitor of germination (Gould, 1969); moreover, many bacterial spore populations require heat activation for maximal germination. These mechanisms ensure that not all spores in a population germinate in response to a single trigger, some remaining dormant in case conditions are not favourable for growth. Such mechanisms are not restricted to bacterial spores, and the response of eukaryotic spores to factors such as light, heat, 'overwintering' or particular chemicals is extensively reviewed by Sussman & Halvorson (1966).

Dispersal

For some organisms, spore dispersal may be more important than survival in time. Successful dispersal may depend upon the number of spores produced; as an extreme example, the giant puff-ball *Calvatia gigantea* produces about 7×10^{12} spores (Buller, 1909). The dispersal of fungal spores over vast distances has been well documented, especially for the wheat rust *Puccinia graminis* (Sussman & Halvorson, 1966). Other fungi have developed more intricate mechanisms for spore dispersal, probably none more curious than that of the genus *Pilobolus*, a common inhabitant of horse and cow dung. The sporangiophores are produced according to a daily sequence (Booth, 1978) and are very phototropic, the sporangium being violently discharged toward light, up to a height of six feet (Buller, 1934).

Sexual spores and genetic recombination: the advantage of variability

Most sexual processes in eukaryotic micro-organisms involve sporulation and usually a resistant spore is formed (Gregory, 1952; Bonner, 1958*b*). Since meiosis and recombination either precedes sporulation directly (as in Ascomycetes, Basidiomycetes, some green algae and commonly among red and brown algae) or follows germination of a resistant spore (Phycomycetes, Zygomycetes and many green algae), an obvious advantage is that the spores or their products have considerable genetic variability and the species has a greater likelihood of survival in a new environment. Not all recombination occurs via a sporulation process since mitotic recombination occurs at low frequency in vegetative diploid organisms (Stern, 1936; James & Lee-Whiting, 1955) and in heterokaryons following rare nuclear fusion events (Pontecarvo, 1956). In the following section the possibility that the evolution of a number of aspects of sexuality *depended* on the capacity of the progenitor to sporulate is discussed. Sussman & Halvorson (1966) have also pointed out that asexual sporulation may also contribute to new genetic combinations, for example both macroconidia and microconidia can give rise to new heterokaryons of different genetic constitution. They have also argued that there may be selective advantage in the periodic purging of disadvantageous mutations that can survive in heterokaryons; the same is true for heterozygous diploids, and in both cases this is achieved by sporulation.

POSSIBLE CONTRIBUTIONS OF SPORULATION TO EVOLUTION

Prokaryotic sporulation: a stage in the evolution of eukaryotes and Gram-negative bacteria?

Gould & Dring (1979) have indicated how an early stage in the possible evolution of sporulation, the Stage III forespore equivalent, is produced by the type of invagination process invoked to account for the evolution of eukaryotes from prokaryotes, according to one of the two general hypotheses currently in contention (Cavalier-Smith, this volume). This is outlined in Fig. 4. Gould & Dring have pointed out that this 'forespore-like' structure would confer a number of important advantages, including the potential to maintain two radically different environments within the cell, and the ability of each compartment to influence the other's activities. One problem with this model which was not discussed is how the initial enclosed forespore (now an organelle) divided and was inherited by the daughter cells on division of the eukaryotic progenitor. One possible explanation is that on restoration of external nutrients, a mutation allowed the 'forespore' to divide faster than the 'mother cell', and led to an initial 'multinucleate' condition from which subsequent, more controlled cell division rapidly evolved. The model has the attraction that the ability to form a Stage III prespore intermediate could have been stably inherited and would have conferred advantages, thereby fulfilling one of the conditions that seems necessary in devising an evolutionary sequence with a fairly large number of inherited changes.

The Stage III prespore may have been an intermediate in another evolutionary pathway – this time a prokaryotic one – leading to the Gram-negative cell. This thought must have occurred to Gould &

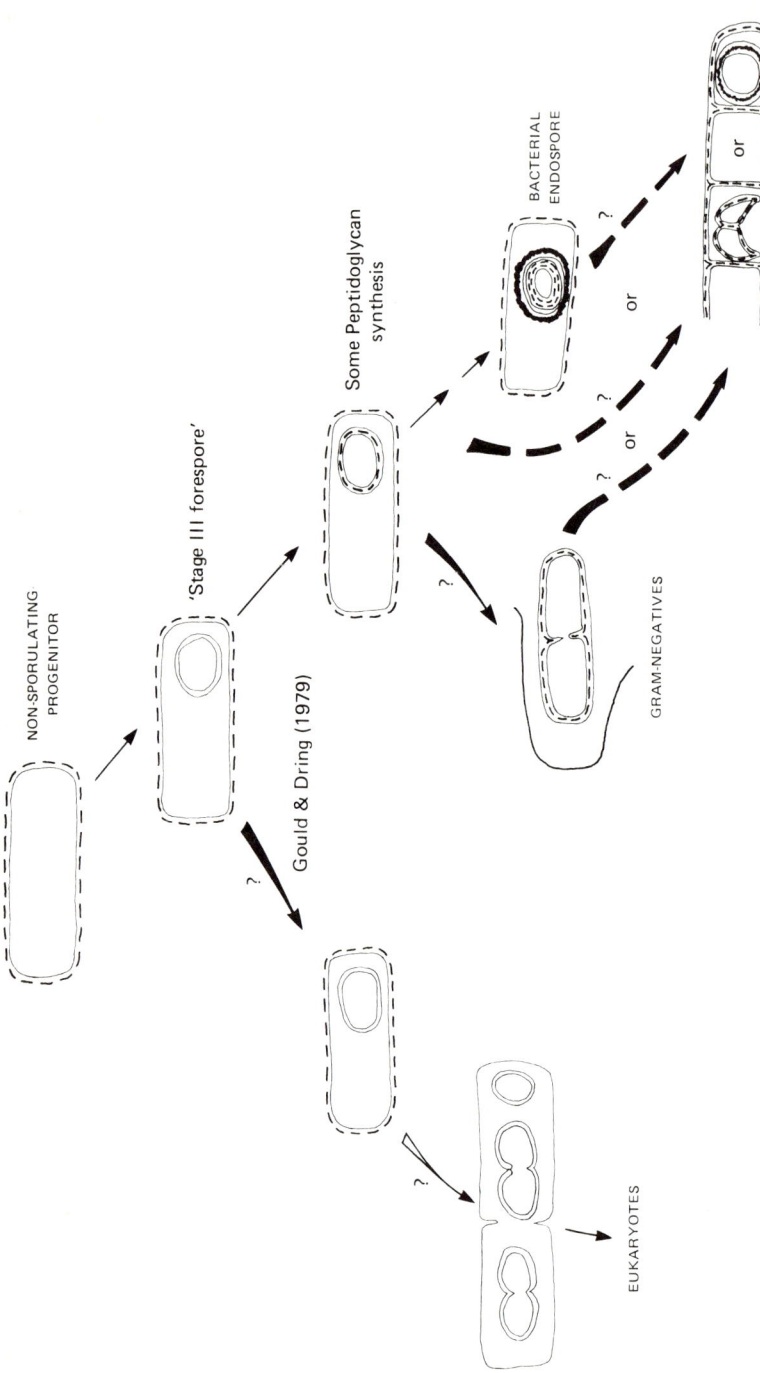

Fig. 4. Scheme for the possible participation of evolving spores in other evolutionary sequences. Peptidoglycan-containing structures are depicted by broken lines, cell membranes by single unbroken lines.

Dring (1979) since they briefly pointed out that forespores resemble Gram-negative bacteria in having two membranes with a simply-structured peptidoglycan layer between them. The similarity is striking when electron micrographs of the two structures are compared. Moreover, the wall composition, while not now identical, is by no means inconsistent with the hypothesis that Gram-negative cells originated in this way. The outer membrane of Gram-negatives is now composed of more chemical species than the inner membrane, but it does contain considerable amounts of phospholipid in the appropriate arrangement (Costerton, Ingram & Cheng, 1974).

Moreover, the peptidoglycan that is common to all Gram-negative bacteria (Chemotype 1; Ghuysen, 1968; Schleifer & Kandler, 1972) is also that which earlier (p. 101) suggested to be formed in the primordial spore before it was modified extensively to form the present cortical structure. The Gram-negative cell peptidoglycan is also characterized by a relative paucity of cross-linking (Rogers, 1970) and this may have been an intermediate condition in the evolution of the spore cortical peptidoglycan. The conversion of a Stage III forespore to a Gram-negative-like cell would have required that under one condition of 'outgrowth' of the prespore, peptidoglycan synthesis continued in the 'sporulation' mode and a mechanism existed for continued insertion of phospholipids into the outer membrane layer. It is interesting in this context to note that sometimes during cell division of Gram-negative bacteria septation proceeds by inner membrane invagination with synthesis of the peptidoglycan layers between them, without the concomitant synthesis of the outer wall layer in the region of the septum (Steed & Murray, 1966; Murray, 1978). That is, the mode of cell division septum formation used by Gram-negatives is not unlike that used by Gram-positives, and there should be no conceptual difficulty in accepting the heritability of the 'initial' Gram-negative state arising from modified outgrowth of a forespore.

What advantages might this Gram-negative cell have had that ensured its survival? One of the main ones may have been the acquisition of resistance to lytic enzymes degrading peptidoglycan. Moreover, this arrangement conferred on the Gram-negative cell the advantages that would accrue from the possession of a periplasmic space – a region for the binding of specific enzymes that may serve a function in scavenging useful molecules from the external medium. In fact, Glenn & Coote (1975) have shown that alkaline phosphatase can be induced between the inner and outer mem-

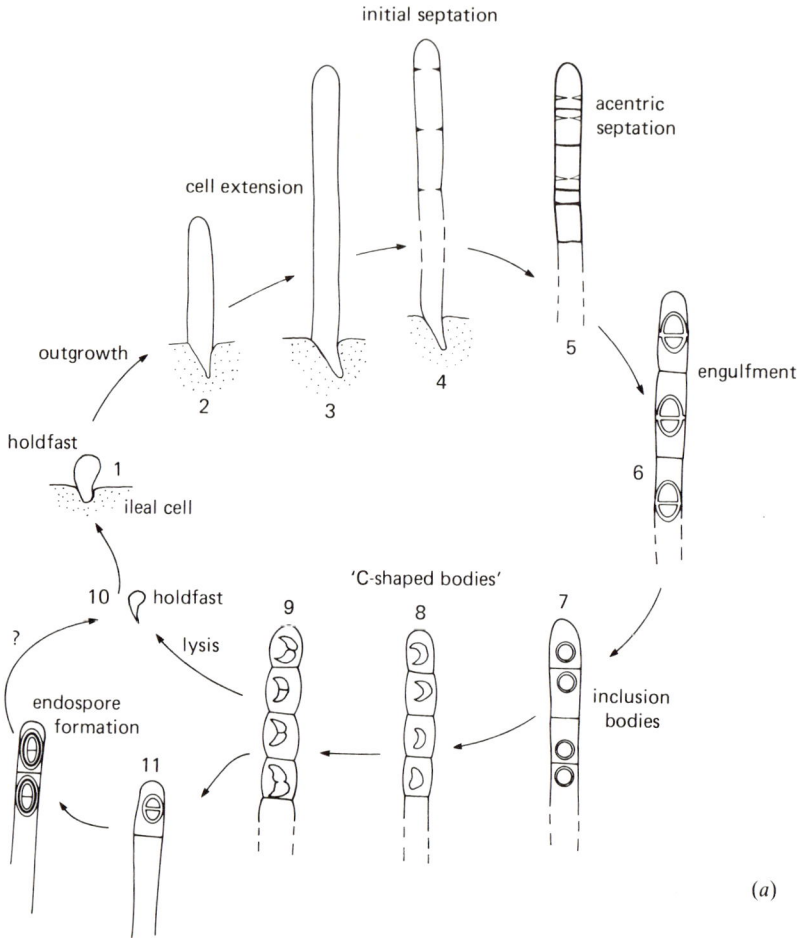

(a)

branes of the developing spore in the same position that it is found in Gram-negatives. Other potential advantages may have been the acquisition of resistance to bacteriophages and some increased resistance of the vegetative cell to starvation.

Sporulation in a filamentous, segmented bacterium from murine ileum

Chase & Erlandsen (1976) have described a remarkable intracellular differentiation in organisms probably belonging to the Arthromictaceae. These filamentous bacteria, comprising up to 90 seg-

SPORULATION IN EVOLUTION 117

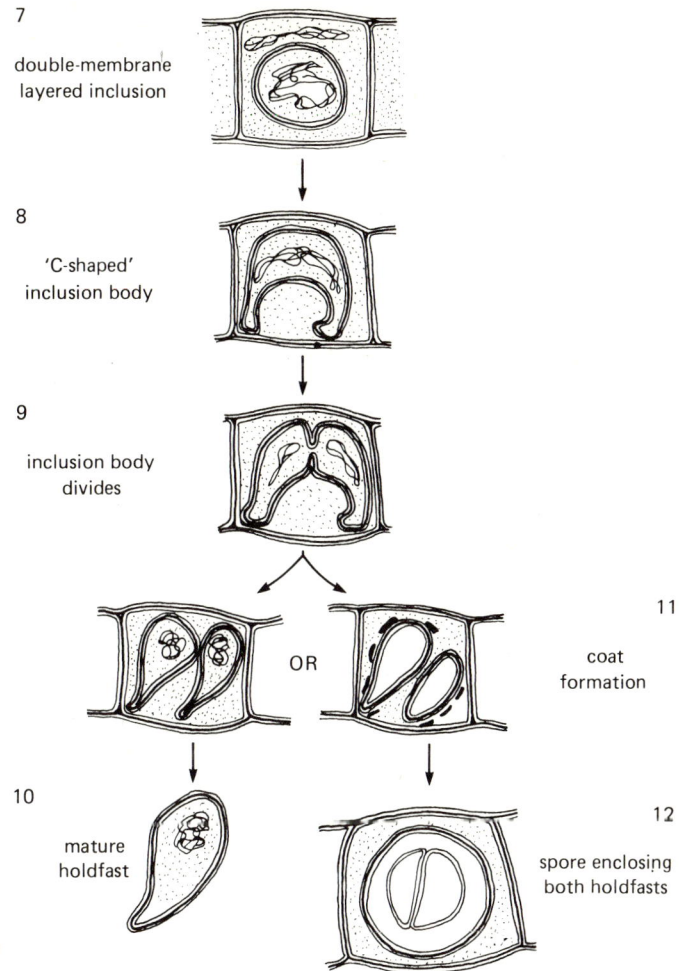

Fig. 5. (*a*) The life cycle of the filamentous, sporulating gut bacteria, after Chase & Erlandsen (1976). (*b*) Detail of holdfast formation and sporulation.

ments, were attached to the epithelial cells by a specialized terminal holdfast segment. Beginning distal to the attachment site, and progressing gradually towards it, segments underwent a process of invagination and engulfment in much the same way that endospores are formed in other bacteria. At this point, however, the processes diverge. The spherical double-membrane enclosed body was converted into a C-shaped inclusion body. The ends of this C-shaped structure matured into holdfasts by becoming pointed and free of ribosomes, and by a septation across the arch of the 'C', two 'holdfasts' were formed per mother cell. On maturation, these were released by the lysis of the mother cell, probably to allow development of new filaments within the same animal. In some filaments, the two newly-formed holdfasts, instead of being released from the filament, were packaged into spores in a manner very similar to that seen during endospore formation in *Bacillus*. Chase & Erlandsen presume this permits survival outside the host and transmission to other rodents. The probable sequence of these events is illustrated in Figs. 5a, b.

This remarkable resemblance to the sporulation process of the clostridia and related organisms seems to be too close to have evolved separately. If it is derived from a modification of the clostridial mechanism, then it has a very direct and important relevance to two of the previous discussions – on the emergence of sporulation and of Gram-negative bacteria. The high quality electron micrographs of Chase & Erlandsen (1976) reveal that the 'holdfast', and the filament that develops from it, have a multilayered wall structure resembling that seen in Gram-negative bacteria. This is what one would expect from the 'outgrowth of a Stage III forespore' as was postulated for the evolution of the Gram-negative cell.

Did this organism then have its origins in one bridging Gram-positives and Gram-negatives? Is it a case of convergent evolution? Or is it a later development, a reiteration of the mechanism whereby Gram-negatives may have arisen? One may incline to the latter possibility, given that the sporulation process bears such a striking resemblance to that in *Bacillus*, etc., and the organism's restricted habitat of relatively recent origin. Clearly it will be of considerable interest to study these bacteria in greater detail: by establishing their phylogenetic relationship to Gram-positive clostridia on the one hand and to Gram-negatives on the other; and by testing for the components characteristic of the Gram-negative cell wall (e.g. lipopolysaccharide) or of the mature *Bacillus* endospore (e.g. DPA, peptidoglycan, heat resistance, Ca^{2+}, etc.).

The involvement of sporulation in the evolution of sexuality

In considering how sexual systems, particularly meiosis, may have arisen, there is a problem in explaining how diploidy, meiosis and conjugation to restore the diploid condition could have arisen in a way that does not presume the acquisition of all three simultaneously. For example, Darlington (1958) has pointed out that a precocity mutation causing meiosis would act on all nuclear divisions in the cell, and meiosis would, therefore, take place as soon as a diploid nucleus had been formed. He has stated that 'the original sexual cycle would be one without any phase of diploid mitosis' and supports this with the argument that the simplest sexual organisms are those in which meiosis is immediately triggered by fusion of the nuclei. Nonetheless, as Darlington also pointed out, 'the second method of making a diploid organism possible is by doing precisely what most fungi seem to have failed to do ... to postpone meiosis by an efficient differentiation in development between mitotic and meiotic divisions of the cell'. In his view, this change prepared the way for the origin of all higher plants and animals.

Therefore, there are two aspects to the evolution of sexuality: how did conjugation, karyogamy and meiosis evolve in a progressive way, and, how could meiosis have become a 'conditional' process such that mitosis of the diploid became a possibility?

Fig. 6 outlines one scheme for the stepwise evolution of meiosis which allows for the formation of stable intermediates with an advantage. This is presented not as a definitive single possibility, nor even as original, but as one scheme which illustrates how possession of the ability to sporulate would have allowed for a progression in the evolution of meiosis and of stable diploidy in a way that takes into account many microbial life cycles still extant.

The stages outlined in Fig. 6 are as follows.

(i) Heterokaryon formation which may have been either the result of the initial evolution of the nucleus, or abnormal cell division, or of a conjugation process. This would have conferred the distinct advantage of allowing complementation of mutational defects (Davis, 1959) and a limited evolution of new genes due to the presence of copies of the non-mutant allele in other nuclei. Conjugation at this stage would greatly increase the opportunity for genetic mixing of nuclei.

(ii) Sporulation evolving as a response to nutrient starvation would confer the additional advantage of increasing the opportunities for genetic mixing in the presence of an efficient conjugation process.

(iii) Diploidy, initially transient due to rare nuclear fusions, would have allowed more extensive recombination of genes (similar to the situation now found in *Aspergillus*; Pontecorvo et al., 1953). The evolution of a mechanism for meiosis would have led to the

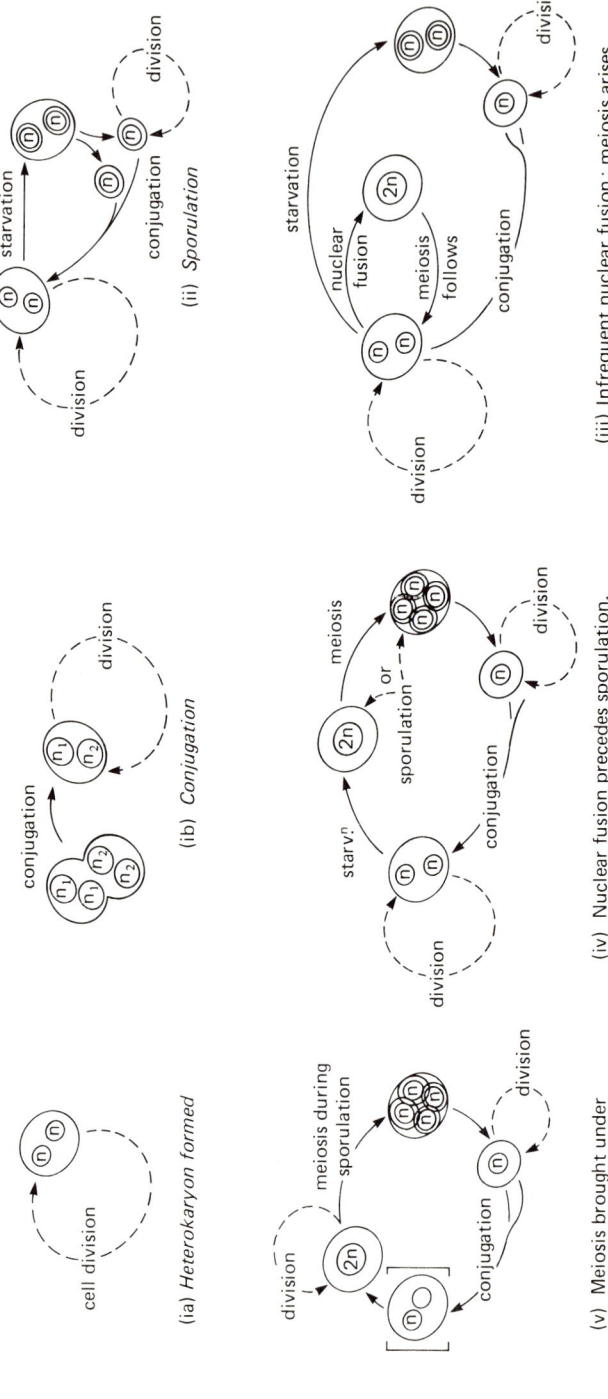

Fig. 6. The possible involvement of sporulation in the evolution of sexuality and the stable diploid state.

possibility of generating new combinations of genes by recombination, but, as pointed out by Darlington (1958), would have led to an immediate return to the haploid, or heterokaryotic, condition. The prior existence of sporulation and conjugation mechanisms would have increased the advantages by allowing for more extensive genetic mixing. Meiosis could have arisen more than once, although there is a basic similarity of mechanisms and structures, which may indicate that it did not (Heywood & Magee, 1976).

(iv) By restricting nuclear fusion to the sporulation phase of the life cycle, the processes of nuclear fusion and meiosis would have become stabilized so that meiosis could follow directly from an organized nuclear fusion. This situation still persists in many fungal and algal life cycles, and would have provided the basis for divergence of those fungi with the dikaryophase as a true alternative to diploidy (Raper & Flexer, 1970).

(v) By a subsequent mutation bringing meiosis under sporulation control, diploid mitosis would have become feasible. This would have led to the formation of diploid spores and, on germination, to a stable diplophase with all its advantages and potential for further rapid evolution (Lewis & Wolpert, 1979).

The stable diplophase has been achieved rarely in fungi, although it is present in a number of yeasts (Fowell, 1969). In one of these, *Saccharomyces cerevisiae,* an interesting observation that highlights the nature and extent of 'sporulation control' over the processes of meiosis and recombination has been made by Klapholz & Esposito (1980). They have shown that the simultaneous presence of just two mutations (*spo12* and *spo13*), each affecting normal sporulation, resulted in a modified 'meiosis' in which meiotic recombination and reductional division were suppressed. Under sporulation conditions asci containing two viable diploid spores were produced, rather than ones with four haploid spores.

In conclusion, sporulation processes have afforded microorganisms the opportunity to segregate important aspects of sexuality to particular phases of their life cycle. Other processes may have enabled this separation, but if one inclines to the view that eukaryotes evolved through microorganisms similar to those currently extant (see Cavalier-Smith, this volume) these, almost without exception, confine their sexual processes to sporulation in one form or another, and it seems quite possible that 'sporulation' played a very important part in the evolution of higher eukaryotes.

And so I turn to the abyss
Of necromancy, try if art
Can voice or power of spirits start,
To do me service and reveal
The things of Nature's secret seal,
And save me from the weary dance
Of holding forth in ignorance.
Then shall I see, with vision clear,
How secret elements cohere,
And what the universe engirds,
And give up hucksteringwith words.

(GOETHE, 'Faust', translation of Philip Wayne)

REFERENCES

ALEXOPOULOS, C. J. (1962). *Introductory Mycology, 2nd edition.* New York & London: John Wiley & Sons.

ARONSON, A. I. & FITZ-JAMES, P. C. (1975). Properties of *Bacillus cereus* spore coat mutants. *Journal of Bacteriology,* **123,** 354–65.

ARONSON, A. I. & FITZ-JAMES, P. C. (1976). Structure and morphogenesis of the bacterial spore coat. *Bacteriological Reviews,* **40,** 360–402.

AUBERT, J.-P., MILLET, J. & SCHAEFFER, P. (1965). Croissance et sporulation de *Bacillus megaterium* en culture continue. *Compte Rendu Hebdomadaire des Seances de l'Academie des Sciences, Paris,* **261,** 2407–9.

BALASSA, G. & YAMAMOTO, T. (1970). Biochemical genetics of bacterial sporulation. III. Correlation between morphological and biochemical properties of sporulation mutants. *Molecular and General Genetics,* **108,** 1–22.

BARGHOORN, E. S. (1971). The oldest fossils. *Scientific American,* **224,** (5), 30–42.

BARTNICKI-GARCIA, S. (1970). Cell wall composition and other biochemical markers in fungal phylogeny. In *Phytochemical Phylogeny,* ed. J. B. Harborne, pp. 81–103. New York & London: Academic Press.

BISSET, K. A. (1950). Evolution in bacteria and the significance of the bacterial spore. *Nature, London,* **166,** 431–2.

BOLD, H. C. & WYNNE, M J. (1978). *Introduction to the Algae. Structure and Reproduction.* New Jersey: Prentice-Hall.

BONNER, J. T. (1958a). *The Evolution of Development.* London: Cambridge University Press.

BONNER, J. T. (1958b). The relation of spore formation to recombination. *American Naturalist,* **92,** 193–200.

BOOTH, C. (1978). Form and function – fungi. In *Essays in Microbiology,* ed. J. R. Norris & M. H. Richmond, pp. 3/1–3/32. Chichester & New York: John Wiley & Sons.

BRILEY, M. S., ILLINGWORTH, R. F., ROSE, A. H. & FISHER, D. J. (1970). Evidence for a surface protein layer on the *Saccharomyces cerevisiae* ascospore. *Journal of Bacteriology,* **104,** 588–9.

BÜCHNER, H. (1890). Ueber die Ursache der Sporenbildung beim Milzbrandbacillus. *Zentralblatt für Bakteriologie, Parasitenkunde, Infektionskrankheiten und Hygiene* (Abteilung I), **8,** 1–6.

BULLER, A. H. R. (1909). *Researches on Fungi*, pp. 153–78. New York: Longmans, Green & Co.
BULLER, A. H. R. (1934). *Researches on Fungi*, volume 6, pp. 1–224. London: Longmans, Green & Co.
BULLOUGH, W. S. (1967). *The Evolution of Differentiation*. London & New York: Academic Press.
BURNETT, J. H. (1968). *Fundamentals of Mycology*. London: Edward Arnold Ltd.
CAMPBELL, L. L. & POSTGATE, J. R. (1965). Classification of the spore-forming sulphate-reducing bacteria. *Bacteriological Reviews*, **29**, 359–63.
CARR, N. G. & BRADLEY, S. (1973). Aspects of development in blue-green algae. In *Microbial Differentiation. Symposia of the Society for General Microbiology* **23**, 161–8, ed. J. M. Ashworth & J. E. Smith. London: Cambridge University Press.
CAVALIER-SMITH, T. (1975). The origin of nuclei and of eukaryotic cells. *Nature, London*, **256**, 463–8.
CHAMBLISS, G. (1979). The molecular biology of *Bacillus subtilis* sporulation. In *Developmental Biology of Prokaryotes*, ed. J. H. Parish, pp. 57–71. Oxford: Blackwell Scientific Publications.
CHASE, D. G. & ERLANDSEN, S. L. (1976). Evidence for a complex life cycle and endospore formation in the attached, filamentous, segmented bacterium from murine ileum. *Journal of Bacteriology*, **127**, 572–83.
CLANCY, M. J., SMITH, L. & MAGEE, P. T. (1980). Regulation of a sporulation-specific enzyme activity. *Abstracts of the 10th International Conference on Yeast Genetics and Molecular Biology, Louvain-la-Neuve*, p. 152.
CLOUD, P. E., LICARI, G. R., WRIGHT, L. A. & TROXEL, B. W. (1969). Proterozoic eukaryotes from eastern California. *Proceedings of the National Academy of Sciences of the U.S.A.*, **62**, 623–30.
COOK, R. P. (1932). Bacterial spores. *Biological Reviews of the Cambridge Philosophical Society*, **7**, 1–23.
COOTE, J. G. (1972). Characterization of oligosporogenous mutants and comparison of their phenotypes with those of asporogenous mutants. *Journal of General Microbiology*, **71**, 1–15.
COOTE, J. G. & MANDELSTAM, J. (1973). Use of constructed double mutants for determining the temporal order of expression of sporulation genes in *Bacillus subtilis*. *Journal of Bacteriology*, **114**, 1254–63.
COSTERTON, J. W., INGRAM, J. M. & CHENG, K.-J. (1974). Structure and function of the cell envelope of Gram-negative bacteria. *Bacteriological Reviews*, **38**, 87–110.
CROSS, T. (1970). The diversity of bacterial spores. *Journal of Applied Bacteriology*, **33**, 95–102.
CROSS, T. & ATWELL, R. W. (1975). Actinomycete spores. In *Spores VI*, ed. P. Gerhardt, R. N. Costilow & H. L. Sadoff, pp. 3–14. Washington D.C.: American Society for Microbiology.
CROSS, T., WALKER, P. D. & GOULD, G. W. (1968). Thermophilic actinomycetes producing resistant endospores. *Nature, London*, **220**, 352–4.
DARLINGTON, C. D. (1958). *The Evolution of Genetic Systems*, 2nd edition. New York: Basic Books Inc.
DAVIS, R. H. (1959). Asexual selection in *Neurospora crassa*. *Genetics, Princeton*, **44**, 1291–308.
DAWES, I. W. (1975). Study of cell development using derepressed mutations. *Nature, London*, **255**, 707–8.
DAWES, I. W. & HARDIE, I. D. (1974). Selective killing of vegetative cells in sporulated yeast cultures by exposure to diethyl ether. *Molecular and General Genetics*, **131**, 281–9.

DAWES, I. W., KAY, D. & MANDELSTAM, J. (1971). Determining effect of growth medium on the shape and position of daughter chromosomes and on sporulation in *Bacillus subtilis*. *Nature, London*, **230**, 567–9.

DAWES, I. W. & MANDELSTAM, J. (1970). Sporulation of *Bacillus subtilis* in continuous culture. *Journal of Bacteriology*, **103**, 529–35.

DAWES, I. W. & THORNLEY, J. H. M. (1970). Sporulation in *Bacillus subtilis*. Theoretical and experimental studies in continuous culture systems. *Journal of General Microbiology*, **62**, 49–66.

DAWES, I. W., WRIGHT, J. F., VEZINHET, F. & AJAM, N. (1980). Separation on Urografin gradients of subpopulations from sporulating *Saccharomyces cerevisiae* cultures. *Journal of General Microbiology*, **119**, 165–71.

DEL REY, F., SANTOS, F., GARCIA-ACHA, I. & NOMBELA, C. (1980). Synthesis of β-glucanases during sporulation in *Saccharomyces cerevisiae:* formation of a new, sporulation-specific, 1,3-β-glucanase. *Journal of Bacteriology*, **143**, 621–7.

DEN DOOREN DE JONG, L. E. (1929). Uber *Bacillus fastidiosus. Zentralblatt für Bakteriologie, Parasitenkunde, Infektionskrankheiten und Hygiene* (Abtelung II), **79**, 344–58.

DOI, R. H. (1977). Role of ribonucleic acid polymerase in gene selection in procaryotes. *Bacteriological Reviews*, **41**, 568–94.

DONACHIE, W. D. (1979). The cell cycle of *Escherichia coli*. In *Developmental Biology of Prokaryotes*, ed. J. H. Parish, pp. 11–35. Oxford: Blackwell Scientific Publications.

DWORKIN, M (1972). The Myxobacteria: new directions in studies of procaryotic development. *Critical Reviews in Microbiology*, **1**, 435–52.

EGOROVA, A. A. & DERYUGINA, Z. P. (1963). The spore-forming thermophilic thiobacterium *Thiobacillus thermophila* Imschenetskii nov. sp. *Mikrobiologiya* (English translation), **32**, 376–81.

ELLAR, D. J. (1978). Spore-specific structures and their function. In *Relations between Structure and Function in the Prokaryotic Cell, Symposia of the Society for General Microbiology*, ed. R. Y. Stanier, H. J. Rogers & J. B. Ward, pp. 295–325. London: Cambridge University Press.

ELMERICH, C. & AUBERT, J.-P. (1975). Involvement of glutamine synthetase and the purine nucleotide pathway in repression of bacterial sporulation. In *Spores VI*, ed. P. Gerhardt, R. N. Costilow & H. L. Sadoff, pp. 385–90. Bethesda: American Society for Microbiology.

ERBEN, K. (1962). Sporulation. In *Physiology and Biochemistry of Algae*, ed. R. A. Lewin, pp. 701–710). New York & London: Academic Press.

ESPOSITO, M. S. & ESPOSITO, R. E. (1975). Mutants of meiosis and ascospore formation. *Methods in Cell Biology*, **9**, 303–26.

ESPOSITO, R. E., FRINK, N., BERNSTEIN, P. & ESPOSITO, M. S. (1972). The genetic control of sporulation in *Saccharomyces*. II. Dominance and complementation of mutants of meiosis and spore formation. *Molecular and General Genetics*, **144**, 241–8.

ESSER, K. (1974). Breeding systems and evolution. In *Evolution in the Microbial World. Symposia of the Society for General Microbiology*, **24**, pp. 87–104. ed. M. J. Carlile & J. J. Skehel. London: Cambridge University Press.

FAY, P., STEWART, W. D. P., WALSBY, A. E. & FOGG, G. E. (1968). Is the heterocyst the site of nitrogen fixation in blue-green algae? *Nature, London*, **220**, 810–12.

FITZ-JAMES, P. C. (1960). Participation of the cytoplasmic membrane in the growth and spore formation of bacilli. *Journal of Biophysical and Biochemical Cytology*, **8**, 507–28.

FITZ-JAMES, P. C. (1964). Sporulation in protoplasts and its dependence on prior forespore development. *Journal of Bacteriology*, **87**, 667–75.

FITZ-JAMES, P. C. (1971). Formation of protoplasts from resting spores. *Journal of Bacteriology*, **105**, 1119–36.
FOSTER, J. W. (1956). Morphogenesis in bacteria: some aspects of spore formation. *Quarterly Reviews of Biology*, **31**, 102–18.
FOWELL, R. R. (1969). Life cycles in yeasts. In *The Yeasts, volume I*, ed. A. H. Rose & J. S. Harrison pp. 461–71. London & New York: Academic Press.
FOWELL, R. R. (1975). Ascospores of Yeasts. In *Spores VI*, ed. P. Gerhardt, R. N. Costilow & H. L. Sadoff, pp. 124–31. Washington, D. C.: American Society for Microbiology.
FOX, G. E., STACKEBRANDT, E., HESPELL, R. B., GIBSON, J., MANIOLOFF, J., DYER, T. A., WOLFE, R. S., BALCH, W. E., TANNER, R. S., MAGRUM, L. J., ZABLEN, L. B., BLAKEMORE, R., GUPTA, R., BONEN, L., LEWIS, B. J., STAHL, D. A., LUEHRSEN, K. R., CHEN, K. N. & WOESE, C. R. (1980). The phylogeny of prokaryotes. *Science*, **209**, 457–63.
GALIZZI, A., GORRINI, F., ROLLIER, A. & POLSINELLI, M. (1973). Mutants of *Bacillus subtilis* temperature-sensitive in the outgrowth phase of spore germination. *Journal of Bacteriology*, **113**, 1482–90.
GHUYSEN, J. M. (1968). Use of bacteriolytic enzymes in determination of wall structure and their role in central metabolism. *Bacteriological Reviews*, **32**, 425–64.
GLENN, A. R. & COOTE, J. G. (1975). Cytochemical studies in alkaline phosphatase production during sporulation in *Bacillus subtilis*. *Biochemical Journal*, **152**, 85–89.
GOODAY, G. W. (1973). Differentiation in the Mucorales. In *Microbial Differentiation. Symposia of the Society for General Microbiology*, **23**, 269–94. Ed. J. M. Ashworth & J. E. Smith. London: Cambridge University Press.
GOULD, G. W. (1969). Germination. In *The Bacterial Spore*, ed. G. W. Gould & A. Hurst, pp. 397–444. London & New York: Academic Press.
GOULD, G. W. (1977). Recent advances in the understanding of resistance and dormancy in bacterial spores. *Journal of Applied Bacteriology*, **42**, 297–309.
GOULD, G. W. & DRING, G. J. (1975). Heat resistance of bacterial endospores and concept of an expanded osmoregulatory cortex. *Nature, London*, **258**, 402–5.
GOULD, G. W. & DRING, G. J. (1979). On a possible relationship between bacterial endospore formation and the origin of eukaryotic cells. *Journal of Theoretical Biology*, **81**, 47–53.
GREGORY, P. H. (1952). Fungus spores. *Transactions of the British Mycological Society*, **35**, 1–18.
GRIFFITHS, D. A. & GRIFFITHS, D. J. (1969). The fine structure of autotrophic and heterotrophic cells of *Chorella vulgaris*. *Plant Cell Physiology*, **10**, 11–19.
GRODZICKER, T., ARDITTI, R. R. & EISEN, H. (1972). Establishment of repression in lambdoid phage in catabolite activator protein and adenylate cyclase mutants of *E. coli*. *Proceedings of the National Academy of Sciences of the United States of America*, **69**, 366–70.
HANSON, R. S. (1979). The physiology and diversity of bacterial endospores. In *Developmental Biology of Prokaryotes*, ed. J. H. Parish, pp. 37–56. Oxford: Blackwell Scientific Publications.
HARTWELL, L. H. (1974). *Saccharomyces cerevisiae* cell cycle. *Bacteriological Reviews*, **38**, 164–98.
HEYWOOD, P. & MAGEE, P. T. (1976). Meiosis in Protists. Some structural and physiological aspects of meiosis in algae, fungi and protozoa. *Bacteriological Reviews*, **40**, 190–240.
HITCHINS, A. D. (1978). Polarity and topology of DNA segregation and septation

in cells and sporangia of the bacilli. *Canadian Journal of Microbiology*, **24**, 1103–34.

HITCHINS, A. D. & SLEPECKY, R. A. (1969). Bacterial sporulation as a modified procaryotic cell division. *Nature, London*, **223**, 804–7.

HRANUELI, D., PIGGOT, P. J. & MANDELSTAM, J. (1974). Statistical estimate of the total number of operons specific for *Bacillus subtilis* sporulation. *Journal of Bacteriology*, **119**, 684–90.

JAMES, A. P. & LEE-WHITING, B. (1955). Radiation-induced segregations in vegetative cells of diploid yeast. *Genetics, Princeton*, **40**, 826–31.

KALAKOUTSKII, L. V. & AGRE, N. S. (1976). Comparative aspects of development and differentiation in actinomycetes. *Bacteriological Reviews*, **40**, 469–524.

KAZMIERCZAK, J. (1979). The eukaryotic nature of Eosphaera ferriferous structures from the Precambrian gunflint iron formation, Canada: a comparative study. *Precambrian Research*, **9**, 1–22.

KEILIN, D. (1958). The problem of anabiosis or latent life. *Proceedings of the Royal Society, London series B*, **150**, 149–91.

KITAHARA, K. & LAI, C. (1967). On the spore formation of *Sporolactobacillus inulinus*. *Journal of General and Applied Microbiology*, **13**, 197–203.

KLAPHOLZ, S. & ESPOSITO, R. E. (1980). Interactions of genes controlling recombination and chromosome segregation in meiosis. *Abstracts of the 10th International Conference on Yeast Genetics and Molecular Biology, Louvain-la-Neuve* p. 153.

KNOLL, A. H. & BARGHOORN, E. S. (1975). Precambrian eukaryotic organisms: a reassessment of the evidence. *Science*, **190**, 52–4.

KRAIG, E., PEARSON, N. & HABER, J. (1980). Changes in the regulation of protein synthesis during sporulation of *Saccharomyces cerevisiae*. *Abstracts of the 10th International Conference on Yeast Genetics and Molecular Biology, Louvain-la-Neuve* p. 153.

LEADBETTER, E. R. & HOLT, S. C. (1968). The fine structure of *Bacillus fastidiosus*. *Journal of General Microbiology*, **52**, 299–307.

LEWIS, J. C. (1969). Dormancy. In *The Bacterial Spore*, ed. G. W. Gould & A. Hurst, pp. 301–358. London & New York: Academic Press.

LEWIS, J. & WOLPERT, L. (1979). Diploidy, evolution and sex. *Journal of Theoretical Biology*, **78**, 425–38.

LICARI, G. R. & CLOUD, P. E. JR. (1968). Reproductive structures and taxonomic affinities of some nanofossils from the Gunflint Iron Formation. *Proceedings of the National Academy of Sciences of the U.S.A.* **59**, 1053–60.

LINN, T. & LOSICK, R. (1976). The program of protein synthesis during sporulation in *Bacillus subtilis*. *Cell*, **8**, 103–14.

LISKE-PETERSON, J. G., KIELLAND-BRANDT, M. C. & NILSSON-TILLGREN, T. (1979). Protein patterns of yeast (*Saccharomyces cerevisiae*) during sporulation. *Carlsberg Research Communications*, **44**, 149–62.

LOPEZ, J. M., MARKS, C. L. & FREESE, E. (1979). The decrease of guanine nucleotides initiates sporulation of *Bacillus subtilis*. *Biochimica et Biophysica Acta*, **587**, 238–52.

LOVETT, J. S. (1975). Growth and differentiation of the water mold *Blastocladiella emersonii*: cytodifferentiation and the role of ribonucleic acid and protein synthesis. *Bacteriological Reviews*, **39**, 345–404.

LYNN, R. R. & MAGEE, P. T. (1970). Development of the spore wall during ascospore formation in *Saccharomyces cerevisiae*. *Journal of Cell Biology*, **44**, 688–92.

MANDELSTAM, J. (1958). Turnover of protein in growing and non-growing populations of *E. coli*. *Biochemical Journal*, **69**, 110–19.

MANDELSTAM, J. (1971). Recurring patterns during development in primitive organisms. In *Control Mechanisms of Growth and Differentiation. Symposia of the Society for Experimental Biology*, **25**, 1–26.

MANDELSTAM, J. (1976). The Leeuwenhoek Lecture, 1975. Bacterial sporulation: a problem in the biochemistry and genetics of a primitive developmental system. *Proceedings of the Royal Society, London; series B*, **193**, 89–106.

MANDELSTAM, J. & HIGGS, S. A. (1974). Induction of sporulation during synchronized chromosome replication in *Bacillus subtilis. Journal of Bacteriology*, **120**, 38–42.

MARGULIS, L. (1970). *Origin of Eukaryotic Cells*. New Haven: Yale University Press.

MICHEL, J. F., CAMI, B. & SCHAEFFER, P. (1968). Sélection de mutants de *Bacillus* bloques au début de la sporulation. II. Sélection par adaptation à une nouvelle source de carbonne et par vieillissement de cultures sporulées. *Annales de l'Institut Pasteur, Paris*, **114**, 21–7.

MILLER, J. J. (1963*a*). The metabolism of yeast sporulation. V. Stimulation and inhibition of sporulation and growth by nitrogen compounds. *Canadian Journal of Microbiology*, **9**, 259–77.

MILLER, J. J. (1963*b*). Determination by ammonium of the manner of yeast nuclear division. *Nature, London*, **198**, 214–15.

MILLS, D. (1972). Effect of pH on adenine and amino acid uptake during sporulation in *Saccharomyces cerevisiae. Journal of Bacteriology*, **112**, 519–26.

MOENS, P. B. (1971). Fine structure of ascospore development in the yeast *Saccharomyces cerevisiae. Canadian Journal of Microbiology*, **17**, 507–10.

MURRAY, R. G. E. (1978). Form and function – Bacteria. In *Essays in Microbiology* ed. J. R. Norris & M. H. Richmond, pp. 2/1–2/31. Chichester: John Wiley & Sons.

NAZINA, T. N. & PIVOVAROVA, T. A. (1979). Submicroscopic organization and spore formation in *Desulfotomaculum nigrificans. Mikrobiologiya*, **48**, 302–6.

NEALE, E. K. & CHAPMAN, G. B. (1970). Effect of low temperature on the growth of *Bacillus subtilis. Journal of Bacteriology*, **104**, 518–28.

NELSON, D. L. & KORNBERG, A. (1970). Biochemical studies of bacterial sporulation and germination XVIII. Free amino acids in spores. *Journal of Biological Chemistry*, **245**, 1128–36.

PIGGOT, P. J. & COOTE, J. G. (1976). Genetic aspects of bacterial endospore formation. *Bacteriological Reviews*, **40**, 908–62.

PITEL, D. W. & GILVARG, C. (1971). Timing of mucopeptide and phospholipid synthesis in sporulating *Bacillus megaterium. Journal of Biological Chemistry*, **246**, 3720–4.

PONTECORVO, G. (1956). The parasexual cycle of fungi. *Annual Reviews of Microbiology*, **1**, 393–400.

PONTECORVO, G., ROPER, J. A., HEMMONS, L. M., MACDONALD, K. D. & BUFTON, A. W. J. (1953). The genetics of *Aspergillus nidulans. Advances in Genetics*, **5**, 141–238.

PRASAD, C., DIESTERHAFT, M. & FREESE, E. (1972). Initiation of spore germination in glycolytic mutants of *Bacillus subtilis. Journal of Bacteriology*, **110**, 321–8.

PUT, H. M. C. & SAND, F. E. M. J. (1975). Some notes on the heat resistance of ascosporogeneous yeasts. *Journal of Applied Bacteriology*, **39**, (3), iii.

RAPER, J. R. & FLEXER, A. S. (1970). The road to diploidy with emphasis on a detour. In *Organization and Control in Prokaryotic and Eukaryotic Cells. Symposia of the Society for General Microbiology*, **20**, 401–32, ed. H. P. Charles & B. C. J. G. Knight. London: Cambridge University Press.

REEVE, J. N., MENDELSON, N. H., COYNE, S. I., HALLOCK, L. L. & COLE, R. M. (1973). Minicells of *Bacillus subtilis*. *Journal of Bacteriology*, **114**, 860–73.

REYSSET, G. & MILLET, J. (1972). Characterization of an intracellular protease in *Bacillus subtilis* during sporulation. *Biochemical and Biophysical Research Communications*, **49**, 328–34.

ROBINOW, C. F. (1960). In *The Bacteria, volume I*, ed. I. C. Gunsalus & R. Y. Stanier, pp. 207–48. New York & London: Academic Press.

ROGERS, H. J. (1970). Bacterial growth and the cell envelope. *Bacteriological Reviews*, **34**, 194–214.

SADOFF, H. L. (1975). Encystment and germination in *Azotobacter vinelandii*. *Bacteriological Reviews*, **39**, 516–39.

SAGER, R. & GRANICK, S. (1954). Nutritional control of sexuality in *Chlamydomonas reinhardi*. *Journal of General Physiology*, **37**, 729–42.

SCHAEFFER, P. (1969). Sporulation and the production of antibiotics, exoenzymes and exotoxins. *Bacteriological Reviews*, **33**, 48–71.

SCHAEFFER, P., MILLET, J. & AUBERT, J.-P. (1965). Catabolic repression of bacterial sporulation. *Proceedings of the National Academy of the U.S.A.*, **54**, 704–11.

SCHIEIFER, K. H. & KANDLER, O. (1972). Peptidoglycan types of bacterial cell walls and their taxonomic implications. *Bacteriological Reviews*, **36**, 407–77.

SCHOPF, J. W. (1974). Paleobiology of the Precambrian: the age of the blue-green algae. In *Evolution Biology, volume I*, ed. T. Dobzhansky, M. K. Hecht & W. C. Steere, pp. 1–43. New York & London: Plenum Press.

SCHOPF, J. W. (1978). The evolution of the earliest cells. *Scientific American* September issue, 84–102.

SCHOPF, J. W. & BLACIC, J. M. (1971). New microorganisms from the Bitter Springs formation (late Precambrian) of the North-Central Amadeus Basin, Australia. *Journal of Paleontology*, **45**, 925–60.

SCHOPF, J. W. & OCHLER, D. Z. (1976). How old are the eukaryotes? *Science*, **193**, 47–49.

SEGALL, J. & LOSICK, R. (1977). Cloned *Bacillus subtilis* containing a gene that is activated early during sporulation. *Cell*, **11**, 751–62.

SETLOW, P. (1977). Protein metabolism during germination of spores of *Bacillus* species. In *Spore Research 1976*, ed. A. N. Barker, J. Wolf, D. J. Ellar, G. J. Dring & G. W. Gould, pp. 661–82. London & New York: Academic Press.

SHILO, V., SIMCHEN, G. & SHILO, B. (1978). Initiation of meiosis in cell cycle initiation mutants of *Saccharomyces cerevisiae*. *Experimental Cell Research*, **112**, 241–8.

SINGH, R. P., SETLOW, B. & SETLOW, P. (1977). Levels of small molecules and enzymes in the mother cell and forespore of sporulating *Bacillus megaterium*. *Journal of Bacteriology*, **130**, 1130–8.

SLEPECKY, R. A. (1972). Ecology of sporeformers. In *Spores V*, ed. H. O. Halvorson, R. S. Hanson & L. L. Campbell, pp. 297–313. Washington D.C.: American Society for Microbiology.

SLEPECKY, R. A. & LEADBETTER, E. R. (1977). The diversity of spore-forming bacteria: some ecological implications. In *Spore Research 1976*, ed. A. N. Barker, J. Wolf, D. J. Ellar, G. J. Dring & G. W. Gould, pp. 869–77. London & New York: Academic Press.

SMITH, D. A., MOIR, A. & LAFFERTY, E. (1978). Progress in genetics of spore germination in *Bacillus subtilis*. In *Spores VII*, ed. G. Chambliss & J. C. Vary, pp. 158–63. Washington D.C.: American Society for Microbiology.

SNIDER, I. J. & MILLER, J. J. (1966). A serological comparison of vegetative cell as ascus walls and the spore coat of *Saccharomyces cerevisiae*. *Canadian Journal of Microbiology*, **12**, 485–8.

SONENSHEIN, A. L. & CAMPBELL, K. M. (1978). Control of gene expression during sporulation. In *Spores VII*, ed. G. Chambliss & J. C. Vary, pp. 179–92. Washington, D.C.: American Society for Microbiology.

STANIER, R. Y. (1970). Some aspects of the biology of cells and their possible evolutionary significance. In *Organization and Control in Prokaryotic and Eukaryotic Cells. Symposia of the Society for General Microbiology* **20**, 1–38, eds. H. P. Charles & B. C. J. G. Knight. London: Cambridge University Press.

STEED, P. & MURRAY, R. G. (1966). The cell wall and cell division of Gram-negative bacteria. *Canadian Journal of Microbiology*, **12**, 263–70.

STERN, C. (1936). Somatic crossing over and segregation in *Drosophila melanogaster*. *Genetics, Princeton*, **21**, 625–730.

STEWART, B. T. & HALVORSON, H. O. (1953). Studies on the spores of aerobic bacteria. I. The occurrence of D-alanine racemase. *Journal of Bacteriology*, **65**, 160–6.

STINE, G. J. & CLARK, A. M. (1967). Synchronous production of conidiophores and conidia of *Neurospora crassa*. *Canadian Journal of Microbiology*, **13**, 447–53.

SUSSMAN, A. S. & HALVORSON, H. O. (1966). *Spores: their dormancy and germination*. New York & London: Harper & Row.

SUTHERLAND, I. W. & MACKENZIE, C. L. (1977). Glucan common to the microcyst wall of cyst-forming bacteria. *Journal of Bacteriology*, **129**, 599–605.

TAGG, J. R., DAJANI, A. S. & WANNAMAKER, L. W. (1976). Bacteriocins of Gram-positive bacteria. *Bacteriological Reviews*, **40**, 722–56.

THOMPSON, R. S. & LEADBETTER, E. R. (1963). On the isolation of dipicolinic acid from endospores of *Sarcina ureae*. *Archiv für Mikrobiologie*, **45**, 27–32.

TIPPER, D. J. & GAUTHIER, J. J. (1972). Structure of the bacterial endospore. In *Spores V*, ed. H. O. Halvorson, R. S. Hanson & L. L. Campbell, pp. 3–12. Washington D.C.: American Society for Microbiology.

TREW, B. J., FRIESEN, J. D. & MOENS, P. B. (1979). Two-dimensional protein patterns during growth and sporulation in *Saccharomyces cerevisiae*. *Journal of Bacteriology*, **138**, 60–9.

TURIAN, G. & BIANCHI, D. E. (1971). Conidiation in *Neurospora*. *Archiv. für Mikrobiologie* **77**, 262–74.

VEZINHET, F., KINNAIRD, J. & DAWES, I. W. (1979). The physiology of mutants derepressed for sporulation in *Saccharomyces cerevisiae*. *Journal of General Microbiology*, **115**, 391–402.

WAITES, W. M., KAY, D., DAWES, I. W., WOOD, D. A., WARREN, S. C. & MANDELSTAM, J. (1970). Sporulation in *Bacillus subtilis*. Correlation of biochemical events with morphological changes in asporogenous mutants. *Biochemical Journal*, **118**, 667–76.

WARTH, A. D. & STROMINGER, J. L. (1969). Structure of the peptidoglycan of bacterial spores: occurrence of the lactam of muramic acid. *Proceedings of the National Academy of Sciences of the U.S.A.* **64**, 528–35.

WATERBURY, J. B. (1979). Developmental patterns of pleurocapsalean cyanobacteria. In *Developmental Biology of Prokaryotes*, ed. J. H. Parish, pp. 203–26. Oxford: Blackwell Scientific Publications.

WATERBURY, J. B. & STANIER, R. Y. (1978). Patterns of growth and cell development in pleurocapsalean cyanobacteria. *Microbiological Reviews*, **42**, 2–44.

WEBER, D. J. & HESS, W. M. (1975). Diverse spores of fungi. In *Spores VI*, ed. P. Gerhardt, R. N. Costilow & H. L. Sadoff, pp. 97–111. Washington D.C.: American Society for Microbiology.

WHITTENBURY, R., DAVIES, S. L. & DAVY, J. F. (1970). Exospores and cysts formed by methane-utilizing bacteria. *Journal of General Microbiology*, **61**, 219–26.

WHITTENBURY, R. & DOW, C. (1977). Morphogenesis and differentiation in *Rhodomicrobium vannielii* and other prosthecate bacteria. *Bacteriological Reviews*, **41**, 754–808.

WOOD, D. A. (1971). Sporulation in *Bacillus* subtilis. The appearance of sulpholactic acid as a marker event for sporulation. *Biochemical Journal*, **123**, 601–5.

WRIGHT, B. E. (1970). On the evolution of substrate control in differentiation. In *Evolutionary Biology, volume 4*, ed. T. Dobzhansky, M. K. Hecht & W. C. Steere, pp. 111–25.

WRIGHT, J. F. & DAWES, I. W. (1979). Sporulation-specific protein changes in yeast. *Federation of European Biochemical Societies Letters*, **104**, 183–6.

ZAKHAROV, I. A. & INGE-VECHTOMOV, S. G. (1964). Ascospore isolation of yeast for genetic analysis without a micromanipulator. *Issledovania po Genetike* (English Translation), **2**, 134–9.

DNA REARRANGEMENTS AND EVOLUTION

JOHN CULLUM AND HEINZ SAEDLER

Max-Planck-Institut für Züchtungsforschung, 5 Köln 30 (Vogelsang), German Federal Republic

INTRODUCTION

In recent years it has been realized that many spontaneous mutations are due to DNA rearrangements of various kinds, rather than to base changes. In this paper we examine the consequences of this fact for evolutionary theory. We first illustrate some of the possible rearrangements seen in *Escherichia coli* K12, the best studied system, and argue that similar events occur in most or all organisms. We then discuss the consequences for evolution at two levels; first, the rôle in producing new enzyme activities and new control pathways and, second, the implications for stability of genome structure and the concept of species. These considerations lead us to the conclusion that DNA rearrangements do, in fact, play an important and distinctive rôle in evolution.

DNA REARRANGEMENTS IN E.COLI K12

Most of the systems we will discuss involve special DNA sequences called 'insertion sequences' (abbreviated IS). When spontaneous Gal$^-$ mutations in the *gal* operon were examined, many were found to arise through the insertion of extra DNA into the bacterial genome (see reviews by Starlinger & Saedler, 1976; Starlinger, 1980). Such insertions have now been observed in many systems. A striking observation was that the insertions examined were usually not random sequences of DNA, but the same sequences were found time and again in independent isolates. The same DNA sequence was often observed to be inserted at many different sites. The insertion sequences have been numbered IS1–IS5 (Table 1). Insertion did not seem to be homology-dependent and occurred in *recA* mutants, where normal homologous recombination was absent.

In this section we discuss some model systems that illustrate the range of DNA rearrangements possible. The subject will be divided into effects on transcription and insertion-sequence-mediated DNA

Table 1. *Insertion sequences of* E. coli *K12*

Name of sequence	Length (base pairs)	Number of copies in E.coli chromosome
IS1	768	ca.7
IS2	1327	5
IS3	ca.1400	5
IS4	ca.1400	1
IS5	ca.1300	?

The data above come from papers quoted in the reviews of Starlinger & Saedler (1976), Starlinger (1980) and Calos & Miller (1980).

rearrangements. This is an arbitrary division, and many systems show both aspects of IS behaviour.

Events affecting transcription

IS insertions into the *gal* operon, giving Gal$^-$ phenotypes, are often very polar – much more polar than classical chain termination mutations in the same region (Jordan & Saedler, 1967; Jordan, Saedler & Starlinger, 1967). At least in the case of IS1 (in both orientations) and IS2 (in orientation I), the polarity is due to *rho*-dependent transcriptional termination signals in the insertion elements (Wetekam & Ehring, 1973; Besemer & Herpers, 1977). Hence, insertions upstream of the first gene of the operon (*galE*), lying between it and the *gal* promoter, are Gal$^-$ and lack all three enzyme activities of the *gal* operon (Fig. 1). An internal deletion of IS1 that removes the transcriptional termination site can restore a Gal$^+$ phenotype (Hans Sommer, personal communication).

When structural genes are not attached to a strong promoter, selection for expression of the genes frequently leads to the isolation of strains in which IS2 has been inserted in orientation II in front of the genes (Saedler, Reif, Hu & Davidson, 1974; Walz, Ratzkin & Carbon, 1978). This is illustrated in Fig. 2 for a *gal* operon that has been disconnected from its promoter by insertion of IS1 between promoter and structural genes, and subsequent deletion of the *gal* promoter. The recovery of gene expression may be due to a promoter situated internally on IS2 (Saedler *et al.*, 1974), although it has also been suggested that the promoter is formed by the new junctions between IS2 and the host sequence on insertion, because some IS2–II insertions show only a weak promoter activity (Boyen

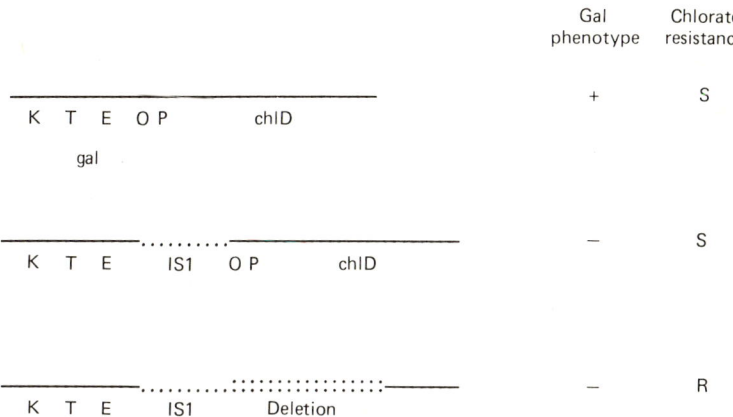

Fig. 1. Insertion of IS1 into the *galOP* region and selection of deletions that inactivate the *chlD* gene.

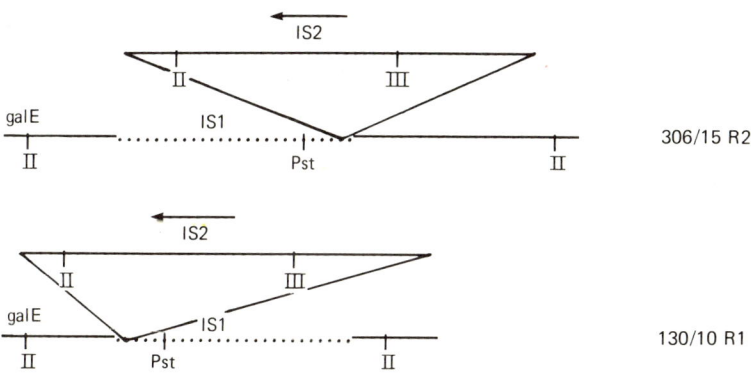

Fig. 2. Insertion of IS2 (orientation II) into IS1. A *gal* operon inactivated by insertion of IS1 into *galOP* and subsequent deletion of *galOP* (bottom line, Fig. 1), was reactivated by insertion of IS2 in front of the structural genes. Restriction sites are shown as follows: Pst = PstI, II = HindII, III = HindIII.

et al., 1978). However, this is still not clear because there are several copies of IS2 in the *E.coli* chromosome, and they may differ in sequence (Saedler & Heiβ, 1973). If only low promoter activity is selected, several other insertion sequences may be found, such as IS2 in orientation I (Boyen *et al.*, 1978).

Promoter activity can also occur as a result of other types of DNA rearrangements. The best studied system makes use of an IS2–I insertion into the *galOP* region (Fig. 3). This produces a Gal⁻ phenotype due to the transcriptional termination site on IS2, and

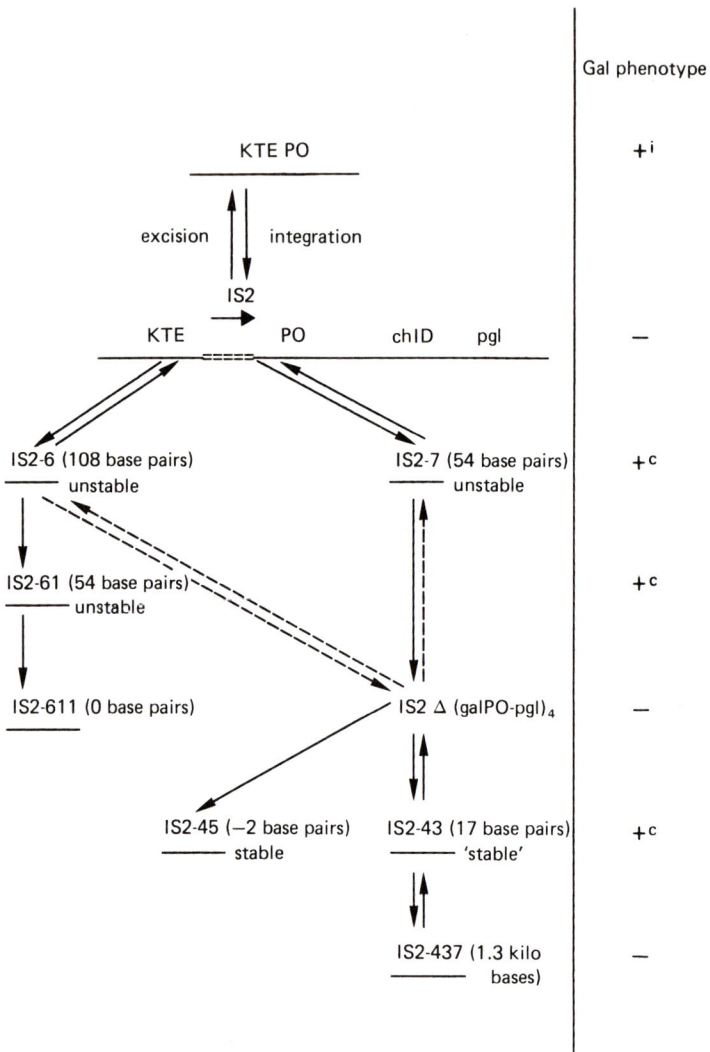

Fig. 3. Alleles of IS2. The figure shows integration of IS2 into the *galOP* region. Subsequent rounds of selection for Gal$^+$ and Gal$^-$ phenotypes produced the range of alleles shown, all of which have been characterized by DNA sequencing. $+^i$ = inducible Gal$^+$ phenotype; $+^c$ = constitutive phenotype.

the Gal$^+$ revertants which may be selected have DNA rearrangements near the end of IS2 that produce new promoters (Ghosal & Saedler, 1977; Sommer, Cullum & Saedler, 1979). A series of such derivatives has been characterized and sequenced (Fig. 3).

The most striking property of the constitutive Gal$^+$ revertants was that the majority were unstable. They could be divided into two

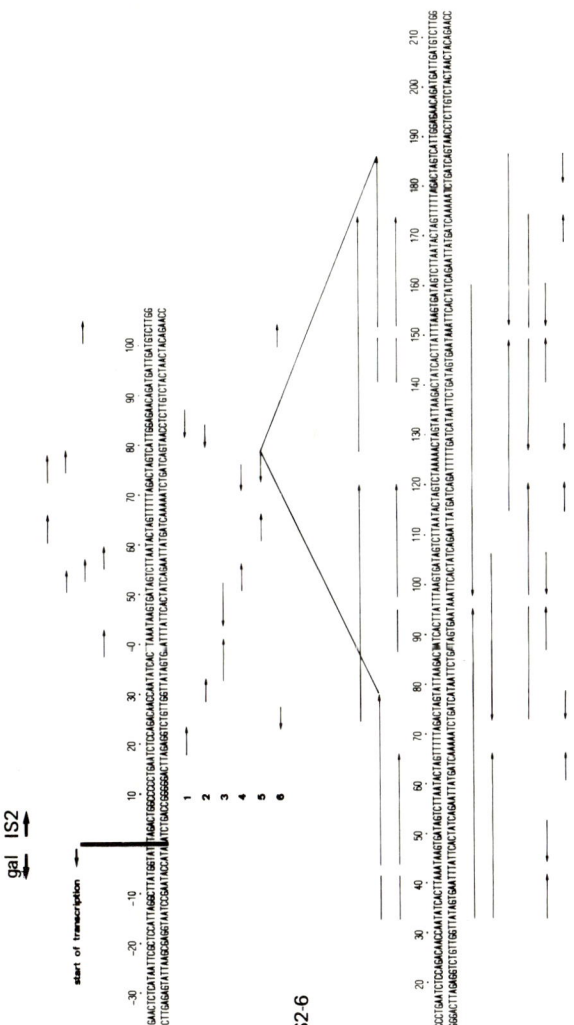

Fig. 4. DNA sequence of the allele IS2-6 compared to the parent IS2. Arrows above the sequences indicate pairs of direct repeats; those below indicate inverted repeats.

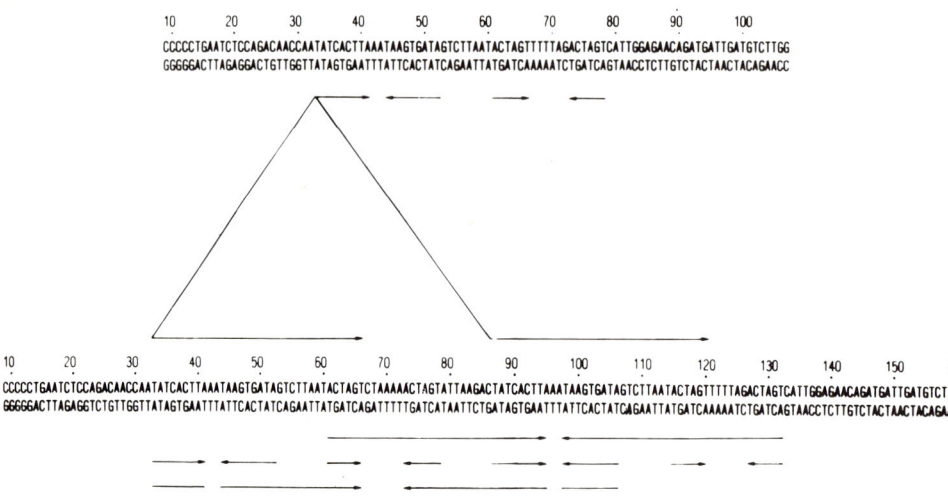

Fig. 5. DNA sequence of allele IS2-7. The upper line shows the sequence of the *galE* proximal end of IS2 and the lower line the sequence of the allele IS2-7.

classes of about equal frequency by their degree of instability; the 'mini-insertions', that produce more than 1% Gal⁻ segregants in overnight cultures grown without selection (Ghosal & Saedler, 1977), and the 'super-minis', that produced about 0.01% segregants (Sommer *et al.*, 1979). The 'mini-insertions' could in turn be divided into two classes; one, represented by the allele IS2-6, was very unstable (giving more than 10% segregants) and the other, represented by IS2-7, was more stable (giving 1% segregants). Sequence analysis showed that they resulted from duplications of 108 base pairs (Ghosal & Saedler, 1978) and 54 base pairs (Ghosal, Gross & Saedler, 1979) respectively (Fig. 4 and 5) within the A + T-rich last 100 base pairs of IS2. The structure of the duplications was rather complicated (Figs. 4 and 5) with both direct repeats and inverted repeats. The structure can be understood with the aid of a model to explain their formation (Ghosal & Saedler, 1978). This assumes (Fig. 6) that, during replication of this section of DNA, the newly synthesized strand attached to the replication complex can unwind. If the DNA then folds back on itself at an inverted repeat, replication could continue using the newly synthesized strand as a template, thus making a duplication. Of course, the polymerase must revert to the original template later in order to complete the round of replication.

The stabler class of the 'super-minis', represented by IS2-43 and

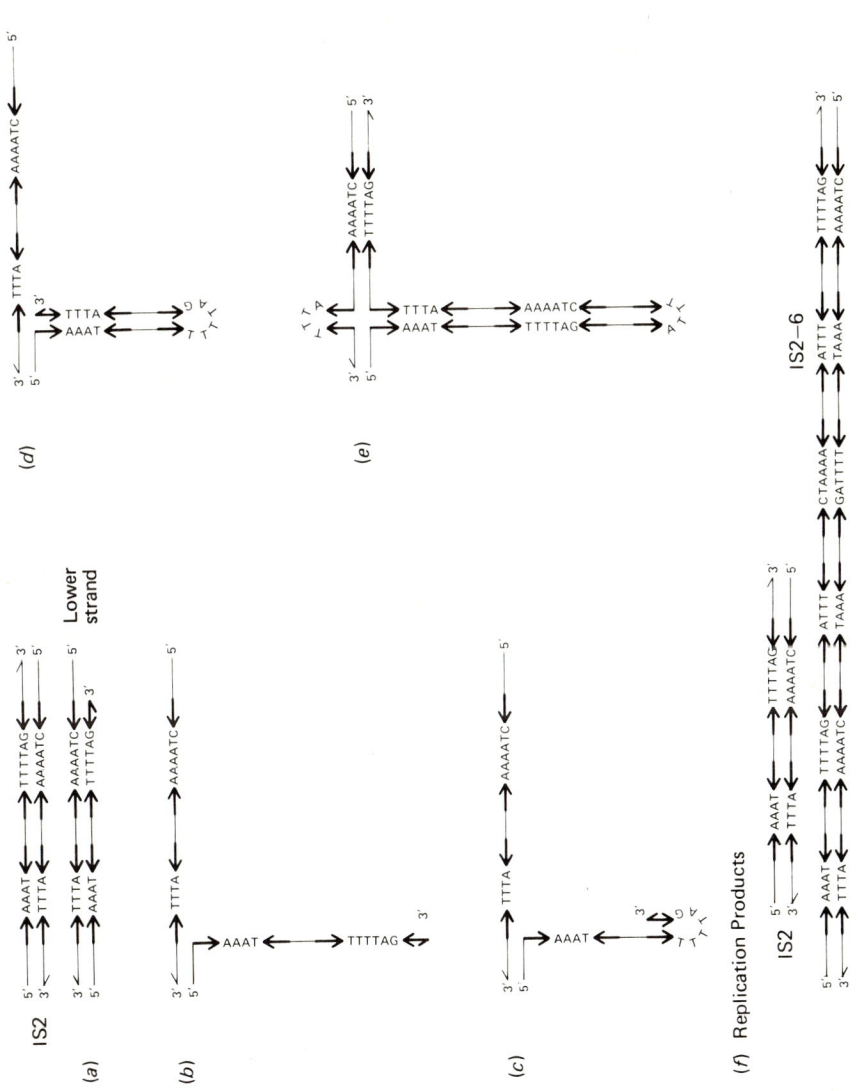

Fig. 6. Model for generation of the 'mini-insertions'. The model is described in the text.

Fig. 7. Sequence of the allele IS2–43 (bottom line) compared to that of IS2 (top line). The arrows above the sequences indicate the duplicated 17 base-pair-long sequence. Below the sequences are marked areas thought to be important for the promoter activity.

IS2-44, carried a 17 base pair tandem duplication within IS2 (Fig. 7; Sommer et al., 1979). This may have been formed by stuttering of the DNA polymerase, similar to that observed *in vitro* with DNA polymerase I (Kornberg, Bertsch, Jackson & Khorana, 1964).

The new promoters described above are unstable. Most Gal⁻ segregants arise through excision of the duplications with restoration of the original sequence (Sommer et al., 1979; Ghosal & Saedler, 1979). In the case of the 108 base pair duplication IS2-6, loss of the promoter can also occur by a two-step process with an intermediate stage (IS2-61), 54 base pairs longer than IS2, which closely resembles IS2-7 in properties (Fig. 8; Ghosal & Saedler, 1979). It is striking that all of these excision events formally involve recombination between short directly repeated sequences (Figs. 7 and 8).

Insertion-sequence-mediated DNA rearrangements

When IS1 is integrated into a DNA sequence it promotes deletion of neighbouring DNA sequences at a high frequency (10^{-4}, Reif & Saedler, 1975). Most of these deletions leave IS1 intact (Reif & Saedler, 1975), and sequence analysis established that one end of the deletions was at nucleotide 1 of IS1 (Ohtsubo & Ohtsubo, 1978; Saedler et al., 1981). Similar deletions occur with IS2, although at a lower frequency (Peterson, Ghosal, Sommer & Saedler, 1979). A convenient system for monitoring deletions is one where IS1 is inserted into the *gal* operon operator–promoter region (Fig. 1). Strains with deletions that run into the neighbouring *chlD* gene can easily be selected as they have a chlorate-resistant phenotype (Reif & Saedler, 1975). Such strains are Gal⁻ and cannot revert to the original *gal*⁺ genotype because the *gal* promoter has been deleted. However, selection for a Gal⁺ phenotype can lead to the isolation of strains with deletions that remove most of IS1 and also neighbouring sequences, fusing the *gal* operon to a new promoter (Saedler et al., 1981). DNA sequence analysis showed that these deletions did not stop exactly at the end of IS1, but, rather, appeared to be due to recombination between short (10 base pair) direct repeats. In certain cases this occurred at a significant frequency even when the repeats were separated by 8 kilobases of DNA.

A further type of event observed with IS1 is inversion. This can most easily be demonstrated in a strain which has a copy of IS1 next to a lactose operon, as found in the lactose transposon Tn951

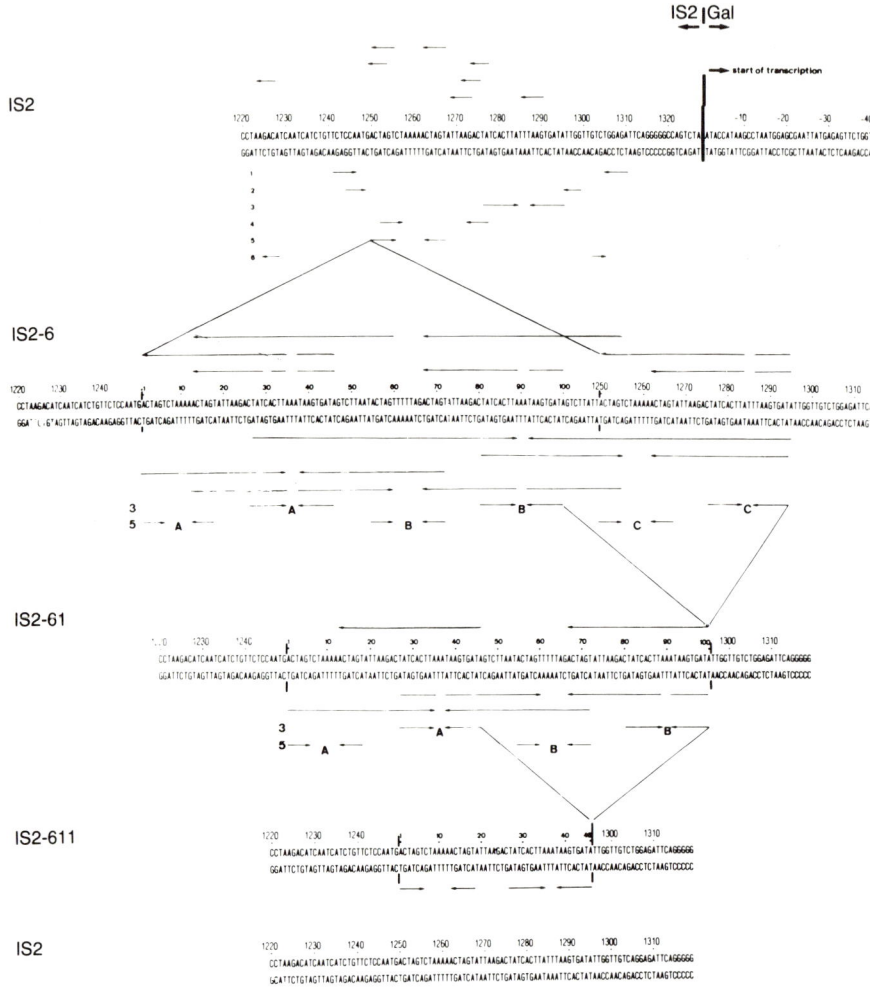

Fig. 8. Sequences of IS2–61 and IS2–611 compared to parent sequences IS2 and IS2–6. Direct and inverted repeats are indicated as in Fig. 4.

(Cornelis & Saedler, 1980). It is possible to select spontaneous LacY⁻ mutants because they are resistant to a toxic lactose analogue, *o*-nitrophenyl-thio-galactoside (Smith & Sadler, 1971). Most such mutants are also LacZ⁻ (Saedler *et al.*, 1981) due to IS1-promoted deletions. However, some of the mutants are LacZ⁺ and thus cannot be explained as deletions, because the Z gene lies between IS1 and the Y gene (Cornelis & Saedler, 1980). Most of these LacZ⁺Y⁻ mutants were shown to arise through insertion of a second copy of IS1 into the *lacY* gene (Saedler *et al.*, 1981). Sometimes, when the second copy of IS1 was inserted in inverse orientation with respect to the resident copy, the DNA sequence between the two copies of IS1 had been inverted.

When two different plasmids are in the same cell, it is sometimes possible to observe their fusion. This is often independent of the host's recombination system, occurring also in *recA* mutants. The best characterized cases of 'replicon fusion' are promoted by transposable elements, and the insertion sequence IS1 can also promote this process (Ohtsubo, Zenilman & Ohtsubo, 1980). When one of the plasmids carried a copy of IS1 and the other did not, the fused replicons had a structure containing all of both plasmids, but with the IS1 region duplicated so that direct repeats of IS1 formed the boundaries between the two plasmids. IS1 was inserted at its usual site in the IS1-carrying plasmid, but seemed to be able to fuse the two plasmids in many ways by interacting at a variety of sites in the second plasmid.

Most transposons in bacteria have repeated sequences at their ends (for review see Calos & Miller, 1980), which are sometimes homologous to known insertion sequences. The best studied case is Tn9 and the related chloramphenicol-resistance transposons that carry a chloramphenicol acetyl transferase gene between direct repeats of IS1 (Chow & Bukhari, 1977; MacHattie and Jackowski, 1977). It appears from experiments where structures were constructed carrying genes between repeats of IS1, that only the IS1 copies, and not the genes in between, are necessary for transposition activity. Therefore, almost any gene could be mobilized to form a new transposon by enclosing it between two copies of IS1 (Rosner & Guyer, 1980). This ability to mobilise genes is probably common to transposable elements.

The events described above are independent of the normal *E.coli* homologous recombination system, since they also occur in *recA* strains. Insertion sequences, of which there may be several copies present per genome (Saedler & Heiβ, 1973), can also serve as regions of homology for the host's recombination system. The best known example is the integration of the F plasmid into the *E.coli* chromosome, to give Hfr strains. This seems to occur by recombination between insertion sequence copies in F and in the chromosome (for a review see Davidson, Deonier, Hu & Ohtsubo, 1975), and is mainly *recA*-dependent (Cullum & Broda, 1979).

A second example occurs when a gene is flanked by two copies of an insertion sequence in the same orientation. Then, unequal crossing over between sister copies of this region can result in tandem duplications of the gene, a process called amplification (Rownd *et al.*, 1978). This process is *recA*-dependent, and the number of repeated copies can change quickly, either increasing on selection for higher levels of the gene product, or decreasing on relaxation of such selection.

Thus, insertion sequences can promote a wide variety of DNA rearrangements, but DNA rearrangements such as deletions also occur in the absence of known insertion sequences albeit at a lower frequency (Farabaugh, Schmeissner, Hofer & Miller, 1978). It must be realised that the type of event observed is very dependent on the selection system used, and the range of events discussed above is therefore probably a very biased sample of those that do occur.

GENERALITY OF DNA REARRANGEMENTS

The data discussed above highlight the importance of insertion sequence elements. The question arises whether such elements are confined to a small group of bacteria, or are more widely distributed. There is evidence for transposable elements in a range of bacteria (for example, see Novick *et al.*, 1979). In eukaryotes, the three systems which have been studied in enough detail all show evidence for the presence of transposable elements. In *Zea mays*, the classical genetical work of McClintock (1965) provided strong evidence for transposable 'controlling elements' with properties that resemble those of *E.coli* insertion sequences (Nevers & Saedler, 1977). In *Saccharomyces cerevisiae* and *Drosophila melanogaster*, recent work has revealed DNA sequences that can change position within the genome and seem to resemble insertion sequences in many ways (Farabaugh & Fink, 1980; Gafner & Philippsen, 1980; Bayev *et al.*, 1980; Dunsmuir, Brorein, Simon & Rubin, 1980). These sequences can also promote various types of DNA rearrangement (Green, 1967; Chaleff & Fink, 1980).

So far, DNA rearrangements of the type that produced new promoters in IS2 have not been observed in other organisms. However, they do occur in other instances in *E.coli* systems; duplications in the *arg* operon were found with a similar structure to the 'mini-insertions' but involving 3.2 kilo bases of DNA rather than only 100 base pairs (Charlier, Crabeel, Cunin & Glansdorff, 1979);

a 9 base pair tandem duplication with promoter activity was shown to occur in the lambda c17 mutation (Rosenberg et al., 1978). It seems likely that such mutations are due to errors of replication and should therefore occur in a wide range of organisms. They would only be detected when a system had been analysed intensively at the molecular level.

It seems likely that DNA rearrangements of the types detected in *E. coli* K12 will also occur in most other organisms, both prokaryotic and eukaryotic. Therefore, any consequences for evolutionary theory of such events will be very widely applicable. In the next section we consider the rôle of rearrangements in the formation of new enzyme functions and control pathways. Later, we consider the consequences for gross genome organization and speciation. The former discussion is equally applicable to prokaryotes and eukaryotes. The latter will point to differences between the two classes of organisms.

EVOLUTION OF NEW FUNCTIONS AND CONTROL PATHWAYS

In this section we discuss two classes of events: first, duplications and divergence of sister genes, and second, gene fusions. Duplications have long been recognized as important in evolution. They can occur by a variety of mechanisms; for example

(i)   Transposition to give a second copy of the gene.
(ii) Tandem duplications due to unequal crossing over.
(iii) 'Mini-insertion'-like duplications.

When a microorganism is selected for growth on a new substrate that it does not normally utilize, duplication may increase the dosage of a gene coding for an enzyme which has a low affinity for the substrate, thus allowing increased utilization. This has been observed for the case of xylitol utilization in *Klebsiella* (Rigby, Burleigh & Hartley, 1974). One copy of the duplicated gene may also mutate to give better utilization of the new substrate, as has been observed in this system (Wu, Lin & Tanaka, 1968). The original copy still survived with retention of the original enzyme activity for use when necessary.

From sequence analysis of proteins, it can be inferred that gene duplications have occurred fairly often in evolution. The structure of the ovalbumin genes, which has been investigated at the molecular level, shows clear evidence of duplicated sequences (Heilig et al., 1980).

Fusion of structural genes to new promoters would modify expression of the genes. The model system discussed above demonstrates the IS1-promoted fusion of the *E.coli gal* operon to a new promoter, which expresses the operon constitutively. Insertion sequences could mobilize genes as new transposons, and move them to different parts of the genome, thus allowing fusions of a variety of systems. It has been suggested that the *E.coli lac* operon arose as a result of a fusion (Cornelis, Ghosal & Saedler, 1978). Of the known *lac* operons, *E.coli* is unusual in having a *lacA* gene (transacetylase) whose *in vivo* rôle is unknown. The operon is homologous to the *lacIZY* region of the *lac* operon of the transposon Tn951, and it seems plausible that the *E.coli* operon could have arisen by insertion of a lactose transposon into the chromosome, followed by fusion events which coupled the *lacA* gene to the operon (Cornelis *et al.*, 1978).

As well as fusing structural genes to new control regions, it is possible to fuse two structural genes. This can allow the formation of hybrid proteins that have new functions which are a combination of the functions of the two parent proteins. An example of a fused protein is provided by strains where the β-galactosidase gene (*lacZ*) has been fused to DNA sequences coding a new N-terminal polypeptide, giving rise to a new protein. If the N-terminal region of the new protein was derived from a membrane protein, then it would carry a signal sequence that allowed the hybrid protein to enter the membrane. It has thus been possible to produce hybrid proteins with β-galactosidase activity that are found in the membrane rather than in the cytoplasm (Silhary *et al.*, 1977).

Finally, transposition of a gene to another place in the genome might affect its expression in the absence of other secondary events. Thus, transposition onto a plasmid or to another place in the chromosome (Masters & Broda, 1971) might affect the copy number of a gene in *E.coli*. In *Drosophila melanogaster*, expression of the *white* gene can be affected by moving it to a different chromosomal location next to heterochromatin ('position effect variegation'; for a review, see Baker, 1968).

GROSS GENOME STRUCTURE AND EVOLUTION

DNA rearrangements such as inversions and transpositions could potentially greatly affect gross genome structure. However, large differences have not been observed between different strains of *E.coli* K12. Moreover, the genetic map of *E.coli* K12 is almost identical with that of *Salmonella typhimurium* LT2 (Riley & Anilionis, 1978); the main differences consist of a series of small deletions/insertions, and one inversion of about 6' map length. Thus, the gross genome structure seems to be fairly stable. This might reflect a very low frequency of major rearrangements, but it is also possible that large-scale rearrangements have a selective disadvantage. This could be due to a disturbance of gene copy number; it has been noticed that many essential genes of *E.coli* are clustered around the origin of replication where they have a relatively high copy number compared to other parts of the chromosome (Bachmann, Brooks Low & Taylor, 1976). It is also possible that DNA topology plays a rôle. The *E.coli* chromosome seems to be organized into up to 80 independent loops (Worcel & Burgi, 1972). The size of a loop, the position in the loop and the particular loop in which a gene occurs might all affect gene expression. A rôle for DNA topology is suggested by the observations of Broda (1967), who isolated a series of Hfr strains with integration sites within a small region of the chromosome. The Hfrs differed from one another in growth rate, and it is difficult to explain this effect by direct influences on transcription or copy number. Integration by recombination between insertion sequences, which are highly polar, should not change the relationship of any genes to their promoters, and the small region of chromosome involved makes copy number effects very unlikely.

Transposition between different replicons and the ability of plasmids to transfer between different bacterial 'species' means that potentially useful genes can be transferred within large groups of bacterial species. The most striking example is the spread of R-factors since the introduction of antibiotics into clinical use (for more information, see books by Falkow, 1975 and Broda, 1979). Here, interspecies transfer of plasmids, coupled with transposition of new drug resistance determinants onto plasmids, has led to the evolution of R-factors conferring resistance to many different drugs.

Many of the characters used to classify bacteria can also be plasmid-borne and mobilised on transposons. For instance, the lactose transposon Tn951 was discovered in a clinical isolate of *Yersinia entercolitica* (Cornelis, Bennett & Grinsted, 1976), which was thus Lac$^+$ rather than the expected Lac$^-$. These considerations make the concept of 'species' in bacteria very different from that in higher organisms, and makes the classification of bacteria potentially very difficult. However, the genome stability noted above may indicate that the concept of bacterial species is not totally useless.

In higher organisms, the meiotic process tends to select against large-scale genome rearrangements. Organisms heterozygous for inversions or translocations will tend to produce unbalanced gametes as a result of crossing over during meiosis, and this will lead to a loss of fertility. The biochemical needs of most higher organisms are probably very similar, and phenotypic differences between related species may reflect mainly differences in control of differentiation. It may not need many gene differences to produce the observed morphological differences. For example, *Zea mays* and teosinte are normally regarded as different species, and differ considerably in morphology. However, they interbreed freely, and analysis of crosses suggested that the differences between them might involve as few as five major genes (Beadle, 1980). It has been suggested that chromosome rearrangements rather than protein differences correlate much better with phenotypic differences between most species (Wilson, Sarich & Maxson, 1974).

Chromosome rearrangements such as inversions or translocations are not, in principle, deleterious if homozygous, only if heterozygous. It would be disadvantageous for subpopulations of a species that differed by such rearrangements to interbreed because of the lowered fertility of the offspring. Therefore, mutations that discourage mating between subpopulations would lead to production of more grandchildren. This would lead to selection of mating barriers, and thus speciation. It would be possible to get 'accidental speciation' driven by a chromosome rearrangement that has no phenotypic effect. The existence of chromosome differences between different subpopulations of a species has been documented; for example, the chromosomes of Sumatran and Bornean orang-utan populations are distinguished by a pericentric inversion (Seuanez, Evans, Martin & Fletcher, 1979).

SUMMARY

We have discussed some of the DNA rearrangements found in *E.coli* K12 and argued that similar events occur in all organisms. These rearrangements could allow the rapid evolution of new functions and control pathways. Gross genome structure is probably fairly stable in both prokaryotes and eukaryotes. In eukaryotes, major rearrangements may trigger speciation.

REFERENCES

BACHMANN, B. J., BROOKS LOW, K. & TAYLOR, A. L. (1976). Recalibrated linkage map of *Escherichia coli* K-12. *Bacteriological Reviews*, **40**, 116–67.

BAKER, W. K. (1968). Position-effect variegation. *Advances in Genetics*, **14**, 133–69.

BAYEV, A. A. JR., KRAYEV, A. S., LYUBOMIRSKAYA, N. V., ILYIN, Y. V., SKRYABIN, K. G. & GEORGIEV, G. P. (1980). The transposable element Mdg3 in *Drosophila melanogaster* is flanked with the perfect direct and mismatched inverted repeats. *Nucleic Acids Research*, **8**, 3263–73.

BEADLE, G. W. (1980). The ancestry of corn. *Scientific American*, **242**, 112–19.

BESEMER, J. & HERPERS, M. (1977). Suppression of polarity of insertion mutations within the *gal* operon of *Escherichia coli*. *Molecular and General Genetics*, **151**, 295–304.

BOYEN, A., CHARLIER, D., CRABEEL, M., CUNIN, R., PALCHAUDHURI, S. & GLANSDORFF, N. (1978). Studies on the control region of the bipolar *argECBH* operon of *Escherichia coli*. I. Effect of regulatory mutations and IS2 insertions. *Molecular and General Genetics*, **161**, 185–96.

BRODA, P. (1967). The formation of Hfr strains in *Escherichia coli* K12. *Genetical Research*, **9**, 35–47.

BRODA, P. (1979). *Plasmids*. W. H. Freeman and Company Ltd.

CALOS, M. P. & MILLER, J. H. (1980). Transposable elements. *Cell*, **20**, 579–95.

CHALEFF, D. T. & FINK, G. R. (1980). Genetic events associated with an insertion mutation in yeast. *Cell*, **21**, 227–37.

CHARLIER, D., CRABEEL, M., CUNIN, R. & GLANSDORFF, N. (1979). Tandem and inverted repeats of arginine genes in *Escherichia coli*. Structural and evolutionary considerations. *Molecular and General Genetics*, **174**, 75–88.

CHOW, L. T. & BUKHARI, A. I. (1977). Bacteriophage Mu genome: Structural studies on Mu DNA and Mu mutants carrying insertions. In *DNA insertion elements, plasmids and episomes*, ed. A. I. Bukhari, J. A. Shapiro & S. L. Adhya pp. 295–306. New York: Cold Spring Harbor Laboratory.

CORNELIS, G., BENNETT, P. M. & GRINSTED, J. (1976). Properties of pGC1, a *lac* plasmid orginating in *Yersinia enterocolitica* 842. *Journal of Bacteriology*, **127**, 1058–62.

CORNELIS, G., GHOSAL, D. & SAEDLER, H. (1978). Tn951: A new transposon-carrying a lactose operon. *Molecular and General Genetics*, **160**, 215–24.

CORNELIS, G. & SAEDLER, H. (1980). Deletions and an inversion induced by a resident IS1 of the lactose transposon Tn951. *Molecular and General Genetics*, **178**, 367–74.

CULLUM, J. & BRODA, P. (1979). Chromosome transfer and Hfr formation by F in rec^+ and $recA$ strains of *Escherichia coli* K12. *Plasmid*, **2**, 358–65.
DAVISON, N., DEONIER, R. C., HU, S. & OHTSUBO, E. (1975). Electron microscope heteroduplex studies of sequence relations among plasmids of *Escherichia coli*. The DNA sequence organization of F and F-primes and the sequences involved in Hfr formation. In *Microbiology 1974*, ed. D. Schlessinger, pp. 56–65. American Society for Microbiology.
DUNSMUIR, P., BROREIN, W. J. JR., SIMON, M. A. & RUBIN, G. M. (1980). Insertion of the Drosophila transposable element *copia* generates a 5 base pair duplication. *Cell*, **21**, 575–9.
FALKOW, S. (1975). *Infectious multiple drug resistance*. Pion Ltd.
FARABAUGH, P. J. & FINK, J. R. (1980). Insertion of the eukaryotic transposable element Ty1 creates a 5bp duplication. *Nature London*, **286**, 352–6.
FARABAUGH, P. J., SCHMEISSNER, U., HOFER, M. & MILLER, J. H. (1978). Genetic studies of the *lac* repressor. VII. On the molecular nature of spontaneous hotspots in the *lacI* gene of *Escherichia coli*. *Journal of Molecular Biology*, **126**, 847–63.
GAFNER, J. & PHILIPPSEN, P. (1980). The yeast transposon Ty1 generates duplications of target DNA on insertion. *Nature London*, **286**, 414–18.
GHOSAL, D., GROSS, J. & SAEDLER, H. (1979). DNA sequence of IS2-7 and generation of mini-insertions by replication of IS2 sequences. *Cold Spring Harbor Symposia on Quantitative Biology*, **43**, 1193–6.
GHOSAL, D. & SAEDLER, H. (1977). Isolation of the mini-insertions IS6 and IS7 of *Escherichia coli*. *Molecular and General Genetics*, **158**, 123–8.
GHOSAL, D. & SAEDLER, H. (1978). DNA sequence of the mini-insertion IS2-6 and its relation to the sequence of IS2. *Nature*, **275**, 611–17.
GHOSAL, D. & SAEDLER, H. (1979). IS2-61 and IS2-611 arise by illegitimate recombination from IS2-6. *Molecular and General Genetics*, **176**, 233–8.
GREEN, M. M. (1967). The genetics of a mutable gene at the *white* locus of *Drosophila melanogaster*. *Genetics*, **56**, 467–82.
HEILIG, R., PERRIN, F., GANNON, F., MANDEL, J. L. & CHAMBON, P. (1980). The ovalbumin gene family: Structure of the X gene and evolution of duplicated split genes. *Cell*, **20**, 625–37.
JORDAN, E. & SAEDLER, H. (1967). Polarity of amber mutations and suppressed amber mutations in the galactose operon of *Excherichia coli*. *Molecular and General Genetics*, **100**, 283–95.
JORDAN, E., SAEDLER, H. & STARLINGER, P. (1967). Strong polar mutations in the transferase gene of the galactose operon in *Escherichia coli*. *Molecular and General Genetics*, **100**, 296–306.
KORNBERG, A., BERTSCH, L. L., JACKSON, J. F. & KHORANA, H. G. (1964). Enzymatic synthesis of deoxyribonucleic acid. XVI Oligonucleotides as templates and the mechanism of their replication. *Proceedings of the National Academy of Sciences, USA*, **51**, 315–23.
MCCLINTOCK, B. (1965). The control of gene action in maize. *Brookhaven Symposia on Biology*, **18**, 162–84.
MACHATTIE, L. A. & JACKOWSKI, J. B. (1977). Physical structure and deletion effects of the chloramphenicol resistance element Tn9 in phage lambda. In *DNA insertion elements, plasmids and episomes*, ed. A. I. Bukhari, J. A. Shapiro & S. L. Adhya, pp. 219–28. New York: Cold Spring Harbor Laboratory.
MASTERS, M. & BRODA, P. (1971). Evidence for the bidirectional replication of the *Escherichia coli* chromosome. *Nature New Biology*, **232**, 137–40.
NEVERS, P. & SAEDLER, H. (1977). Transposable genetic elements as agents

of gene instability and chromosomal rearrangements. *Nature London*, **268**, 109–15.
NOVICK, R. P., EDELMAN, I., SCHWESINGER, M. D., GRUSS, A. D., SWANSON, E. C. & PATTEE, P. A. (1979). Genetic translocation in *Staphylococcus aureus*. *Proceedings of the National Academy of Sciences, USA*, **76**, 400–4.
OHTSUBO, H. & OHTSUBO, E. (1978). Nucleotide sequence of an insertion element, IS1. *Proceedings of the National Academy of Sciences, USA*, **75**, 615–19.
OHTSUBO, E., ZENILMAN, M. & OHTSUBO, H. (1980). Plasmids containing insertion elements are potential transposons. *Proceedings of the National Academy of Sciences, USA*, **77**, 750–4.
PETERSON, P. A., GHOSAL, D., SOMMER, H. & SAEDLER, H. (1979). Development of a system useful for studying formation of unstable alleles of IS2. *Molecular and General Genetics*, **173**, 15–21.
REIF, H. J. & SAEDLER, H. (1975). IS1 is involved in deletion formation in the *gal* region of *E.coli* K12. *Molecular and General Genetics*, **137**, 17–28.
RIGBY, P. W. J. BURLEIGH, B. D. JR. & HARTLEY B. S. (1974). Gene duplication in experimental enzyme evolution. *Nature*, **251**, 200–4.
RILEY, M. & ANILIONIS, A. (1978). Evolution of the bacterial genome. *Annual Reviews of Microbiology*, **32**, 519–60.
ROSENBERG, M., COURT, D., SHIMATAKE, H., BRADY, C. & WULFF, D. C. (1978). The relationship between function and DNA sequence in an intercistronic regulatory region in phage lambda. *Nature London*, **272**, 414–23.
ROSNER, J. L. & GUYER, M. S. (1980). Transposition of IS1-lambdaB10-IS1 from a bacteriophage lambda derivative carrying the IS1-*cat*-IS1 transposon (Tn9). *Molecular and General Genetics*, **178**, 111–20.
ROWND, R., MIKI, T., APPELBAUM, E. R., MILLER, J. R., FINKELSTEIN, M. & BARTOCK, C. R. (1978). Dissociation, amplification, and reassociation of composite R-plasmid DNA. In *Microbiology 1978*, ed. D. Schlessinger, pp. 138–46. American Society for Microbiology.
SAEDLER, H., CORNELIS, G., CULLUM, J., SCHUMACHER, B. & SOMMER, H. (1981). IS1-mediated DNA rearrangements. *Cold Spring Harbor Symposia on Quantitative Biology*, **45**, in press.
SAEDLER, H. & HEIβ, B. (1973). Multiple copies of the insertion-DNA sequences IS1 and IS2 in the chromosome of *Escherichia coli* K12. *Molecular and General Genetics*, **122**, 267–77.
SAEDLER, H., REIF, H. J., HU, S. & DAVIDSON, N. (1974). IS2, a genetic element for turn-off and turn-on of gene activity in *E.coli*. *Molecular and General Genetics*, **132**, 265–89.
SEUANEZ, H. N., EVANS, H. J., MARTIN, D. E. & FLETCHER, J. (1979). An inversion of chromosome 2 that distinguishes between Bornean and Sumatran orangutans. *Cytogenetics and Cell Genetics*, **23**, 137–140.
SILHARY, T. J., SHUMAN, H. A., BECKWITH, J. & SCHWARTZ, M. (1977). Use of gene fusions to study outer membrane protein localization in *Escherichia coli*. *Proceeding of the National Academy of Sciences, USA*, **74**, 5411–5.
SMITH, T. F. & SADLER, J. R. (1971). The nature of lactose operator constitutive mutations. *Journal of Molecular Biology*, **59**, 273–305.
SOMMER, H., CULLUM, J. & SAEDLER, H. (1979). IS2-43 and IS2-44: New alleles of the insertion sequence IS2 which have promoter activity. *Molecular and General Genetics*, **175**, 53–6.
STARLINGER, P. (1980). IS elements and transposons. *Plasmid*, **3**, 241–59.
STARLINGER, P. & SAEDLER, H. (1976). IS-elements in microorganisms. *Current Topics in Microbiology and Immunology*, **75**, 111–52.
WALZ, A., RATZKIN, B. & CARBON, J. (1978). Control of expression of cloned yeast

(*Saccharomyces cerevisiae*) gene (*trp5*) by a bacterial insertion element (IS2). *Proceedings of the National Academy of Sciences, USA*, **75,** 6172–6.

WETEKAM, W. & EHRING, R. (1973). A role for the product of gene *suA* in restoration of polarity *in vitro*. *Molecular and General Genetics*, **124,** 345–58.

WILSON, A. C., SARICH, V. M. & MAXSON, L. R. (1974). The importance of gene rearrangements in evolution: Evidence from studies on rates of chromosomal, protein, and anatomical evolution. *Proceedings of the National Academy of Sciences, USA*, **71,** 3028–30.

WORCEL, A. & BURGI, E. (1972). On the structure of the folded chromosome of *Escherichia coli*. *Journal of Molecular Biology*, **71,** 127–47.

WU, T. T., LIN, E. C. C. & TANAKA, S. (1968). Mutants of *Aerobacter aerogenes* capable of utilizing xylitol as a novel carbon. *Journal of Bacteriology*, **96,** 447–56.

tRNA SPLICING IN YEAST

J. ABELSON, G. KNAPP, C. L. PEEBLES, R. C. OGDEN, P. F. JOHNSON AND J. D. JOHNSON

Department of Chemistry, University of California, San Diego, La Jolla, California, 92093, USA

INTRODUCTION

One dramatic discovery in the initial encounters with eukaryotic genes was that they are often not colinear with their products. Instead many genes contain intervening sequences (or introns) which interrupt the continuity of the genetic information. As a result, a eukaryotic gene coding for a protein may be much larger than required for the simple coding of the amino-acid sequence. Intervening sequences have been found in a wide range of eukaryotic organisms, from yeast to man. The β-haemoglobin gene in mouse, for example, contains two intervening sequences of 116 and 646 base pairs (Konkel, Tilghman & Leder, 1978). The coding sequence is 432 base pairs. This is hardly an extreme example. $\alpha 2$ (I) collagen in chicken is coded for by a mRNA which is 5000 nucleotides in length; however, the gene contains more than 50 introns and has a length of 38 000 base pairs (Okhubo *et al.*, 1980; Vogeli *et al.*, 1980). Genes for structural RNA molecules, rRNA and tRNA, often contain intervening sequences (for a review see Abelson, 1979). In addition, extra-nuclear genes in mitochondria (Halbreich *et al.*, 1980) and chloroplasts (Rochaix & Malnoe, 1978; Allet & Rochaix, 1979) have been shown to contain intervening sequences.

Expression of these genes depends on a newly discovered enzymatic reaction, splicing. The entire gene is transcribed as a long RNA precursor; the intervening sequences are excised and the internal termini rejoined to yield the uninterrupted coding sequence as the functional RNA. Our research, reviewed in this article, has been concerned with the mechanism of RNA splicing in the expression of yeast tRNA genes containing intervening sequences. We have also created, and studied the expression of, altered tRNA genes in order to understand the function of the intervening sequences.

INTERVENING SEQUENCES IN YEAST tRNA GENES

The first example of an intervening sequence in a tRNA gene was described by Goodman, Olson & Hall (1977) who discovered that the tyrosine tRNA genes in yeast contain an intervening sequence of 14 base pairs. Subsequently, intervening sequences of 18 or 19 base pairs were demonstrated in three phenylalanine tRNA genes (Valenzuela et al., 1978). In both sets of genes, the insertion occurs at a position on the 3' side of the anticodon. The intervening sequences are closely related within each set of tRNA genes but the tyrosine- and phenylalanine-specific sequences are dissimilar.

Further investigation of yeast tRNA genes containing intervening sequences was made easier by the discovery by Hopper, Banks & Evangelidis (1978) that tRNA precursors accumulate in a mutant strain of yeast (ts136). This mutant, isolated by Hutchison, Hartwell & McLaughlin (1969), defines the *rna*1 locus. It is presumed to be defective in a step in RNA transport from nucleus to cytoplasm. At the nonpermissive temperature, the 35S rRNA precursor accumulates (Hopper et al., 1978), the appearance of mRNA in the cytoplasm is halted, poly(A)-containing RNA accumulates in the nucleus (Shiokawa & Pogo, 1974), and a particular subset of tRNA precursors accumulates (Knapp et al., 1978).

The separation of the tRNA precursors that accumulate in ts136 has been accomplished by two-dimensional polyacrylamide gel electrophoresis. A typical two-dimensional separation is shown in Fig. 1. Originally the precursor-specific spots were identified by hybridization of the RNA to a set of *Escherichia coli* recombinant plasmid clones, each of which carries one or more yeast tRNA genes (Beckmann, Johnson, Abelson & Fuhrman, 1977*b*). Five of the RNAs (indicated in Fig. 1) hybridized to clones that have been identified as containing genes for $tRNA^{Tyr}$, $tRNA^{Phe}$, $tRNA_3^{Leu}$, $tRNA_{UCG}^{Ser}$ and $tRNA^{Trp}$. The nucleotide sequences of the genes for $tRNA^{Tyr}$ (Goodman et al., 1977), $tRNA^{Phe}$ (Valenzuela et al., 1978), and $tRNA^{Trp}$ and $tRNA_3^{Leu}$ (Kang, Ogden & Abelson, 1980) have been determined. More recently the nucleotide sequence of the gene for the 'Spot 19' precursor has been determined (Ogden, Harrell & Abelson, unpublished) and it is a gene for an isoleucine tRNA.

The RNA sequences of precursors isolated as described in Fig. 1 have been compared with the sequences of the genes (Knapp et al., 1978; Kang et al., 1980). In all cases they were homologous.

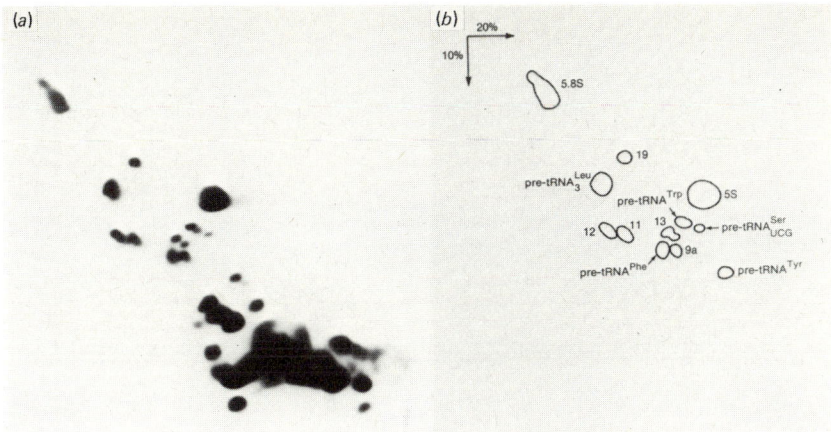

Fig. 1. Two-dimensional polyacrylamide gel electrophoresis of ^{32}P-RNA isolated from the yeast *rna*1 mutant. ^{32}P-RNA was prepared using a temperature-sensitive diploid, M304, that is homozygous for the *rna*1 locus (as described in Ogden *et al.*, 1979). A typical preparation yields approximately 1.0–2.0 mCi of labelled RNA which is purified by electrophoresis in a 10% polyacrylamide-4 M urea gel, followed by electrophoresis in a second dimension using a 20% polyacrylamide-4 M urea gel. Autoradiography revealed the positions of the different pre-tRNAs. The yields of individual ^{32}P-pre-tRNAs recovered by elution were of the order of 2 to 25 × 10^6 cpm. (*a*) Autoradiogram of the polyacrylamide gel separations. (*b*) Diagrammatic representation and identification of the ^{32}P-pre-tRNAs. Spot 19 is a precursor of a tRNA containing an isoleucine anticodon.

Sequences for five pre-tRNAs are shown in Fig. 2. The 5' and 3' ends of the precursors are homologous with the 5' and 3' ends of the mature tRNA. The 3' terminal CCA sequence is present and must have been added post-transcriptionally since it is not encoded in any of these genes. Each precursor contains an intervening sequence corresponding to the sequence in the gene. Many of the modified nucleotides present in the mature tRNA are present in the precursor although some modifications are notably absent. A detailed catalogue of these modifications has been published (Kang *et al.*, 1979). The general conclusion is that nucleosides in the anticodon region including the 'hypermodified' nucleosides adjacent to the anticodon are unmodified in precursors, whereas nucleoside modifications in the mature tRNA portion of the precursors, for example in the D and the TψC loops, are frequently present. Thus, the precursors which accumulate at the non-permissive temperature in ts136 all contain intervening sequences. They are partially matured but final maturation requires the removal of the intervening sequence and the modification of certain nucleosides.

Other tRNA genes in yeast have been sequenced which do not

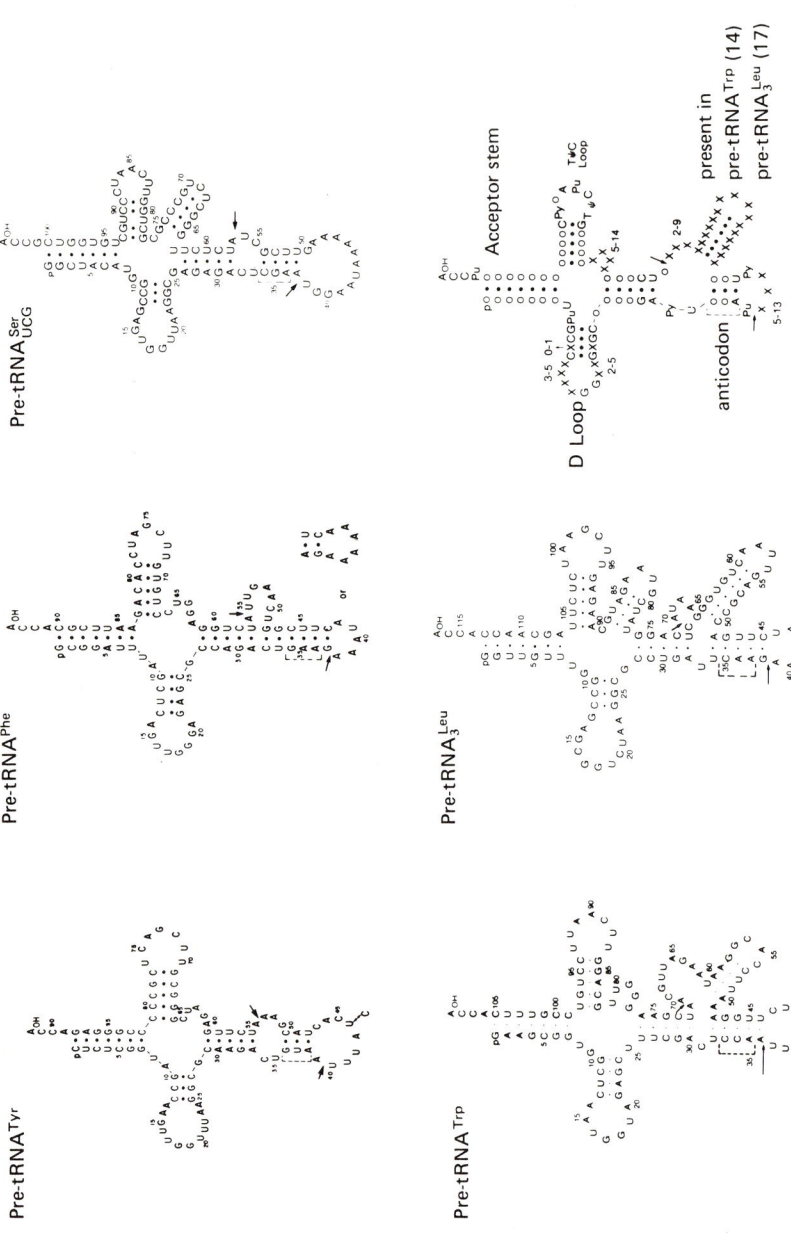

Fig. 2. Nucleotide sequences of five yeast tRNA precursors. The nucleotide sequences of the precursors have all been arranged in secondary structures similar to those derived for pre-tRNA^{Tyr} and pre-tRNA^{Phe} (Knapp et al., 1978). In the cases of pre-tRNA^{Trp} and pre-tRNA$_3^{Leu}$ additional hairpin helices in the intervening sequence contribute to the favourable free energy of the secondary structures. The arrows indicate the splice points. A composite secondary structure is also shown which summarizes the constant and variable positions of nucleotides. Variable nucleotides of the precursor are denoted by open circles. Positions with variable numbers of nucleotides are denoted by crosses. The numbers of nucleotides in these are indicated adjacent to the structure.

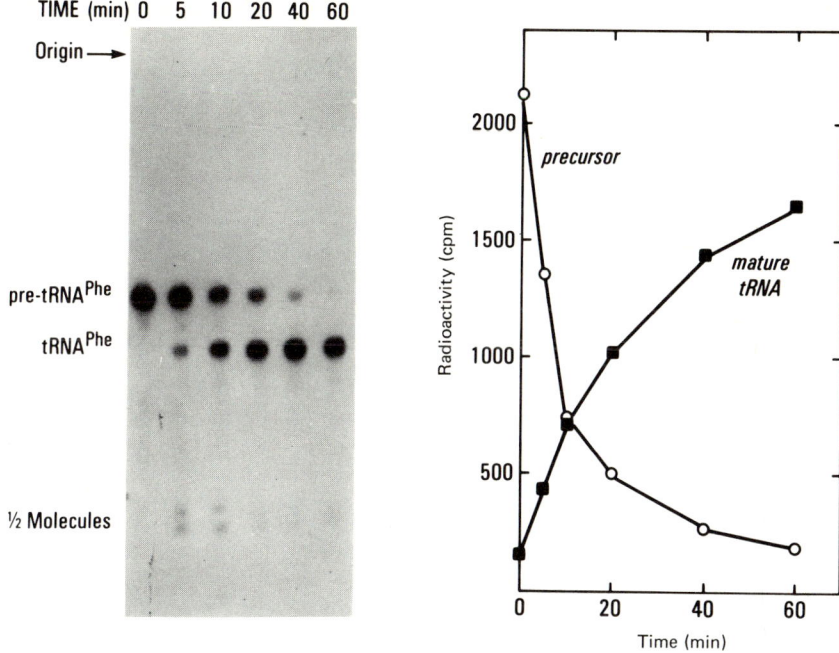

Fig. 3. Time course of the *in vitro* splicing reaction. Aliquots (1 μl) from an *in vitro* splicing reaction (10 μl) containing ^{32}P-labelled pre-tRNA^{Phe} (105 000 cpm) were removed after various times of incubation at 30 °C and analysed by polyacrylamide gel electrophoresis (reaction conditions have been described in Peebles *et al.*, 1979). The autoradiogram, shown at left, served as a guide for slicing the gel. Fractions were counted without scintillant (40% efficiency) and cpm plotted directly. Identification of pre-tRNA^{Phe}, tRNA^{Phe}, and half-tRNA sized RNA (1/2 molecules) has been confirmed by fingerprint analysis of the products.

contain intervening sequences. These include genes for tRNA$_3^{Arg}$ and tRNA^{Asp}, which are sometimes clustered in the yeast genome (Beckmann *et al.*, 1977*b*; Schmidt *et al.*, 1980). Precursors for these tRNAs do not accumulate in ts136, suggesting that only those tRNAs which require a splicing step in maturation accumulate as precursors in ts136. Sequences of additional tRNA genes will be required to substantiate this theory.

In vitro SPLICING OF tRNA PRECURSORS

All of the tRNA precursors that accumulate at the nonpermissive temperature in ts136 are substrates for splicing, a processing reaction in which the intervening sequence is removed intact and the ends of the 'half-tRNA-sized molecules' are joined to form a mature tRNA molecule. Although some progress towards the purification of the enzymes involved in RNA splicing has been made, we have been able to characterize many of the properties of the splicing system using a yeast ribosomal wash fraction (prepared as described by Peebles, Ogden, Knapp & Abelson, 1979).

Fig. 3 illustrates the time course of *in vitro* splicing for the precursor of yeast tRNA^{Phe}. Concomitantly with the disappearance of the precursor, mature-sized tRNA is formed. At early times the transient appearance of smaller RNA products is seen. These RNAs migrate in the polyacrylamide gel with the mobility of half-tRNA-sized molecules. The size and time course of appearance of these smaller RNAs suggested that they were likely to be the halves generated by excision of the intervening sequence. As evidence presented below will show, it is probable that these halves are the true intermediates in the splicing reaction. Appearance in excellent yield of mature tRNA implies that nearly all of the precursor was processed and that, even in the crude extract, neither random degradation nor abortive splicing consume a significant fraction of the RNA precursor.

An investigation of the role of ATP in the splicing reaction is shown in Fig. 4. The RNA substrate used in this experiment has been designated '19' (see Fig. 1) and contains the anticodon assigned to isoleucine. It is both the largest tRNA precursor yet identified and also one of the more abundant. In the presence of 1 mM ATP (Fig. 4), the results of incubation with the yeast ribosomal wash fraction are similar to those shown for pre-tRNA^{Phe} in Fig. 3. The precursor is rapidly converted to the tRNA product, and a second band, migrating slightly faster than the mature-sized tRNA, is observed. This RNA is the excised intervening sequence and is 60 nucleotides in length. In the absence of ATP (Fig. 4) the precursor is utilized at a similar rate; however, the pattern of product formation is very different. No tRNA product is formed. Instead, half-tRNA-sized products are formed along with the 60 nucleotide-long molecule. At intermediate levels of ATP a mixture of mature-sized tRNA and halves appears.

This experiment suggests that the endonucleolytic stage of the *in vitro* splicing reaction is not dependent on ATP and can be separated from the ATP-dependent ligation step. Fig. 5 demonstrates that half-tRNA-sized molecules can be ligated in the presence of ATP and that both halves must be present for the reaction to yield tRNA products. Characterization of the product by fingerprint analysis has shown that the halves are indeed covalently joined

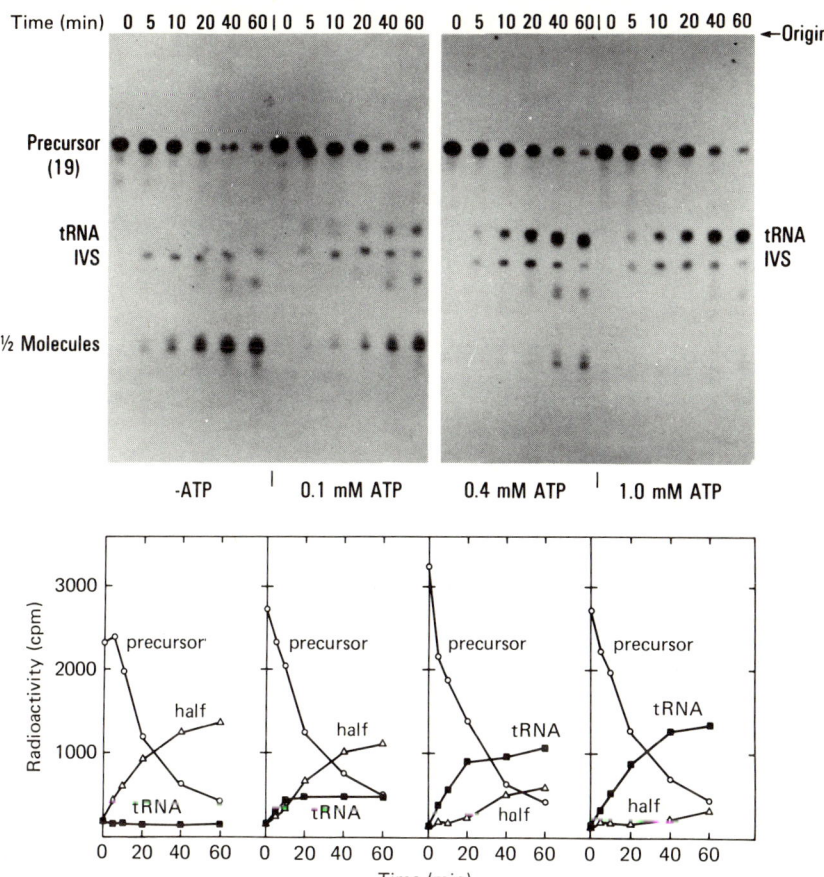

Fig. 4. ATP is required for the *in vitro* splicing reaction. The reaction mixture (10 μl) contained precursor-tRNA ('19', 92000 cpm) and the indicated concentrations of ATP (reaction conditions have been described in Peebles *et al.*, 1979). Aliquots (1 μl) were taken after various periods of incubation at 30 °C, subjected to gel electrophoresis and autoradiography, and analysed quantitatively as for Fig. 3. Identification of intervening sequence (IVS) and half-tRNA sized molecules (1/2 molecules) depends on mobility and minor nucleotide analysis.

as in the mature tRNA. Thus, it has been demonstrated that the splicing reaction can be divided operationally into two reactions: (1) an endonucleolytic step in which the precursor is cleaved twice in an ATP-independent reaction to yield the intervening sequence and half-tRNA-sized molecules and (2) a ligation step that requires ATP and produces the mature-sized tRNA product.

Characterization of the half-tRNA-sized molecules and intervening sequences has proved to be equally important to an understanding of the mechanism of the splicing reaction. There are several

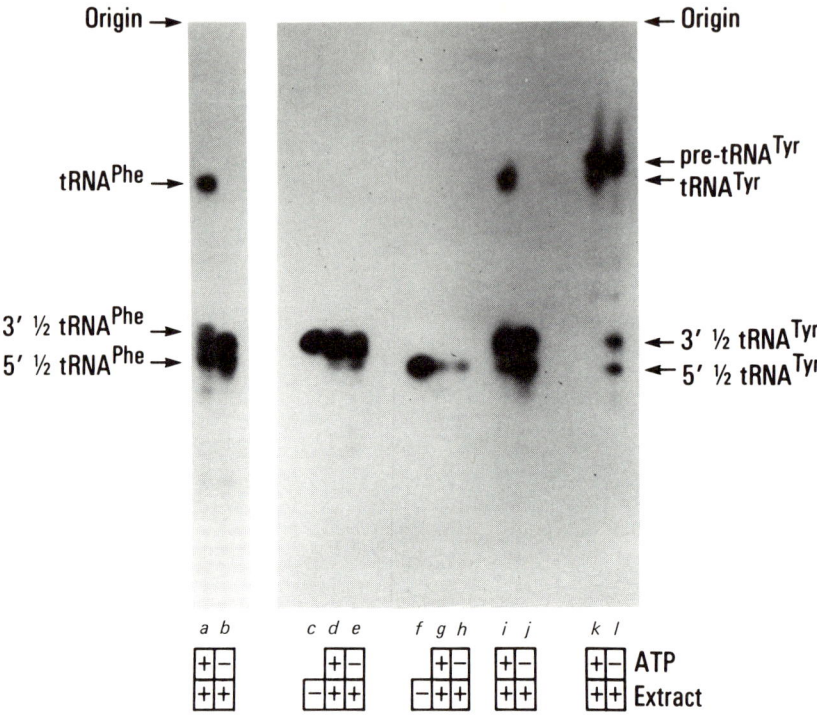

Fig. 5. Half-tRNA molecules are substrates for ligation. Reaction mixtures (1.5 μl) contained either the 3′-half of tRNA^{Phe} (4 000 cpm) plus the 5′-half of tRNA^{Phe} (4 000 cpm; lanes a and b); the 3′-half of tRNA^{Tyr} (5 000 cpm; lanes c, d and e); the 5′-half of tRNA^{Tyr} (5 000 cpm; lanes f, g and h); the 3′-half of tRNA^{Tyr} (5 000 cpm) plus the 5′-half of tRNA^{Tyr} (5 000 cpm; lanes i and j); or pre-tRNA^{Tyr} (5 700 cpm; lanes k and l). Pre-incubation at 42 °C and 37 °C (5 min each) of the reaction mixture without added extract was to allow formation of correct secondary structures. Extract was replaced by buffer A (see Peebles et al., 1979) plus 10 mM MgCl$_2$ and 0.3 M KCl in lanes c and f; ATP was omitted from lanes b, e, h, j and l. Incubation was stopped after 40 min at 30 °C, and the samples analysed by polyacrylamide gel electrophoresis and autoradiography. Half-tRNA molecules were prepared from pre-tRNA molecules in a large scale reaction similar to lane l, and products identified by fingerprint analysis.

ways in which the excision of the intron could occur. The structure of the excised intervening sequence should reflect the way in which it has been removed. Possible structures for the excised intervening sequence from pre-tRNA^{Tyr} are shown in Fig. 6. First, the reaction might proceed *via* a concerted mechanism, so that the intervening sequence forms a circle. This could happen if the splicing occurred with a reciprocal exchange at the two points of cleavage. Alternatively, the product could be a unique linear molecule. There are two possible classes of linear products: those with 5′-phosphate and 3′-OH termini and those with 5′-OH and 3′-phosphate termini. In

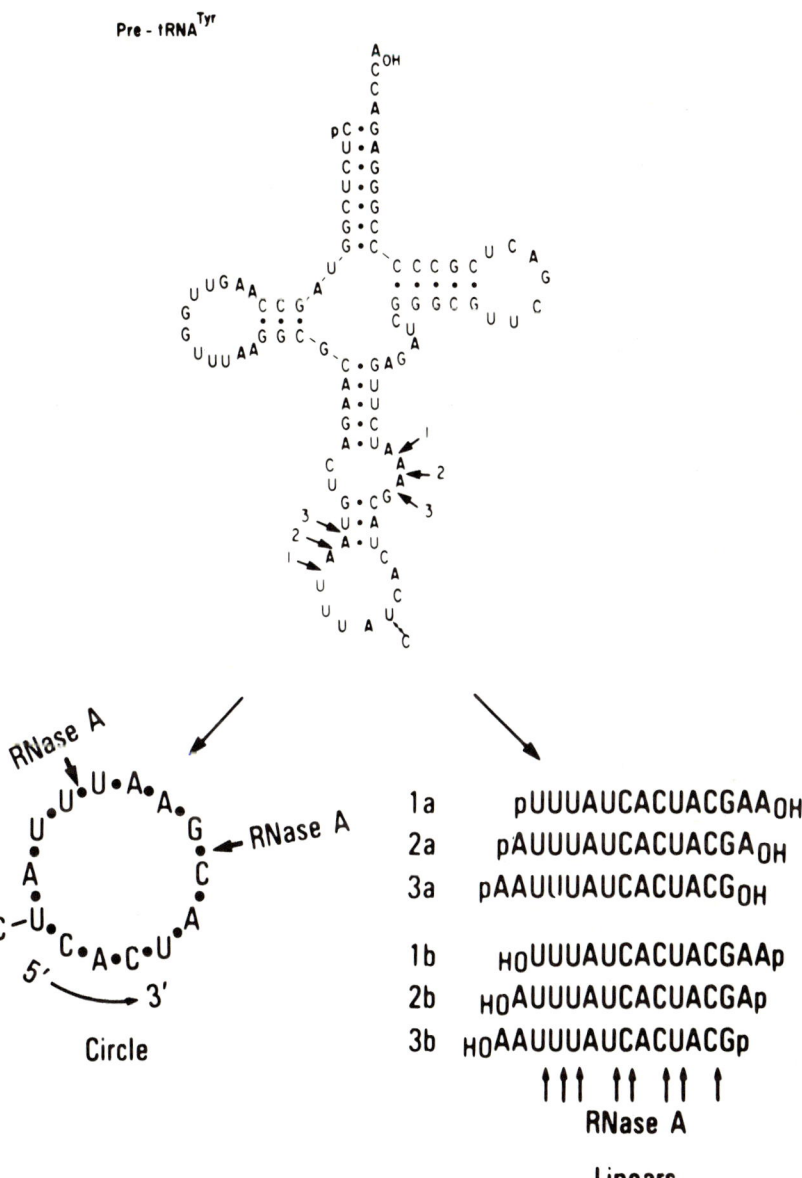

Fig. 6. Possible structures of the excision product. Numbered arrows indicate the three possible excision points to which the linear sequence permutations correspond. Unnumbered arrows indicate RNase A cleavage sites.

both of these classes, there are three possible sequence permutations in the case of pre-tRNA^{Tyr}, because of the repeated sequence at the cleavage sites. Another possibility is that a mixture of all three linear sequences would be produced in the splicing reaction. Investigation of the intervening sequence excised from pre-tRNA^{Tyr} revealed that it is not a circle (whether produced in the presence or absence of ATP) but is a linear sequence designated 1b in Fig. 6. This product has 5'-OH and 3'-phosphate termini.

If the cleavage reaction is a simple scission of two phosphodiester bonds as the previous data imply, the half-tRNA-sized molecules produced in the ATP-independent reaction should also have 3'-phosphate and 5'-OH terminals. Fingerprint analyses of the halves produced by cleavage of five precursors demonstrated that each 5'-half molecule has a 3'-phosphate end and each 3'-half molecule has a 5'-OH end. In these five precursors, the splice point is adjacent to the nucleotide at the 3'-side of the anticodon (the nucleotide that is often hypermodified in the mature tRNA). These results have strong implications concerning the mechanism of precursor cleavage and subsequent ligation. Several features of one mechanism that may be constructed are novel when compared with the mechanisms of other processing enzymes. First, in all endonucleolytic RNA processing-enzymes previously described, the scission of the phosphodiester chain produces 5'-phosphate and 3'-OH ends. This, of course, is not a chemical necessity; there are many endonucleases, e.g., RNase A and RNase T1, that produce 3'-phosphate termini. However, these enzymes are generally considered to be degradative enzymes. The position of the terminal phosphate left by the splicing endonuclease is also surprising when the ligation step is considered. The ligase must join a 3'-phosphate to a 5'-OH to produce the tRNA. This reaction is not a feature of the T4 RNA ligase or of the known DNA ligases, which specifically join 5'-phosphate to 3'-OH terminals. We have shown that the 3'-phosphate is required for the ligase reaction (Knapp, Ogden, Peebles & Abelson, 1979). 5'-half molecules from which the 3'-phosphate has been removed are no longer substrates in the ATP-dependent joining reaction.

We have not yet been able to determine the role of ATP in the ligase reaction. If activation of the tRNA halves occurs in a manner analogous to activation of the 5'-phosphate with ATP by T4 RNA ligase, the 3'-phosphate on the 5'-half would be adenylylated and AMP released on formation of the phosphodiester bond.

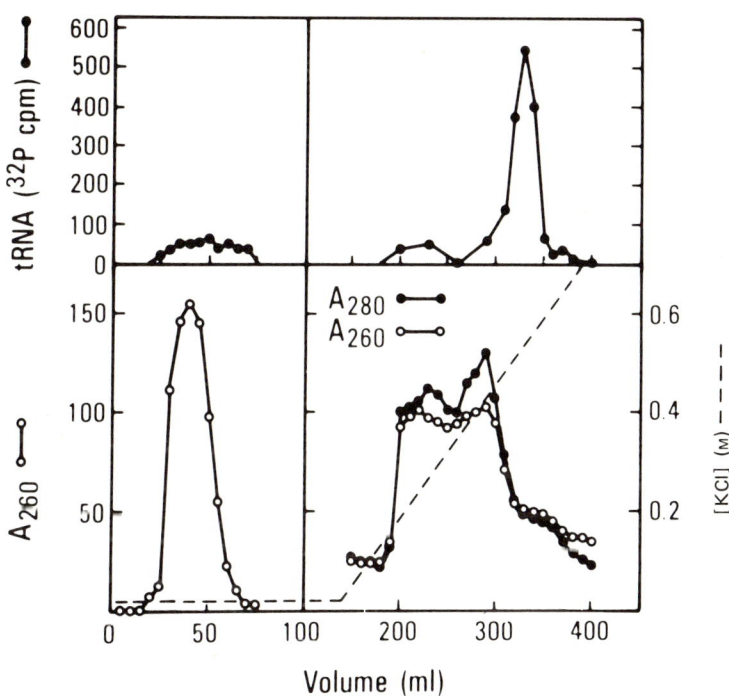

Fig. 7. Partial purification of tRNA ligase on an A15M heparin–agarose column. Wild-type yeast cells (10g) were disrupted by passage through an Eaton pressure cell and debris was removed by centrifugation. The extract was desalted on a Sephadex G25 column equilibrated in Buffer A (20 mM Tris–HCl, pH 8.0, 20 mM KCl, 10 mM 2-mercaptoethanol, 0.5 mM EDTA, 10% glycerol, 0.5 mM phenylmethylsulphonyl fluoride). The fractions containing material absorbing at 280 nm were combined and applied to a 40-ml heparin-agarose column which had been equilibrated in Buffer A. The column was washed with 80 ml Buffer A and eluted with a linear KCl gradient (400 ml Buffer A and 400 ml Buffer A plus 1 M KCl). The absorbances at 280 nm (filled circles) and 260 nm (open circles) are shown in the lower panel of the figure. Splicing ligase assays (autoradiogram and top panel on graph) were performed as described in Peebles *et al.* (1979).

Purification of the splicing activity has provided evidence that individual enzyme activities which catalyze these two steps can be physically separated. Extracts of yeast are prepared by passing frozen pellets of yeast through an Eaton press to disrupt the cells.

The debris is removed by low speed centrifugation and a high speed supernatant fraction is prepared. This fraction contains most of the ligase activity and has been used as a starting material for its purification. The first step is chromatography on heparin–agarose and the column profile of a typical fractionation is shown in Fig. 7. About 90% of the soluble protein passes through the column while most of the ligase is recovered later by a salt step or linear gradient. We have determined that the splicing endonuclease passes through this column and is thus quantitatively separated from the ligase in the first step. Mixing the material passing through the column with the ligase reconstitutes total splicing activity. The ligase has been further purified by chromatography on DEAE-sephacel and by affinity chromatography on tRNA-agarose.

We have characterized some of the properties of the partially-purified ligase fraction. The reaction proceeds at a linear rate for at least an hour at 30 °C, and at a reduced but substantial rate at 23 °C or 35 °C. A similar extent of reaction is approached after several hours at 15 °C, or more slowly at 5 °C. The apparent Km for ATP is about 42 μM; other natural ribonucleoside triphosphates are ineffective. Magnesium is required and the optimum concentration is 15 mM $MgCl_2$. The optimum salt concentration is 40 mM KCl. At this stage of purification there are about ten major polypeptides evident in SDS polyacrylamide gel electrophoresis. We cannot at this point assign to the ligase a native molecular weight or peptide chain composition.

THE ROLE OF THE INTERVENING SEQUENCE IN THE EXPRESSION OF THE tRNA GENE

We have carried out two sets of experiments designed to test the rôle of intervening sequences in the expression of tRNA genes which contain them. It is not yet known whether splicing is an essential step in the synthesis of yeast tRNA. Could a tRNA gene with an altered intervening sequence be expressed? Not all tRNA genes contain intervening sequences (in yeast we estimate that perhaps one-tenth of the yeast tRNA genes contain intervening sequences); thus, splicing is not an obligatory step in the synthesis of all tRNAs. The presence of an intervening sequence in a precursor gives that RNA molecule a distinctive structure so that one could invoke an essential role for the intervening sequence in processing,

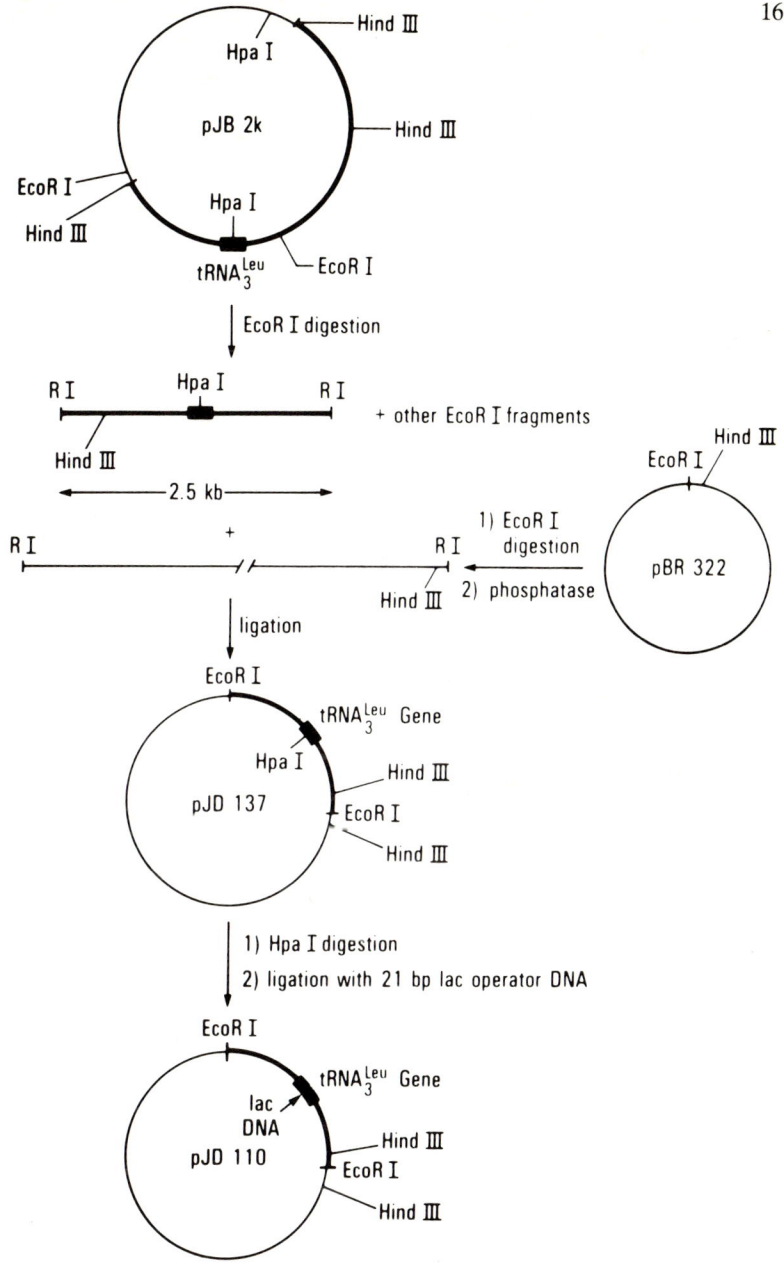

Fig. 8. Construction of an altered $tRNA_3^{Leu}$ gene. *Hind*III fragments of yeast genomic DNA were inserted in the vector pBR313 and screened by hybridization to purified yeast tRNAs (Beckmann, Johnson & Abelson, 1977a). A clone hybridizing to $tRNA_3^{Leu}$, pJB2k, was identified. An *Eco*RI fragment that contains the $tRNA_3^{Leu}$ gene from pJB2k was subcloned into pBR322. The orientation of the yeast fragment in pJD137 was deduced from the observation that, after *Hind*III digestion, the $tRNA_3^{Leu}$ gene was found in a fragment that co-migrates with linear plasmid DNA. This subclone, pJD137, was made linear by digestion with *Hpa* I and then ligated to a 21-base pair DNA molecule with a sequence corresponding to *lac* operator, to produce the plasmid pJD110.

base modification or transport from nucleus to cytoplasm. Alternatively, the intron could be part of the transcriptional recognition site. This latter possibility is raised from the results of Brown and his colleagues, who have shown that transcriptional initiation of the *Xenopus* 5S gene by RNA polymerase III requires sequences in the middle of the gene (Bogenhagen, Sakonju & Brown, 1980; Sakonju, Bogenhagen & Brown, 1980). Eukaryotic tRNA genes are also transcribed by RNA polymerase III.

The first set of experiments (Johnson *et al.*, 1980) has utilized a cloned tRNA$_3^{Leu}$ gene of yeast. This gene contains an intervening sequence of 32 base pairs that includes an Hpa I restriction endonuclease site near the middle of the intervening sequence (see Fig. 8). We have altered the intervening sequence in this gene by insertion of the 21 base pair synthetic *lac* operator DNA (Bahl *et al.*, 1976) into the Hpa I site. The parent and modified plasmids were then transcribed in a *Xenopus* germinal vesicle extract. RNA fingerprint analysis of the transcription products revealed that both the tRNA$_3^{Leu}$ gene and its modified counterpart were accurately transcribed. Transcription products corresponding to mature tRNA$_3^{Leu}$ and pre-tRNA$_3^{Leu}$ with the normal and *lac*-containing intervening sequences were identified (Fig. 9). Precursors extended at their 5' and 3' ends were also present. Both parent and modified genes were transcribed efficiently and the various products accumulated in similar amounts, indicating that no deleterious effects on transcriptional competence, stability of the transcripts, or processing result from insertion of the 21 base-pair *lac* operator DNA. Further experiments showed that precursors which were synthesized in the germinal vesicle extracts (bands 3 and *lac* 3, Fig. 9) can be correctly spliced by yeast extracts. Thus, the insertion of 21 nucleotides into the intervening sequence does not affect transcription in the *Xenopus* system and it does not affect splicing in *Xenopus* or yeast. We are led to the conclusion that there may be a different mechanism for the recognition of tRNA and 5S genes by RNA polymerase III.

It is also of interest that insertion of the 21 base-pair *lac* sequence into the tRNA$_3^{Leu}$ gene does not affect the processing of the precursor in either the *Xenopus* or yeast systems. The correct 5' and 3' ends are produced in precursors containing the modified intron. It would thus seem that the structure of the intervening sequence is not a crucial factor in the recognition of pre-tRNA molecules by either the nuclease(s) responsible for end-trimming or the

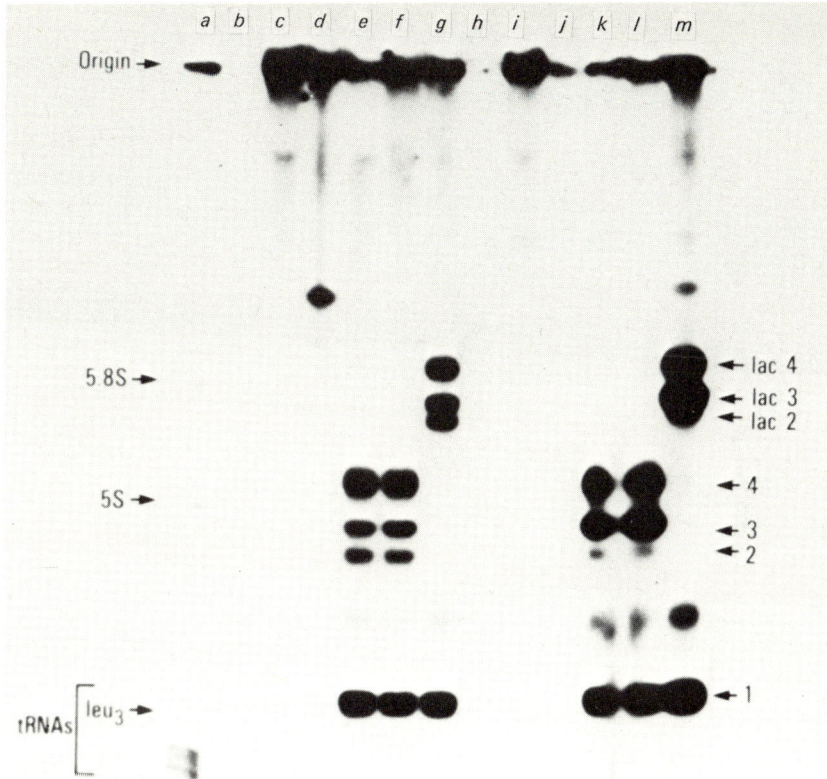

Fig. 9. Autoradiography of the [^{32}P]RNA products of transcription of cloned yeast tRNA$_3^{Leu}$ genes in *Xenopus* germinal vesicle extracts. Equivalent amounts of DNA from the vectors and recombinant plasmids were incubated with an extract of *X. laevis* stage V and VI oocytes. The reaction mixtures contained either [α-^{32}P]ATP (lanes $b - g$) or [α-^{32}P]UTP (lanes $h - m$). The reaction mixtures were then subjected to a protease digestion followed by phenol extraction and ethanol precipitation. The RNA was then electrophoresed on a 10% polyacrylamide–4 M urea gel. ^{32}P-Labelled yeast 4S, 5S and 5.8S RNAs were included as markers (lane *a*). Lanes *b* and *h*, no DNA; lanes *c* and *i*, pBR322; lanes *d* and *j*, pBR313; lanes *e* and *k*, pJB2k, a pBR313 derivative containing the yeast tRNA$_3^{Leu}$ gene; lanes *f* and *l*, pJD137; lanes *g* and *m*, pJD110.

nucleotidyl transferase. Furthermore, the endonuclease responsible for excising the intervening sequence is able to recognize the modified substrate and cleave it at precisely the same sites as the unmodified pre-tRNA$_3^{Leu}$. Thus, the positions of cutting for excision of the intron are not determined by the size of the base-paired region in the central part of the intervening sequence.

In a more crucial test of the rôle of the intervening sequence in tRNA gene expression, we have engineered the precise deletion of the intron from the tRNA^{Tyr} SUP6 ochre suppressor gene (Wallace et al., 1980). In *Saccharomyces cerevisiae* there are eight tRNA^{Tyr}

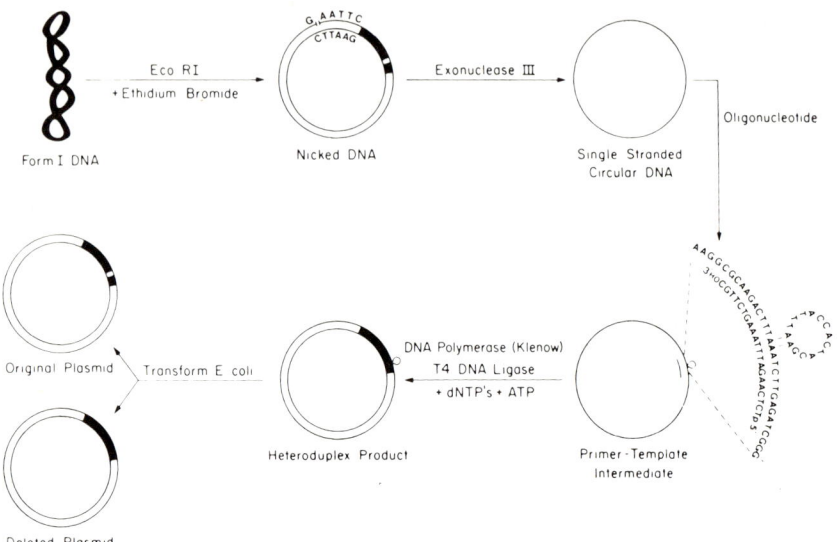

Fig. 10. Construction of an altered tRNA^{Tyr} gene. The 14 base-pair intervening sequence was deleted from the yeast SUP6 gene cloned in pBR322 as shown in this diagram. The experimental procedure has been described in Wallace et al. (1980).

genes (Olson et al., 1977) and each can be converted to either an ochre or amber suppressor, a dominant phenotype (Gilmore, Stewart & Sherman, 1971). The ochre suppressor gene can be introduced into yeast by transformation (Hinnen, Hicks & Fink, 1978), so that the phenotype of the altered gene can be compared to that of the parent.

The strategy for constructing the gene with the intervening sequence deleted is shown in Fig. 10. A synthetic oligonucleotide complementary to residues 28–48 of the mature yeast ochre suppressor tRNA^{Tyr} was synthesized using the triester method (Hirose, Crea & Itakura, 1978). This oligonucleotide was used as a primer for DNA synthesis on a single-stranded plasmid DNA containing the tRNA^{Tyr} SUP6 gene. Completion of the double-stranded circle by DNA synthesis and ligation yields a heteroduplex. This DNA is transformed into E. coli and the heteroduplex is resolved into parent and mutant by replication. This technique allowed us to successfully isolate a mutant of the tRNA^{Tyr} SUP6 gene containing a precise deletion of the intervening sequence.

In order to determine whether the SUP6 gene, lacking the intervening sequence, is still functional, we introduced the altered gene into yeast by transformation. DNA fragments containing the parental SUP6 allele or the mutant with the intron deletion were ligated into the yeast transformation vector CV7 (Hicks, Hinnen & Fink, 1978). This pBR322 derivative, which carries the wild type yeast *leu2* gene an the 1.4 megadalton *Eco*RI fragment of the endogenous 2 micron yeast plasmid pScl, replicates in yeast as an episomal element with about ten copies per cell (Struhl, Stinchcomb, Scherer & Davis, 1979; Petes, 1980). The recipient yeast strain carries a double mutation at the *leu2* locus and contains three ochre nonsense mutations (*lys*1–1, *trp*5–48, *ade*2–1). This strain was transformed with CV7, or CV7 containing either the parental SUP6 gene or the deleted gene. Leu$^+$ transformants were selected and then tested for the simultaneous suppression of the three ochre loci. As expected, leu$^+$ transformants from the vector alone do not exhibit ochre suppression, whereas every transformant from CV7 containing the SUP6 allele shows the suppressor character. Moreover cells harboring the SUP6 intervening sequence deletion plasmids also suppress ochre mutations. Further experiments verified that the SUP6 gene is indeed linked to *leu2* in the transformants.

The results of this experiment indicate that the intervening sequence in the tRNA^{Tyr} SUP6 gene is not essential for suppressor function. It remains to be seen whether or not the gene functions optimally although it should be straightforward to test the level of gene expression. Since not all yeast tRNA genes contain intervening sequences, it is not surprising that processing pathways exist for the correct maturation of the suppressor. We do not yet know whether the suppressor tRNA produced by the gene lacking the intervening sequence is normally modified.

We have provided evidence that an intervening sequence is not essential for the expression of a yeast tRNA suppressor gene. Analogous experiments in which the introns are deleted from eukaryotic genes have been performed with genes from SV40 and with the mouse β-globin gene (Gruss, Lai, Dhar & Khoury, 1979; Hamer *et al.*, 1979; Goff & Berg, 1979; Mulligan, Howard & Berg, 1979; Hamer & Leder, 1979; Lai & Khoury, 1979). The results are different from those we have obtained with the yeast tRNA genes. An intervening sequence must be present in the transcriptional unit for synthesis of stable mRNA and/or transport of mRNA from nucleus to cytoplasm. This leads to a central hypothesis concerning the role of intervening sequences and, indeed, of RNA processing in the expression of eukaryotic genes. In eukaryotes, RNA is transcribed in the nucleus and must pass, probably through nuclear pores, to the cytoplasm. The RNA-processing reactions, the modification of the 5' terminus (capping), poly(A) addition and splicing may play a role in, or be coincident with, the process of RNA transport from the nucleus. For some classes of RNA that do not contain intervening sequences, there must be alternative modes of transport. A fascinating and important area for future research will be the localization of RNA processing activities within the nucleus.

We have not yet dealt with what are certainly the most intriguing questions regarding intervening sequences: How did they enter the gene in the first place and what is their role in evolution? Gilbert has hypothesized that intervening sequences have been a potent agent in the evolutionary assembly of genes (Gilbert, 1978). In this model, genes containing functional domains may recombine in an imprecise manner. The joint between the two genes becomes an intervening sequence and is removed by RNA splicing. This theory certainly receives strong support from the structure of immunoglobulin genes where each domain in the light and heavy chain genes is separated from its neighbour by intervening sequences. In the insulin gene, the regions coding for the A and B chains are separated by an intron in the C region. The theory is less successful in explaining the intervening sequences in tRNA genes where it seems likely that the archetypal tRNA structure is the one found in all organisms. In that structure the halves do not constitute separate functional domains but must conserve complementary sequences for the formation of the base-paired helices of the anticodon and acceptor stems.

PERSPECTIVES

The outlines of a mechanism of tRNA splicing are beginning to emerge. In all tRNA genes which contain an intervening sequence, it is located in exactly the same place, one base to the 3' side of the anticodon. In the precursor the ligation sites are therefore held in conjunction by the secondary and tertiary interactions that are common to tRNA structure. Although neither activity involved in tRNA splicing has been completely purified, we have, as yet, found no evidence for multiple splicing systems. It seems likely that a single pair of enzymes acts on the entire set of yeast tRNA precursors, locating the splice sites by recognizing common features

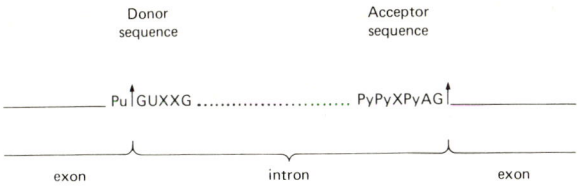

Fig. 11. Sequences found at the left-hand splice junction (donor sequence) and right-hand splice junction (acceptor sequence) of the intron, illustrating the Chambon rule. Pu and Py indicate positions where purines or pyrimidines, respectively, are found. X indicates a variable nucleotide.

of the RNA structure. In this regard, tRNA splicing is similar to a number of RNA processing reactions that have been characterized in prokaryotes (Abelson, 1979), in which a single enzyme, e.g., RNase III or RNase P, is able to recognize a large number of substrates by detecting common features.

One is led to the conclusion that mRNA splicing must proceed by a somewhat different mechanism. A large number of eukaryotic genes has now been sequenced and one can say with certainty that there is a rule – the Chambon rule – for the sequences at the intron-exon boundary (Fig. 11; Breathnach et al., 1978). These sequences are present at the borders of virtually every intervening sequence though it is doubtful if splice sites could be identified unambiguously by examining sequences. The Chambon rule holds among vertebrates from fish to man and apparently in all eukaryotes, since the actin gene in yeast contains a single intervening sequence whose boundaries also obey the Chambon rule (Ng & Abelson, 1980). A number of *in vitro* genetic studies, notably those of Berg and his colleagues, allow the conclusion that sequences in the middle of an intron can be deleted without affecting splicing, but alterations to splice junction sequences abolish splicing (Gruss et al., 1979; Mulligan et al., 1979; Hamer & Leder, 1979). Berg and his colleagues have also shown that it is possible to construct functional hybrid introns in which, for example, the donor site is derived from mouse β-haemoglobin gene and the acceptor from the SV40 T-antigen gene (P. Berg, personal communication). Thus one is led to the conclusion that a short sequence at each of the splice junctions provides the specificity for the splicing reaction. How does a donor recognize the appropriate acceptor? We will not know the answer to this fascinating question until an *in vitro* assay for pre-mRNA splicing has been developed. The absence of any common secondary structure which base-pairs the donor to the acceptor and the adherence of sequences at the boundary to the Chambon rule have led to the suggestion that a separate adaptor RNA holds the edges of the intron together, providing the specificity for splicing (Lerner *et al.*, 1980; Rogers & Wall, 1980). J. Steitz and her collaborators have suggested that this RNA (Lerner *et al.*, 1980) is the ubiquitous small nuclear RNA (snRNA), U1 (Reddy, Ro-Choi, Henning & Busch, 1974). U1 has intriguing sequence complementarity to the intron boundaries. In the nucleus, U1 is found as an snRNP, a complex of U1 snRNA with several polypeptides (Lerner & Steitz, 1979). Antibodies to snRNP found in patients suffering from lupus inhibit splicing of some early adenovirus RNA in intact virus-infected nuclei (J. Flint & J. Steitz, personal communication). Thus, there is evidence encouraging the belief that the snRNP's are involved in splicing, but an *in vitro* assay will be required to prove this.

Although few details are as yet known about the biochemistry of mRNA splicing, it is not possible to explain both mRNA and tRNA splicing in terms of a single splicing system. The difference in the possible structures that can be predicted for the two classes of precursors and the probable role of an adaptor RNA in mRNA splicing suggest that the mechanisms of recognition by the specific endonucleases may differ. However, it is our belief that the mechanisms will share fundamental features. The termini produced by excision of the intervening sequence from both classes of RNA precursors may be the same. Thus, the unique joining of a 3'-phosphate to a 5'-OH in an ATP-dependent reaction may be common to all splicing reactions.

REFERENCES

ABELSON, J. (1979). RNA processing and the intervening sequence problem. *Annual Reviews of Biochemistry*, **48**, 1035–69.

ALLET, B. & ROCHAIX, J.-D. (1979). Structure analysis at the ends of the intervening DNA sequences in the chloroplast 23S ribosomal genes of *C. reinhardii*. *Cell*, **18**, 55–60.

BAHL, C. P., WU, R., ITAKURA, K., KATAGUI, N. & NARANG, S. A. (1976). Chemical and enzymatic synthesis of lactose operator of *Escherichia coli* and its binding to lactose repressor. *Proceedings of the National Academy of Sciences USA*, **73**, 91–4.

BECKMANN, J. S., JOHNSON, P. F. & ABELSON, J. (1977*a*). Cloning of yeast transfer RNA genes in *Escherichia coli. Science*, **196**, 205–8.

BECKMANN, J, S., JOHNSON, P. F., ABELSON, J. & FUHRMAN, S. A. (1977b). Isolation and characterization of *Escherichia coli* clones containing genes for the stable RNA species. In *Molecular Approaches to Eukaryotic Systems*, ed. G. Wilcox, J. Abelson & C. F. Fox, pp. 213–26. New York: Academic Press.

BOGENHAGEN, D. F., SAKONJU, S. & BROWN, D. D. (1980). A control region in the center of the 5S RNA gene directs specific initiation of transcription: II. The 3' border of the region. *Cell,* **19,** 27–35.

BREATHNACH, R., BENOIST, C., O'HARE, K., GANNON, F. & CHAMBON, P. (1978). Ovalbumin gene: Evidence for a leader sequence in mRNA and DNA sequences at the exon-intron boundaries. *Proceedings of the National Academy of Sciences USA,* **75,** 4853–7.

GILBERT, W. (1978). Why genes in pieces? *Nature,* **271,** 501.

GILMORE, R. A., STEWART, J. W. & SHERMAN, F. (1971). Amino acid replacements resulting from super-suppression of nonsense mutants of iso-1-cytochrome *c* from yeast. *Journal of Molecular Biology,* **61,** 157–73.

GOFF, S. P. & BERG, P. (1979). Construction, propagation and expression of Simian virus 40 recombinant genomes containing the *Escherichia coli* gene for thymidine kinase and a *Saccharomyces cerevisiae* gene for tyrosine transfer RNA. *Journal of Molecular Biology,* **133,** 359–83.

GOODMAN, H. M., OLSON, M. V., & HALL, B. D. (1977). Nucleotide sequence of a mutant eukaryotic gene: the tyrosine-inserting ochre suppressor SUP4-0. *Proceedings of the National Academy of Sciences USA,* **74,** 5453–7.

GRUSS, P., LAI, C.-J., DHAR, R. & KHOURY, G. (1979). Splicing as a requirement for functional 16S mRNA of Simian virus 40. *Proceedings of the National Academy of Sciences USA,* **76,** 4317–21.

HALBREICH, A., PAJOT, P., FOUCHER, M., GRANDCHAMP, C. & SLONIMSKI, P. (1980). A pathway of cytochrome *b* mRNA processing in yeast mitochondria: specific splicing steps and intron-derived circular RNA. *Cell,* **19,** 321–9.

HAMER, D. H. & LEDER, P. (1979). Splicing and the formation of stable RNA. *Cell,* **18,** 1299–302.

HAMER, D. H., SMITH, K. O., BOYER, S. H. & LEDER, P. (1979). SV40 recombinants carrying rabbit β-globin gene coding sequences. *Cell,* **17,** 725–35.

HICKS, J. B., HINNEN, A. & FINK, G. R. (1978). Properties of yeast transformation. In *Cold Spring Harbor Symposium on Quantitative Biology,* Vol. **43,** pp. 1305–13. Cold Spring Harbor: Cold Spring Harbor Laboratory.

HINNEN, A., HICKS, J. & FINK, G. R. (1978). Transformation of yeast. *Proceedings of the National Academy of Sciences USA,* **75,** 1929–33.

HIROSE T., CREA, R. & ITAKURA, K. (1978). Rapid synthesis of trideoxyribonucleotide block. *Tetrahedron Letters,* 2449–52.

HOPPER, A. K., BANKS, F. & EVANGELIDIS, V. (1978). A yeast mutant which accumulates precursor tRNAs. *Cell,* **14,** 211–19.

HUTCHISON, H. T., HARTWELL, L. H., & MCLAUGHLIN, C. S. (1969). Temperature-sensitive yeast mutant defective in ribonucleic acid production. *Journal of Bacteriology,* **99,** 807–14.

JOHNSON, J. D., OGDEN, R., JOHNSON, P., ABELSON, J., DEMBECK, P. & ITAKURA, K. (1980). Transcription and processing of a yeast tRNA gene containing a modified intervening sequence. *Proceedings of the National Academy of Sciences USA,* **77,** 2564–8.

KANG, H. S., OGDEN, R. C. & ABELSON, J. (1980). Two yeast tRNA genes containing intervening sequences. In *Mobilization and reassembly of genetic information,* ed. W. A. Scott, R. Werner, D. R. Joseph & J. Schultz. New York: Academic Press.

KANG, H. S., OGDEN, R. C. KNAPP, G., PEEBLES, C. L. & ABELSON, J. (1979). Structure of yeast tRNA precursors containing intervening sequences. In *Eukaryotic Gene Regulation*, ed. R. Axel, T. Maniatis & C. F. Fox, pp. 69–84. New York: Academic Press.

KNAPP, G., BECKMANN, J. S., JOHNSON, P. F., FUHRMAN, S. A. & ABELSON, J. (1978). Transcription and processing of intervening sequences in yeast tRNA genes. *Cell*, **14**, 221–36.

KNAPP, G., OGDEN, R. C., PEEBLES, C. L. & ABELSON, J. (1979). Splicing of yeast tRNA precursors: structure of the reaction intermediates. *Cell*, **18**, 37–45.

KONKEL D. A., TILGHMAN, S. M. & LEDER, P. (1978). The sequence of the chromosomal mouse β-globin major gene: homologies in capping, splicing and poly(A) sites. *Cell*, **15**, 1125–32.

LAI, C.-J. & KHOURY, G. (1979). Deletion mutants of Simian virus 40 defective in biosynthesis of late viral mRNA. *Proceedings of the National Academy of Sciences USA*, **76**, 71–5.

LERNER, M. R., BOYLE, J. A., MOUNT, M. S., WOLIN, L. S. & STEITZ, J. A. (1980). Are snRNPs involved in splicing? *Nature, London*, **283**, 220–4.

LERNER, M. R. & STEITZ, J. A. (1979). Antibodies to small nuclear RNAs complexed with proteins are produced by patients with systemic lupus erythematosus. *Proceedings of the National Academy of Sciences USA*, **76**, 5495–9.

MULLIGAN, R. C., HOWARD, B. H. & BERG, P. (1979). Synthesis of rabbit β-globin in cultured monkey kidney cells following infection with a SV40 β-globin recombinant genome. *Nature London*, **277**, 108–14.

NG, R. & ABELSON, J. (1980). Isolation and sequence of the gene for actin in *Saccharomyces cerevisiae*. *Proceedings of the National Academy of Sciences USA*, **77**, 3912–16.

OGDEN, R. C., BECKMANN, J. S., KANG, H. S., ABELSON, J., SÖLL, D. & SCHMIDT, O. (1979). *In Vitro* transcription and processing of a yeast tRNA gene containing an intervening sequence. *Cell*, **17**, 399–406.

OGDEN, R. C., KNAPP, G., PEEBLES, C. L., KANG, H. S., BECKMANN, J. S., JOHNSON, P. F., FUHRMAN, S. A. & ABELSON, J. (1980). Enzymatic removal of intervening sequences in the synthesis of yeast tRNAs. In *Transfer RNA: Biological Aspects*, ed. D. Söll, J. Abelson & P. R. Schimmel, pp. 173–90. Cold Spring Harbor: Cold Spring Harbor Laboratory.

OKHUBO, H., VOGELI, G., MUDRYJ, M., AVVEDIMENTO, V. E., SULLIVAN, M., PASTAN, I. & DE CROMBRUGGHE, B. (1980). Isolation and characterization of overlapping genomic clones covering the chick α2(I) collagen gene. *Proceedings of the National Academy of Sciences USA*, in press.

PEEBLES, C. L., OGDEN, R. C., KNAPP, G. & ABELSON, J. (1979). Splicing of yeast tRNA precursors: a two-stage reaction. *Cell*, **18**, 27–35.

PETES, T. D. (1980). Molecular genetics of yeast. *Annual Reviews of Biochemistry*, **49**, 845–76.

REDDY, R., RO-CHOI, T. S., HENNING, D. & BUSCH, H. (1974). Primary sequence of U-1 nuclear ribonucleic acid of Novikoff hepatoma ascites cells. *Journal of Biological Chemistry*, **249**, 6486–94.

ROCHAIX, J.-D. & MALNOE, P. (1978). Anatomy of the chloroplast ribosomal DNA of *Chlamydomonas reinhardii*. *Cell*, **15**, 661–70.

ROGERS, J. & WALL, R. (1980). A mechanism for RNA splicing. *Proceedings of the National Academy of Sciences USA*, **77**, 1877–9.

SAKONJU, S., BOGENHAGEN, D. F. & BROWN, D. D. (1980). A control region in the center of the 5S RNA gene directs specific initiation of transcription: I. The 5' border of the region. *Cell*, **19**, 13–25.

SCHMIDT, O., MAO, J., OGDEN, R., BECKMANN, J., SAKANO, H., ABELSON, J. & SÖLL, D. (1980). Dimeric tRNA precursors in yeast. *Nature, London* **287**, 750–2.

SHIOKAWA, K. & POGO, A. O. (1974). The role of cytoplasmic membranes in controlling the transport of nuclear messenger RNA and initiation of protein synthesis. *Proceedings of the National Academy of Sciences USA*, **71**, 2658–62.

STRUHL, K., STINCHCOMB, D. T., SCHERER, S. & DAVIS, R. W. (1979). High-frequency transformation of yeast: autonomous replication of hybrid DNA molecules. *Proceedings of the National Academy of Sciences USA*, **76**, 1035–9.

VALENZUELA, P., VENEGAS, A., WEINBERG, F., BISHOP, R. & RUTTER, W. J. (1978). Structure of yeast phenylalanine-tRNA genes: an intervening DNA segment within the region coding for the tRNA. *Proceedings of the National Academy of Sciences USA*, **75**, 190–4.

VOGELI, G., AVVEDIMENTO, V. E., SULLIVAN, M., MAIZEL, J. V., JR., LOZANO, G., ADAMS, S. L., PASTAN, I. & DE CROMBRUGGHE, B. (1980). Isolation and characterization of genomic DNA coding for $\alpha 2$ Type I collagen. *Nucleic Acids Research*, **8**, 1823–37.

WALLACE, R. B., JOHNSON, P. F., TANAKA, S., SCHÖLD, M., ITAKURA, K. & ABELSON, J. (1980). Directed deletion of a yeast transfer RNA intervening sequence. *Science*, **209**, 1396–400.

THE EVOLUTION OF FERREDOXIN AND SUPEROXIDE DISMUTASE IN MICROORGANISMS

K. K. RAO AND R. CAMMACK

Department of Plant Sciences, University of London King's College, 68 Half Moon Lane, London SE24 9JF, UK

INTRODUCTION

Although fossil evidence for the existence of microorganisms has been observed in some of the earliest rocks on earth, 3.4–3.5×10^9 years old (Walter, Buick & Dunlop, 1980), this does not tell us how they lived or how they originated. The only way we can tell this is by examining the microorganisms of the present day. This will tell us which types of metabolism are possible and which are not. Moreover, present-day microorganisms retain a record of their ancestors in their genes, although these have gone through a process of mutation and evolutionary change. It is possible, at least in principle, to determine what the ancestors of these organisms were like, and how they evolved, by comparing the base sequences of DNA in different organisms. At present this information is not available although methods to obtain it have recently been developed. Another approach is to compare sequences of amino acids in proteins, and this has been successfully applied to a number of proteins, notably haemoglobin and cytochrome *c*. For ease of determination the method works best with widely distributed proteins which are readily extracted and of fairly low molecular weight. When the sequences of a protein from two species are compared, differences are seen, due to 'neutral' mutations where an amino-acid substitution does not affect the efficiency of a protein significantly and is therefore not selected for or against in the course of evolution. These mutations are usually at positions remote from the active site, and the frequency with which they occur varies from one protein to another. It is assumed that the more differences there are between the two proteins, the longer the period of time that has elapsed since they evolved from a common ancestor. In this way a 'natural' classification of the animals, for example, can be derived which agrees well with the fossil record. Some anomalies will inevitably arise, because of the random nature of the mutations, and because even within a species there are genetic variations between individuals. Such anomalies can be smoothed out by determining more sequences of more proteins.

The amino-acid sequence technique can be extended further back to the time when only microorganisms were present and the fossil record is uninformative. For example, extrapolation of protein and nucleic-acid sequence data suggests that the eukaryotic ancestors of animals and plants diverged about 1.2×10^9 years ago. It is potentially even more useful in determining how the prokaryotes evolved and how life originated. There are however, several difficulties in applying the method to microorganisms.

The great metabolic diversity of the bacteria means that the same protein may have different functions in different organisms. Also, the fact that they have been evolving for a very long time means that the proteins from even closely related species may be quite different. In some cases it is difficult to decide which proteins should be compared. Two proteins with the same active group, such as a haem or iron–sulphur cluster, may show some similarities because they have evolved from a common ancestor, or because they have evolved convergently to accommodate the same type of group. In the latter case it is necessary to observe those features of the protein molecule that are not directly related to its function. For example, the cytochrome c of mitochondria, the cytochrome c_2 of purple photosynthetic bacteria and cytochrome f of algae and plants show little evidence of homology in their amino-acid sequences and differ considerably in their molecular weights. However comparison of their three dimensional structures shows that they all fold around the haem group in the same way and that the differences are principally confined to loops of amino acids on the surface of the protein distant from the haem (Dickerson, Timkovich & Amassy, 1976). When the amino acids in conserved positions in the structure are compared, they show a clear homology between these proteins.

A more serious difficulty in determining the course of evolution of microorganisms is the possibility that genetic material can be transferred from one species to another. When one considers the enzymes involved in a process (such as nitrogen fixation, ATP synthesis, and the many other diverse metabolic pathways) in most cases they are of very similar properties and activity, whichever organisms they occur in. It seems that once the system had been developed and optimized, it was incorporated into all organisms that needed to use that process; they did not have to evolve it separately. As an extreme view one could imagine a 'pool' of genetic material, so that each prokaryote is a collection of genes, selected to specify the particular metabolic processes associated with its own environment and way of life. In that case, the concept of an evolutionary 'tree' is clearly meaningless. However it seems that although there has obviously been a considerable amount of genetic transfer from one species to another, the study of a particular protein in a series of related organisms can show a significant evolutionary trend. This trend may often conflict with present methods of classification, which are based more on the need to identify microorganisms than on evolutionary principles. For example Ambler *et al.* (1979), considering the sequences of cytochrome c_2, find that the differences between them do not reflect the present classification of the photosynthetic bacteria. Dickerson (1980) considers it would be more logical to reclassify the bacteria on the basis of data such as these amino-acid sequences.

In this paper we will discuss the available information about the sequence and structure of two other types of protein, the ferredoxins and superoxide dismutases. They are widely distributed in prokaryotes and eukaryotes, and they are small proteins with a defined function: the first to transfer electrons, the second to protect against oxygen inactivation (Hall, 1977). They show contrasting distribution across the boundary between the prokaryotes and eukaryotes. The two-iron ferredoxins occur in the prokaryotic cyanobacteria and halobacteria, and also in the eukaryotic algae and plants. The Cu/Zn type of superoxide dismutase appears to have developed with the eukaryotic cytoplasm, while the Fe and Mn superoxide dismutases of prokaryotes were transferred to intracellular organelles. The study of these distributions, and the exceptions to them, give an instructive insight into the course of microbial evolution.

FERREDOXINS

Ferredoxins have a good claim to have been among the first proteins to have appeared on earth. They are a group of metalloproteins generally designated as iron–sulphur proteins. The iron–sulphur proteins contain non-haem iron bonded to sulphur ligands, both cysteinyl residues from the protein and, except in the case of rubredoxins, inorganic 'labile' sulphur (Orme-Johnson, 1973; Palmer, 1975; Rao & Hall, 1977; Yoch & Carithers, 1979; Cammack,

1979). Iron–sulphur proteins are ubiquitous and have been isolated from diverse groups of bacteria, algae, plants and animals. For classification purposes they are subdivided into simple iron–sulphur proteins (ferredoxins, rubredoxins and hydrogenase) and conjugated iron–sulphur proteins, containing other prosthetic groups (e.g. succinate dehydrogenase, NADH dehydrogenase, nitrogenase, nitrate reductase, nitrite reductase, sulphite reductase, xanthine oxidase).

The iron–sulphur proteins function in a wide variety of biological reactions such as hydrogen metabolism, nitrogen fixation, oxidative and photosynthetic phosphorylation, mitochondrial hydroxylation, and nitrite and sulphite reductions (Hall, Lumsden & Tel-Or, 1977).

Ferredoxins are simple iron–sulphur proteins with relatively low molecular weights, from 6 000 to 25 000 Daltons and by definition, their sole function is to transfer electrons. They contain one or more iron–sulphur clusters covalently linked to a monomeric protein chain and each Fe–S cluster can accept or donate one electron at a time. Their common biological function is in electron transport between soluble or membrane-bound oxidoreductases (Table 1).

The function of a ferredoxin is principally determined by the midpoint reduction (redox) potential of its iron–sulphur cluster. Ferredoxins have midpoint potentials between -600 mV and $+350$ mV relative to the hydrogen electrode (Table 2). The potentials of the iron–sulphur clusters are controlled by the protein. Such control is well-known for cytochromes and flavoproteins where mechanisms such as variation of the exposure of the haem or flavin to water and the formation and breaking of internal hydrogen bonds have been suggested (Stellwagen, 1978; Adman, 1979). In the ferredoxins, in addition to these mechanisms, the protein can influence the type of iron–sulphur cluster that is formed, and in the case of the [4Fe–4S] clusters, the type of redox change that it can undergo.

The influence of the protein can be assessed by comparing their properties with those of the analogue compounds which have been synthesized with [4Fe–4S] and [2Fe–2S] clusters bound to simple thiolate ligands (Holm, 1977). The [2Fe–2S] clusters require a more definite steric arrangement of the thiolate ligands than the [4Fe–4S] clusters. When iron, thiolate and sulphide are mixed together, [4Fe-4S] compounds are readily formed. The redox potentials of these are not greatly different from those in the clostridial ferredoxins, if correction is made for the dielectric properties of the solvent (Hill, Renaud, Holm & Mortenson, 1977). Therefore the

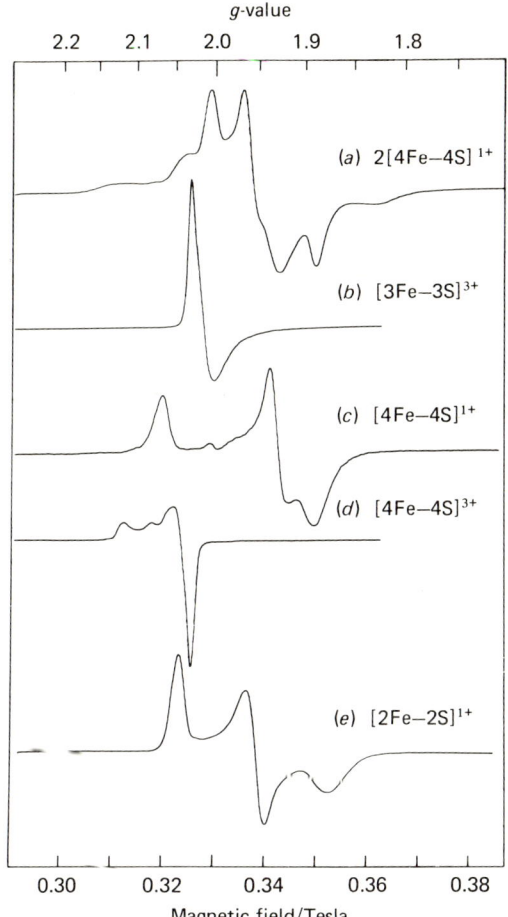

Fig. 1. Low temperature EPR Spectra of typical ferredoxins. (a) Reduced *Clostridium pasteurianum* ferredoxin; the complex shape is due to spin coupling between the two [4Fe–4S] clusters. (b) Oxidized *Desulfovibrio gigas* 4[3Fe–3S]$^{3+}$ ferredoxin. (c) Reduced *Bacillus stearothermophilus* ferredoxin. (d) Oxidized *Chromatium vinosum* HiPIP. (e) Reduced *Mastigocladus laminosus* ferredoxin.

clostridial ferredoxins can be simple molecules since the iron–sulphur clusters are themselves very stable structures. However the analogue compounds are sensitive to oxidation of the sulphur by oxygen, and to hydrolysis by water and therefore require the presence of excess thiol ligand. These effects are mitigated by wrapping the clusters in a polypeptide to which the thiolate sulphurs are attached.

Analogue compounds have been formed in which the ligands are

Table 1. *Electron transfer reactions catalysed by ferredoxins. Modified from Rao, Hall & Cammack, 1981*

Type of active centre and source of ferredoxin	Reaction	Other components required
I [2Fe–2S] (a) Algae and plants	1. Non cyclic photophosphorylation and NADP reduction $NADP^+ + H_2O + 2ADP + 2Pi \rightarrow NADPH_2 + \frac{1}{2}O_2 = 2ATP$	Ferredoxin-NADP$^+$ oxidoreductase Illuminated chloroplast membranes
	2. Cyclic photophosphorylation $ADP + Pi \rightarrow ATP$	Illuminated chloroplast electron transport chain
	3. Sulphite reduction $SO_3^{2-} \rightarrow S^{2-}$	Sulphite reductase
	4. Nitrite reduction $NO_2^- \rightarrow NH_3$	Nitrite reductase
	5. Fatty acid desaturation	Fatty acid desaturase NADPH oxidase
	6. Pyruvate decarboxylation $Pyruvate + CoA \rightleftharpoons Acetyl\ CoA + CO_2$	Pyruvate ferredoxin oxidoreductase ATP. In blue-green algae only
	7. Hydrogen metabolism $H_2 \rightleftharpoons 2H^+ + 2e^-$	Blue green algal membranes
(b) Animal mitochondria (adrenodoxin) and bacteria (putidaredoxin)	8. Hydroxylation $R-H + O_2 + NAD(P)H \rightarrow ROH + H_2O + NAD(P)$	Pyridine nucleotide oxidoreductases Cytochrome P-450

II [4Fe–4S] *Rhodospirillum rubrum* *Bacillus polymyxa* *Desulfovibrio gigas*	9. Nitrogen fixation $N_2 + 3H_2 \rightleftharpoons 2NH_3$	Nitrogenase complex
	10. Sulphite reduction Hydrogen metabolism	Sulphite reductase Hydrogenase, cytochrome c_3
III 2[4Fe–4S] or 8Fe–8S Fermentative and Photosynthetic bacteria	11. Phosphoroclastic reaction Pyruvate + Pi $\xrightarrow{CoA}$ Acetyl phosphate + CO_2	Pyruvate dehydrogenase
	12. CO_2 fixation (synthesis of α ketoacids) Acetyl CoA + $CO_2 \rightarrow$ Pyruvate + CoA Succinyl CoA + $CO_2 \rightarrow \alpha$-oxoglutarate + CoA Propionyl CoA + $CO_2 \rightarrow \alpha$-oxobutyrate + CoA	Various α-ketoacid synthetases, TPP
	13. One-carbon metabolism $CO_2 \rightleftharpoons HCO_2^-$	CO_2 reductase, formate dehydrogenase
	14. Hydrogen metabolism	Hydrogenase
	15. Nitrogen fixation	Nitrogenase
	16. NAD reduction $NAD^+ + 2H^+ \rightarrow NADH$	*Chlorobium* membranes + light

Table 2. *Properties of representative types of microbial ferredoxins. After Cammack, 1979*

Protein	Typical source	RMM[a] $\times 10^{-3}$	Fe-S	$E_m{}^b$ (mV)
2-Fe ferredoxin (Plants and eukaryotic algae)	*Scenedesmus obliquus*	10.5	[2Fe–2S]	−420
2-Fe ferredoxin (cyanobacteria)	*Spirulina platensis*	10.5	[2Fe–2S]	−390
2-Fe ferredoxin	*H. halobium*	15	[2Fe–2S]	−345
2-Fe ferredoxin (monooxygenase)	*Ps. putida*	12.5	[2Fe–2S]	−240
2-Fe ferredoxin	*C. pasteurianum*	25	[2Fe–2S]	−300
Fe–S protein I	*Azotobacter vinelandii*	21	[2Fe–2S]	−350
Fe–S protein II	*A. vinelandii*	24	[2Fe–2S]	−225
HiPIP	*Chromatium vinosum*	9.5	[4Fe–4S]$^{2+; 3+}$	+350
4-Fe ferredoxin	*Bacillus stearothermophilus*	8.5	[4Fe–4S]$^{1+; 2+}$	−280
	Desulfovibrio desulfuricans	6	[4Fe–4S]$^{1+; 2+}$	−370
	Rhodospirillum rubrum	14.5	[4Fe–4S]$^{1+; 2+}$	
6-7-Fe ferredoxin	*Thermus thermophilus*	9	[4Fe–4S] and [2Fe–2S] or [3Fe–3S]	−530 −250
7-Fe ferredoxin I (Fe–S protein III)	*A. vinelandii*	14.5	[4Fe–4S]$^{2+; 3+}$ [3Fe–3S]	+340 −420
Ferredoxin IV	*R. rubrum*	14	8Fe, 8S	+355 −380
8-Fe ferredoxin	*C. pasteurianum*	6	2[4Fe–4S]$^{1+; 2+}$	−400
8-Fe ferredoxin	*Chr. vinosum*	10	2[4Fe–4S]$^{1+; 2+}$	−490
Ferredoxin II (Fe–S protein IV)	*A. vinelandii*			−460
Ferredoxin I	*D. gigas*	18	3[4Fe–4S]$^{1+; 2+}$	−455
Ferredoxin II	*D. gigas*	24	4[3Fe–3S]$^{2+; 3+}$	−130

[a] RMM = Relative molecular mass (Molecular weight).
[b] E_m = mid point reduction potential, relative to the standard hydrogen electrode.

polypeptides of cysteine and glycine (Que *et al.*, 1974) or partial ferredoxin sequences (Gunter, Ridge, Rydon & Sharpe, 1979). These usually accommodate the more stable [4Fe–4S] clusters, but in some cases the [2Fe–2S] clusters may be produced. However, in these compounds it is likely that each cluster binds to cysteines of more than one polypeptide. It appears that in the ferredoxins protein conformation plays an important part in controlling the conformation and type of cluster.

Identification of cluster types

The [Fe–S] cluster imparts a colour to the ferredoxins which reflects the type of cluster that is present. [2Fe–2S] proteins are red, and

[4Fe–4S] proteins are brownish black. Therefore the optical absorption spectra are characteristic of the type of cluster [see Hall, Cammack & Rao, 1974a]. The electron paramagnetic resonance (EPR) spectra of the clusters (Fig. 1) in different oxidation states are prominent and readily distinguished (Orme-Johnson & Sands, 1973) and this is one of the common methods of detecting and

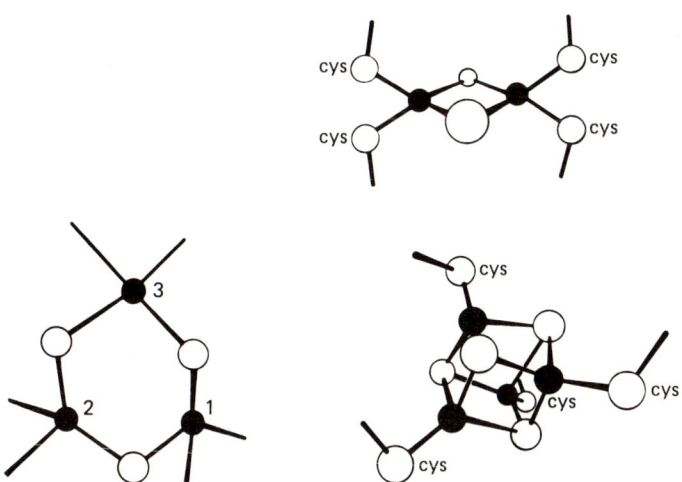

Fig. 2. Structures of the two, three and four iron clusters in ferredoxins. The three-iron structure is derived from the model of Stout *et al.* (1980) for the second cluster of *Azotobacter vinelandii* ferredoxin 1. The ligands to this cluster have not yet been determined, but probably are not all cysteine sulphurs. Filled circles, Fe; open circles, S.

characterizing ferredoxins. Mössbauer spectroscopy (Cammack *et al.*, 1977a; Emptage *et al.*, 1980) and magnetic circular dichroism measurements (Stephens *et al.*, 1978) are also helpful in characterizing the cluster. Three types of iron–sulphur clusters have been identified in ferredoxins by the application of the above mentioned techniques and by X-ray crystallography (Adman, Sieker & Jensen, 1973; Carter, 1977; Fukuyama *et al.*, 1980; Stout, Ghosh, Pattabhi & Robbins, 1980). These three cluster types are illustrated in Fig. 2.

Various methods of distinguishing between the types of iron–sulphur clusters have been developed based on the displacement of the clusters from the protein (Orme-Johnson & Holm, 1978) or EPR spectroscopy (Cammack, 1975).

Ferredoxins with two [4Fe–4S] clusters

The first non-haem iron protein to be isolated and named 'ferredoxin' was that from *Clostridium pasteurianum* (Mortenson, Valentine

& Carnahan, 1962). Later, ferredoxins with similar properties and sequences were isolated from a number of clostridia and other anaerobic bacteria (Yasunobu & Tanaka, 1973). All of these ferredoxins were found to contain eight iron and eight sulphide atoms per molecule of 54 to 56 amino-acid residues. X-ray structure determination of the eight-iron ferredoxin from *Peptococcus aerogenes* (Adman, Sieker & Jensen, 1973) revealed (Fig. 3) that

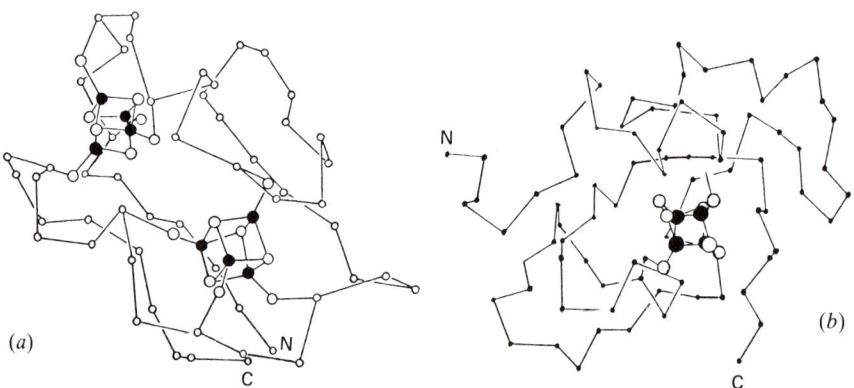

Fig. 3. Structures determined by X-ray crystallography. (*a*) *Peptoccus aerogenes* eight-iron ferredoxin (Adman *et al.*, 1973) and (*b*) *Chromatium vinosum* HiPIP from Carter (1977). Filled circles, Fe open circles, S.

the iron and sulphur atoms are in two identical clusters, 1.2 nm apart, each cluster having four iron, four inorganic sulphur and four cysteinyl sulphur atoms. The electronic state of the iron atoms in these clusters was investigated by Mössbauer studies on *C. pasteurianum* ferredoxin (Thompson *et al.*, 1974). They appear to be covalent, mixed-valence structures in which the spins of the iron atoms are antiferromagnetically coupled together.

Eight-iron ferredoxins have been purified from diverse bacteria including *Chlorobium, Chromatium, Veillonella alcalescens* (Zubieta & Dalton, 1973) and *Rhizobium japonicum* bacteroids (Carter *et al.*, 1980) *Ruminococcus albus* (Tewes & Thauer, 1979) and *Desulfuromonas acetoxydans* (Probst *et al.*, 1980). This distribution is probably due to their relative instability in the presence of oxygen. Under anaerobic conditions they are stable, efficient electron carriers. However the reduced form is reoxidized by oxygen, forming a superoxide radical (Orme-Johnson & Beinert, 1969) which is presumably the cause of their destruction.

The eight-iron ferredoxins so far isolated all operate at low redox potentials of about −400 to −500 mV. They are involved in a number of different electron transfer systems where their ability to transfer electrons either one or two at a time may be an advantage (Fig. 4). In view of their low molecular weight they can be considered as effectively a coenzyme, rather like NAD. It is of

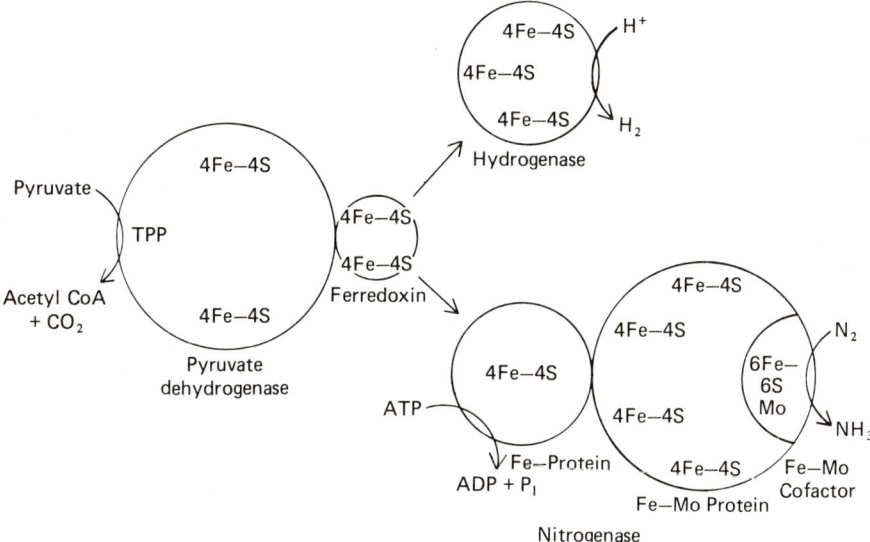

Fig. 4. Interactions of the 2[4Fe–4S] ferredoxin of *Clostridium pasteurianum* with conjugated iron–sulphur proteins in fermentations. The relative sizes of the enzymes are drawn on the assumption that they are spherical molecules.

interest that the redox potential of the ferredoxins is lower than that of NAD, so that photosynthetic bacteria can fix carbon dioxide by a series of reactions which is essentially the citric acid cycle driven in reverse (Evans, Buchanan & Arnon, 1966). This may have been the predecessor of the citric acid cycle.

Ferredoxins containing a single [4Fe–4S] cluster

Low-potential ferredoxins of this type have been isolated from *Bacillus* spp., photosynthetic bacteria and sulphate-reducing bacteria. The various ferredoxins differ in molecular weight and reactivity with different enzymic systems (Yoch & Carithers, 1979). The [4Fe–4S] cluster in *B. stearothermophilus* has been well characterized by physico-chemical techniques and appears to be very similar to those in the eight-iron ferredoxins (Cammack *et al.*, 1977a).

Table 3. *Oxidation levels of iron-sulphur clusters*

Type of cluster	Type of protein	Oxidation level	Formal states of Fe	Typical mid point potentials
[4Fe–4S]	4Fe and 8Fe ferredoxin	↑ +1 ↓	$3Fe^{2+} + 1Fe^{3+}$	−280 to −550 mV
[4Fe–4S]	HiPIP A. vinelandii Fd. I, cluster I etc.	↑ +2 ↓ +3	$2Fe^{2+} + 2Fe^{3+}$ $1Fe^{2+} + 3Fe^{3+}$	+350 to +200 mV
[3Fe–3S]	A. vinelandii Fd I, cluster II	↑ +2 ↓ +3	$1Fe^{2+} + 2Fe^{3+}$ $3Fe^{3+}$	−50 to −420 mV
[2Fe–2S]	2Fe ferredoxins	↑ +1 ↓ +2	$1Fe^{2+} + 1Fe^{3+}$ $2Fe^{3+}$	−180 to −450 mV

The oxidation level is derived by adding the charge on the iron atoms and −2 for each sulphide, ignoring the thiolate ligands. This formal ionic description does not give an accurate picture of the individual iron atoms, particularly in the [4Fe–4S] clusters where the atoms are essentially indistinguishable because of the effects of covalency.

High-potential iron-sulphur protein [HiPIP] is another type of four-iron ferredoxin. Bartsch (1963) isolated it from *Chromatium* cells in which it is present in high concentration. Unlike the clostridial ferredoxins which have very negative redox potentials the *Chromatium* non-haem iron protein was found to have a positive mid-point potential of +350 mV. Although it is now classified as a ferredoxin the name HiPIP is retained for this type of protein. Since then several more high-potential iron–sulphur proteins have been isolated, mostly from photosynthetic bacteria, but one from a halophilic *Paracoccus* sp. Though the reduction of HiPIP by photosynthetic electron transport in *Chromatium* was demonstrated by EPR spectroscopy (Evans, Lord & Reeves, 1974), the physiological role for the protein was unknown. Fukumori & Yamanaka (1979) have reported that HiPIP in *Chromatium vinosum* acts as an electron acceptor for thiosulphate in the presence of a thiosulphate oxidizing enzyme.

The eight-iron ferredoxin from *Peptococcus aerogenes* (E_m − 400 mV) and HiPIP from *Chromatium* (E_m + 350 mV) were found to possess the same type of [4Fe–4S] cluster by X-ray structure determination (Carter, 1977) (Fig. 3). To explain the considerable difference in redox potential between the two proteins, Carter *et al.* (1972) proposed the [4Fe–4S] cluster can have different oxidation levels (with different potentials) in the two proteins. This 'three state hypothesis' is summarized in Table 3. Thus *P. aerogenes* and the clostridial ferredoxins are diamagnetic in the oxidized state (as isolated) whereas HiPIP is diamagnetic in the reduced state (as isolated). This state in both proteins is now called the +2 oxidation level, and was the C state of Carter et al. (1972). In ferredoxin the clusters can be reduced to a +1 level (C^- state) which is paramagnetic, with an EPR signal centred around g = 1.96. In HiPIP the cluster can be oxidized to a +3 level (C^+ state) which gives an EPR signal with g-values greater than 2.

The difference in redox potentials can therefore be explained by the different changes in oxidation level which the proteins can undergo. These differences in turn are the result of the structures of the proteins (Sweeney & Rabinowitz, 1980). Thus the cluster in HiPIP can be reduced chemically to the +1 level if the protein is denatured first (Cammack, 1973), and oxidation of the clusters in *C. pasteurianum* ferredoxin to the +3 level by $K_3Fe(CN)_6$ has been reported by Sweeney, Bearden & Rabinowitz (1974).

Ferredoxins with mixed types of clusters

Ferredoxin I from *Azotobacter vinelandii* contains two iron–sulphur clusters, both giving EPR signals in the oxidized state, like HiPIP, but with remarkably different redox potentials, +320 mV and −420 mV (Sweeney & Rabinowitz, 1980). X-ray crystallography (Stout et al., 1980) indicates that one of these clusters is a [4Fe–4S] type as in HiPIP, while the other appears to be a novel three-iron cluster (Fig. 2). The reason why a protein should require such widely differing potentials is unknown. Similar ferredoxins have been isolated from a number of other species including *Pseudomonas ovalis* and *Mycobacterium smegmatis* (Hase et al., 1979).

Another type of ferredoxin from *Thermus thermophilus* contains 6–7 iron atoms and appears to have two clusters, one with an EPR signal in the oxidized state and a redox potential of −260 mV, and one with a signal in the reduced state and a potential of −520 mV. Ohnishi et al. (1980) interpreted this as due to one HiPIP-type [4Fe–4S] cluster and a [2Fe–2S] cluster, but a [3Fe–3S] cluster and a [4Fe–4S] cluster as in clostridial ferredoxin are also possibilities (J. Fee, personal communication).

Another remarkable case is provided by ferredoxins I and II of *Desulfovibrio gigas* (Cammack et al., 1977b). Both are polymers of the same polypeptide of molecular weight approx. 6 000. Ferredoxin I appears to be a trimer, with [4Fe–4S] clusters in each subunit while ferredoxin II is a tetramer, with [3Fe–3S] clusters in each subunit

(Huynh et al., 1980). The former has a midpoint potential of −455 mV, the latter −130 mV. Thus the same polypeptide might even function at two points in the same electron-transport chain (Moura, Xavier, Hatchikian & Le Gall, 1978).

Amino-acid sequences of the [4Fe–4S] and [3Fe–3S] ferredoxins

With the exception of the HiPIPs (Fig. 5) the available sequences of the ferredoxins with [4Fe–4S] and [3Fe–3S] clusters can be aligned with a considerable degree of homology (Fig. 6), indicating that they all belong to the same family. For each cluster there are four cysteine residues in a conserved position.

In the sequence of *Peptococcus aerogenes* eight-iron ferredoxin there are eight cysteines in two groups of four. The two halves of the molecule are so similar that this is one of the most widely-cited cases for gene duplication. It should, however, be noted that the two domains of the sequence do not correspond with the cysteines that bind the two clusters. It can be seen from the structure of this molecule (Fig. 3) that one cluster binds to cysteines 8, 11, 14 and 45 and the other to cysteines 35, 38, 41 and 18 as the polypeptide wraps itself twice around the clusters. When the two halves of the ferredoxin from *Clostridium acidi-urici* were enzymatically cleaved and then reconstituted it was found that the protein again produced dimers with two clusters (Orme-Johnson, 1972). Furthermore in the four-iron ferredoxins from *Desulfovibrio gigas* and *Bacillus stearothermophilus* the cysteines that bind the cluster are three closely-spaced and one distant residue. There may be steric restrictions on placing four closely-spaced cysteines around one cluster. All this suggests that if a 28-residue polypeptide with four cysteines and one [4Fe–4S] cluster was an evolutionary precursor of the bacterial ferredoxins, the molecule was probably dimeric.

The evolution of the ferredoxins can be considered as starting from the simple eight-iron ferredoxins such as *Clostridium butyricum* and *Peptococcus aerogenes*, and appears to have developed in several directions, by addition of extra 'loops' of amino acids which can probably be accommodated on the surface of the molecule. In Fig. 6 these have been incorporated by inserting spaces in the shorter sequences.

The two ferredoxins of the green photosynthetic bacterium *Chlorobium limicola* and that of *Chlorobium thiosulfatophilum* show, when compared with *Clostridium* ferredoxin, an insertion

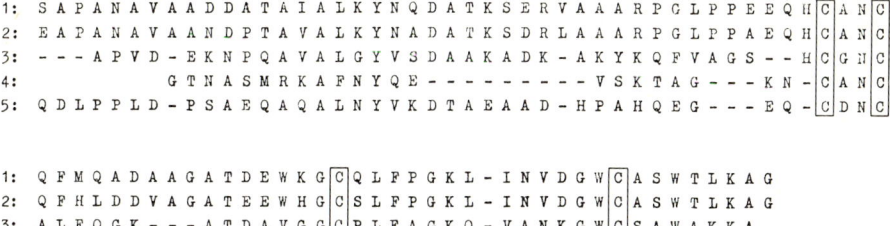

Fig. 5. Amino-acid sequences of high-potential iron-sulphur proteins. (1) *Chromatium vinosum;* (2) *Thiocapsa pfennigii;* (3) *Rhodopseudomonas gelatinosa;* (4) *Rhodospirillum tenue;* (5) *Paracoccus* sp. Sequences are in one letter code and are rearranged so as to align the cysteines (C) that bind the [4Fe–4S] cluster. Modified from Tedro, Meyer & Kamen (1979).

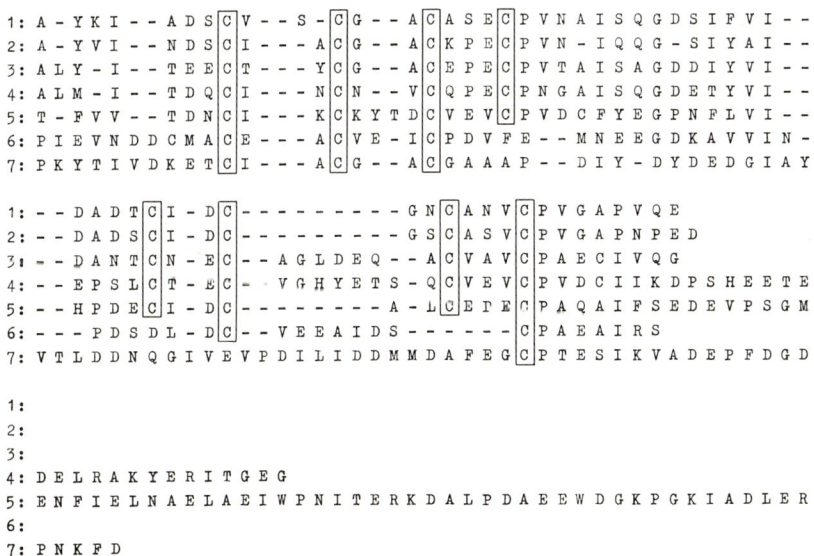

Fig. 6. Amino-acid sequences of bacterial ferredoxins. (1) *Clostridium pasteurianum;* (2) *Peptococcus aerogenes;* (3) *Chlorobium limicola* (Fd I); (4) *Chromatium vinosum;* (5) *Pseudomonas ovalis;* (6) *Desulfovibrio gigas;* (7) *Bacillus stearothermophilus.* Sequences are in one letter code and are arranged so as to align the cysteines (C) that probably bind the iron–sulphur clusters. Data from Yasunobu & Tanaka (1973), Tanaka *et al.* (1975), Hase *et al.* (1977), Bruschi (1979), Matsubara *et al.* (1980).

between residues 39 and 40. This insertion is also present in the sequence of ferredoxin from the purple photosynthetic bacterium *Chromatium vinosum* which also has an extra sequence at the C terminal end. These additions may be required for the proteins to interact with additional enzymic systems. They suggest a sequence of evolution: anaerobic fermentative bacteria → green sulphur bacteria → purple sulphur bacteria. No sequences of ferredoxins from non-sulphur bacteria are known, but those from *Rhodospirillum rubrum* appear to have higher molecular weights (Yoch, Carithers & Arnon, 1977).

The four-iron ferredoxins of *Desulfovibrio* and *Bacillus* spp. can be considered as evolving from *Clostridium* ferredoxin by loss of four cysteines that bind one [4Fe–4S] cluster. The ferredoxins of *Pseudomonas ovalis* and *Thermus thermophilus* which may contain [3Fe–3S] and HiPIP-type [4F3–4S]$^{2+;\ 3+}$ clusters appear to be yet another line of development, with a greatly extended polypeptide at the C-terminus. The factors that influence the formation of these different types of cluster are not evident from the sequences, and full elucidation of the structure of *Azotobacter vinelandii* ferredoxin I (Stout *et al.*, 1980) is awaited with interest.

Two-iron ferredoxins

The soluble ferredoxins isolated from cyanobacteria (blue-green algae), red and green algae and chloroplasts of plants all contain a [2Fe–2S] cluster linked to a protein chain of 95 to 98 amino-acid residues. These ferredoxins function at the reducing side of photosystem I of these organisms in the transfer of electrons from water to NADP$^+$. Being soluble proteins they can also donate electrons to other systems and are involved in the photosynthetic reduction of nitrate, nitrite and sulphite and in the cyanobacteria in dinitrogen fixation (Rao & Hall, 1977). In addition they play an important role in regulating the activity of many enzymes of the carbon dioxide reduction pathway, through the thioredoxin/thioredoxin reductase system (Buchanan, Wolosiuk & Schurmann, 1979).

The three-dimensional structure of *Spirulina platensis* ferredoxin, determined by X-ray crystallography at 2.5Å resolution, (Fig. 7) shows that the [Fe–S] cluster is located near the molecular surface and is mainly surrounded by hydrophobic residues (Fukuyama *et al.*, 1980). One of the iron atoms is coordinated to Cys 41 and Cys 46 and the other to Cys 49 and Cys 79 of the protein chain – these cysteines are invariant in all algal and plant ferredoxins so far sequenced. This structure bears no obvious resemblance to the bacterial ferredoxins.

This X-ray determination confirms the structure for the [2Fe–2S] cluster predicted from spectroscopic measurements (Palmer, 1973).

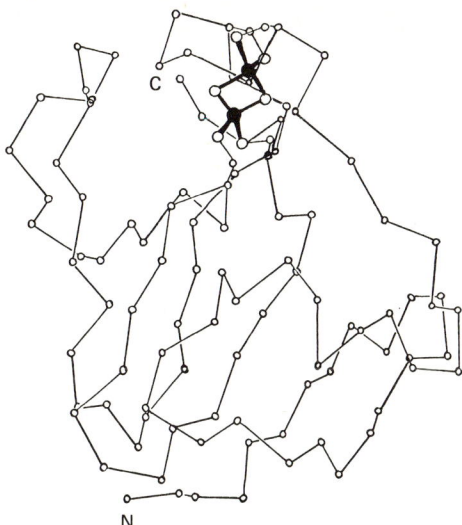

Fig. 7. X-ray structure of the [2Fe–2S] ferredoxin from *Spirulina platensis*. Adapted from Fukuyama *et al.* (1980).

The two iron atoms are antiferromagnetically coupled together and in the oxidized state both are Fe (III). On reduction, one iron atom becomes Fe (II) so these proteins are one-electron transfer agents.

A similar ferredoxin has been found to be present in high concentrations in *Halobacterium halobium* (Kerscher, Oesterhelt, Cammack & Hall, 1976), despite the fact that this organism has no photosynthetic electron transport. Ferredoxins from halobacteria have been implicated in α-oxoacid metabolism (Kersher & Oesterhelt, 1977) and nitrite reduction (Werber & Mevarech, 1978). The sequence homology of this ferredoxin with those of the cyanobacteria is remarkable, since it has been suggested that the halobacteria are taxonomically distinct from most other prokaryotes including cyanobacteria (Magrum, Luehrsen & Woese, 1978). The distribution of this ferredoxin suggests that gene transfer has occurred.

[2Fe–2S] ferredoxins have been isolated from photosynthetic bacteria such as *Chlorobium thiosulfatophilum, Chromatium vinosum* and *Rhodospirillum rubrum,* and from some anaerobes including *Clostridium pasterianum*. The function of these proteins is unknown. A [2Fe–2S] ferredoxin from *Azotobacter vinelandii* is involved in the stabilization of nitrogenase against oxygen inactivation (Scherings, Haaker & Veeger, 1978). In *Pseudomonas putida* a [2Fe–2S] ferredoxin is involved in a well-characterized camphor monooxygenase system which also includes a flavoprotein and cytochrome P-450 (Estabrook *et al.*, 1973).

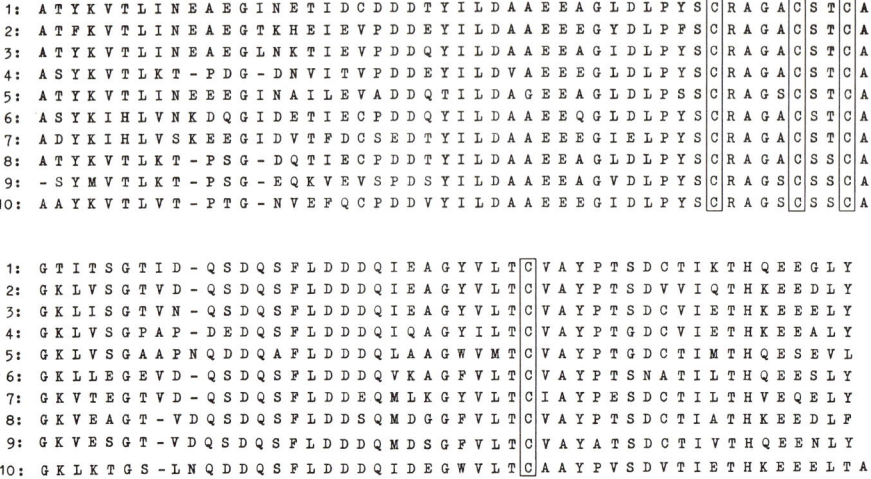

Fig. 8. Amino-acid sequences of [2Fe–2S] ferredoxins. (1) *Spirulina platensis:* (2) *Nostoc muscorum* Fd I; (3) *Mastigocladus laminosus;* (4) *Aphanothece sacrum* Fd I; (5) *A. sacrum* Fd II, (6) *Cyanidium caldarium;* (7) *Porphyra umbilicalis;* (8) *Scenedesmus quadricauda;* (9) *Dunaliella salina* Fd I; (10) *Spinacia oleracea.* The invariant cysteines that bind the [Fe–S] cluster are boxed. Data from Matsubara *et al.* (1980).

A very similar type of monooxygenase system, in this case for steroids, occurs in mitochondria of the mammalian adrenal cortex although the ferredoxins of the two systems are not interchangeable. It may be of evolutionary significance that whereas the P-450 monooxygenase systems of mitochondria involve a [2Fe–2S] ferredoxin, those of microsomes employ FMN in the same place. The [2Fe–2S] ferredoxins of *Escherichia coli* and *Agrobacterium tumefaciens* may also be involved in P-450 systems, though so far the only evidence is a similarity in EPR spectra (Hall *et al.*, 1974*b*).

Amino-acid sequences of cyanobacterial and algal ferredoxins

The two-iron ferredoxins involved in oxygen-evolving photosynthesis all show a considerable degree of homology and can therefore be used as taxonomic markers, (Fig. 8). However, many species contain two types of ferredoxin with significantly different sequences. This could be interpreted as showing that the two proteins diverged earlier in the history of the organism, or, as seems likely, that the two proteins have different functions. The two ferredoxins of *Nostoc* strain MAC differ in activity with enzyme systems, and in redox potential (Hutson *et al.*, 1978).

From a matrix comparison of the sequences Matsubara, Hase, Wakabayashi & Wada (1980) have constructed a phylogenetic tree for algal ferredoxins. From this tree they have found that the algal ferredoxins are evolutionarily a very diverse group and diverged early in evolution. The relative positions of the cyanobacterial ferredoxins are in accordance with their cellular morphology. The tree shows that ferredoxin from the red alga, *Porphyra umbilicalis*, diverged from the prokaryotic algae; a relationship which agrees with the distribution of phycobiliproteins in these two algal types. *Cyanidium caldarium* is a unicellular alga of uncertain relationship. The sequence of *Cy. caldarium* ferredoxin is more homologous to *Porphyra umbilicalis* ferredoxin than to that of any other algal ferredoxin and hence *Cy. caldarium* may be closely related to the red algae. The amino-acid sequence of the heat-stable ferredoxin from the thermophilic alga *Mastigocladus laminosus* did not show any special features and was therefore of little help in predicting the origin of thermophily. Two ferredoxins were isolated and sequenced from the halophyte *Dunaliella salina* (Hase *et al.*, 1980). These ferredoxins were homologous to other algal ferredoxins and did not show any special resemblance to halobacterial ferredoxins. The longer chain length (128 amino acids) of *Halobacterium halobium* ferredoxin may reflect the high intracellular concentration (3M) of KCl and its effects on protein–protein interactions (Lanyi, 1980). *Dunaliella*, by contrast, maintains its osmotic balance by a high concentration of glycerol.

SUPEROXIDE DISMUTASES

Superoxide dismutase is an enzyme that catalyses the elimination of the superoxide radical O_2^- with the formation of oxygen and hydrogen peroxide.

$$O_2^- + O_2^- + 2H_+^+ \to H_2O_2 + O_2$$

The generation of O_2^- as a respiratory intermediate of aerobic organisms was discovered by McCord & Fridovich (1968) and later they observed that the protein hemocuprein whose function had been unknown for 30 years is involved in the dismutation of this radical (McCord & Fridovich, 1969). They named the enzyme superoxide dismutase and suggested that the enzyme provided a defence against oxygen toxicity in aerobic cells. During the past ten

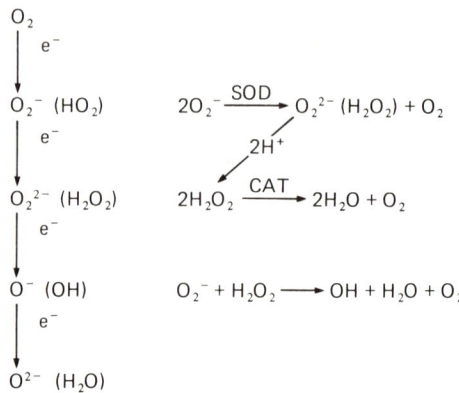

Fig. 9. Reduced species derived from oxygen. From Hall (1977).

years the superoxide radical and dismutase have been very active areas of research and reviews and symposium volumes have appeared describing the physical, biochemical and evolutionary aspects of the enzyme (Fridovich, 1975, 1976; Michelson, McCord & Fridovich, 1977; Bannister & Hill, 1980).

Superoxide is the initial product in the univalent pathway of reduction of molecular oxygen to water. Fig. 9 shows the products of the reaction of oxygen with electrons and the enzymes involved in breaking down superoxide to peroxide, hydroxyl radical and water. A number of biological reactions have been shown to generate O_2^- as an intermediate. These include the autoxidation of ferredoxin, flavins, haem proteins, quinols, etc. (Fridovich, 1975; Hill, 1978). The catalytic action of enzymes such as xanthine oxidase, aldehyde oxidase, and dihydroorotic dehydrogenase also produce O_2^-. The superoxide formed by accidental autoxidation of respiratory proteins can therefore damage the enzymes themselves, and other cell constituents. Superoxide formation has been suggested as one of the causes for the *in vitro* deactivation of *Alcaligenes eutrophus* hydrogenase, a protein containing iron–sulphur centres and flavin (Schneider & Schlegel, 1981).

Superoxide has been implicated in the oxygen-dependent destruction of proteins, nucleic acids, carbohydrates and membrane lipids in reactions that are inhibited by superoxide dismutase. The mechanism of inactivation is still not clearly understood. O_2^- itself is a fairly innocuous species and its damaging effects must therefore be due to a more reactive species derived from it. Current evidence favours the hydroxyl radical generated from O_2^- and H_2O_2 in the presence of traces of transition metal ions. It has also been suggested that O_2^- might give rise to singlet oxygen, but this is hotly disputed.

It appears that superoxide dismutase is an important part of the cell's defences against oxygen toxicity which also include catalase and reductants such as glutathione and ascorbate. Almost all aerobic organisms examined have been found to contain a constant, high level of at least one type of superoxide dismutase. The only exceptions found so far are certain strains of two pathogenic bacteria; *Mycoplasma pneumoniae* and *Neisseria gonorrhoeae* (Norrod & Morse, 1979). Mutants of *E. coli* with a high sensitivity to oxygen have been found to lack superoxide dismutase activity (Fridovich, 1976). However they also lacked other enzymes; some lacked catalases, others peroxidases.

The O_2^- radical is a short-lived species in water, as it dismutates spontaneously with a half-life of a few milliseconds at pH 7.4.

$$2O_2^- + 2H^+ = H_2O_2 + O_2 \quad k = 2 \times 10^5 \text{ M}^{-1}\text{s}^{-1}$$

From the high concentration of superoxide dismutase in the cell, it is inferred that even small amounts of O_2^- would be damaging. Transition metal ions which can undergo redox changes are themselves efficient catalysts of superoxide dismutation (Pasternack & Halliwell, 1979). The Cu^{2+} ion in water is more efficient than Cu/Zn-superoxide dismutase. However, it is unlikely that free metal ions could exist in the cell, as they would be sequestered by many chelating molecules, including proteins, especially as they can themselves give rise to oxygen radicals. The function of the superoxide dismutase protein therefore, is to hold the metal ion (Cu^{2+}, Mn^{3+} or Fe^{3+}) in an environment where it is accessible to O_2^- in solution, and where the appropriate electron and hydrogen transfers can take place. As will be noted later, the sequences of the superoxide dismutases have been highly conserved in evolution. It appears that the efficient action of these enzymes places severe structural constraints on the proteins. Indeed other copper proteins react only sluggishly with superoxide if at all.

Occurrence and types of superoxide dismutases

Since the biological function of superoxide dismutase is to scavenge the O_2^- radical generated during oxygen metabolism, early investigations on the location of its activity were confined to aerobic and

Table 4. *Distribution and properties of superoxide dismutases*[a]

Enzyme Type	Metal content g atom/mole	Molecular Weight	Subunit structure
Cu/Zn type			
Many eukaryotes including mammalian tissues, yeast, *Neurospora* and green plants	2Cu, 2Zn	32000	$\alpha 2$
Photobacterium leiognathi	1Cu, 2Zn	33,100	$\alpha\beta$
Paracoccus denitrificans	n.d.[b]	n.d.	n.d.
Mn type			
E. coli (cell matrix)	1.2	40000	$\alpha 2$
Bacillus stearothermophilus	1	40000	$\alpha 2$
Streptococcus mutans	1.2	40000	$\alpha 2$
Rhodopseudomonas spheroides	1.1	37000	$\alpha 2$
Thermus aquaticus	2.1	80000	$\alpha 4$
Mycobacterium sp	1.7	62000	$\alpha 3$
Chicken liver mitochondria	2.3	80000	$\alpha 4$
Yeast mitochondria	4	100000	$\alpha 4$
Pleurotus olearius (fungus), two types	2.1	76000	$\alpha 2\ \beta 2$
	1.9	78000	$\alpha 4$
Paracoccus denitrificans	2	41000	$\alpha 2$
Fe type			
Brassica campestris	1.6	41000	$\alpha 2$
Porphyridium cruentum (Red alga)	1.0	40000	$\alpha 2$
Spirulina platensis (Cyanobacteria)	1.0	37000	$\alpha 2$
Plectonema boryanum (Cyanobacteria)	0.9	36500	$\alpha 2$
Desulfovibrio desulfuricans	1.6	43000	$\alpha 2$
E. coli (Periplasmic space)	1–1.8	39000	$\alpha 2$
Thiobacillus denitrificans	1.35	43000	$\alpha 2$
Anacystis nidulans	1	37500	$\alpha 2$
Chromatium vinosum	2	41000	$\alpha 2$
Euglena gracilis	1	57500	n.d.

[a] For references to original data see Hall *et al.* (1977), Michelson *et al.* (1977), Hill (1978) and Bannister & Hill (1980).
[b] Not determined.

aerotolerant anaerobic organisms. However, since 1975 it has been shown that superoxide dismutase occurs also in obligate anaerobic bacteria, such as *Clostridium, Desulfovibrio, Chlorobium* and *Chromatium* (Hewitt & Morris, 1975; Lumsden & Hall, 1975; Asada *et al.*, 1975), though usually at lower levels than in aerobes.

Three different types of superoxide dismutases have now been isolated and characterized. Cytosols of eukaryotic cells contain an enzyme that has a molecular weight of 32 000, is made up of two identical subunits and contains one Cu^{2+} and one Zn^{2+} per sub unit. The Cu^{2+} is essential for catalytic activity; the Zn^{2+}, which preserves the conformation of the molecule, may be replaced by Co^{2+}, Hg^{2+} or Cd^{2+} without loss of enzymic activity.

The superoxide dismutases in the clostridia, sulphate-reducing bacteria and photosynthetic anaerobes are iron-containing enzymes, usually containing 1 Fe atom per molecule of approximately 40 000 Daltons. Fe-superoxide dismutases are also found in facultative and aerobic bacteria, in prokaryotic and eukaryotic algae and in protozoa (Asada *et al.*, 1980). From electron paramagnetic resonance the valence of iron in the enzyme has been identified as high spin Fe^{3+}.

The third type of superoxide dismutase contains manganese [Mn^{3+}] as the metal component. Mn-superoxide dismutases with molecular weights of 40 000 and 80 000 containing 1 or 2 atoms of manganese per molecule respectively have been isolated. The Mn-enzyme occurs in facultative and aerobic bacteria, the cyanobacteria, algae, protozoa, fungi and mitochondria. A summary of the occurrence and properties of superoxide dismutases from various sources is given in Table 4.

Functional and structural comparison of the three types of superoxide dismutases

The three types of superoxide dismutases can be distinguished in cell extracts by selective inhibition or inactivation. Thus cyanide inhibits the Cu/Zu enzyme but not the Mn or Fe enzyme. Hydrogen peroxide inactivates both the Cu/Zn and Fe superoxide dismutase but not the Mn enzyme. Azide inhibits these enzymes in the order Fe > Mn > Cu/Zn. However, irrespective of their metal content and structural differences (see below) the specific activities of these three enzymes in the pure state are approximately the same – about 3 000 units per mg protein (Kirby, Blum, Kahane & Fridovitch, 1980). This implies that there is no immediate evolutionary advantage in converting from one type to another – indeed both types can coexist in the same cell.

The metals present in all three types of superoxide dismutases have been reversibly removed by treatment with protein denaturating agents followed by dialysis. With each inactive apoenzyme thus obtained, the enzyme activity can be restored only by the addition of the particular metal or metals found in the holoenzyme. Kirby *et al.* (1980) have used this to distinguish between Mn and Fe containing enzymes in crude extracts of prokaryotic cells. They produced the apoenzymes by exposing the extracts to low pH in the presence of guanidinium hydrochloride and then regenerated the active holoenzyme by the addition of Mn or Fe.

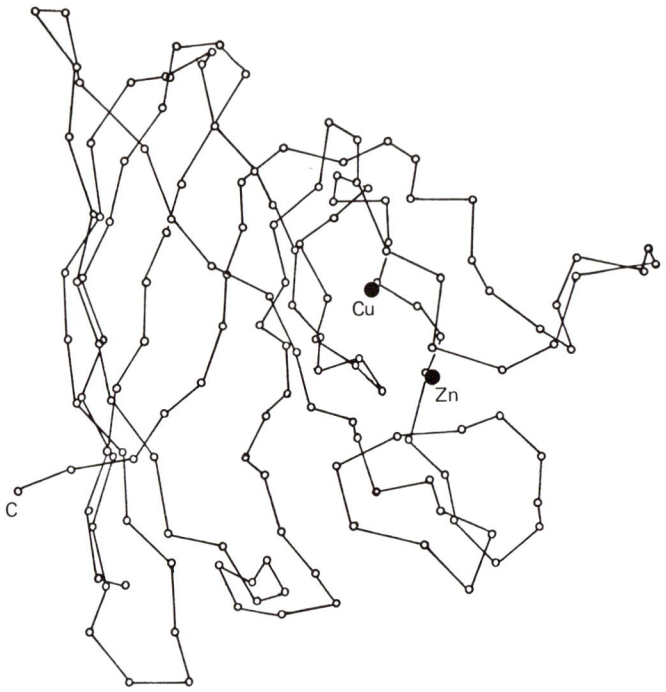

Fig. 10. Structure of the Cu/Zn superoxide dismutase of bovine erythrocytes. After Richardson (1977).

Amino-acid sequences

The complete amino acid sequences of Cu/Zu superoxide dismutase from bovine erythrocytes (Steinman, Naik, Abernethy & Hill, 1974) and *Saccharomyces cerevisiae* (Johansen *et al.*, 1979) have been determined. The two proteins are highly homologous with 55% identity in the sequences. X-ray diffraction analysis (Fig. 10) of the bovine erythrocyte enzyme (Richardson, 1977) indicates that the Cu and Zn are joined in the enzyme by a common ligand, the imidazole ring of histidine 61. The other ligands to the Cu are His_{44}, His_{46} and His_{118} while the other Zn ligands are His_{69}, His_{78} and Asp_{81}. The positions of all the Cu and Zn ligands are identical in the sequences of the bovine erythrocyte and *S. cerevisiae* superoxide dismutases (Fig. 11).

McLachlan (1980) from a study of the amino-acid sequence and X-ray structure of bovine Cu/Zn superoxide dismutase, has proposed a detailed hypothesis for the evolution of this enzyme. The X-ray structure (Richardson, 1977) shows two paired subdomains

```
           10                  20                  30
1: - V Q A V A V L K G D A G V S G V V K F E Q A S E S E P T T V S Y E I A G N S
2: A T K A V C V L K G D G P V E G T I H F E A K G D T V V V T G S - - I T G - L
         40                  50                  60                  70
1: P N A E R G F[H]I[H]E F G D A T N G C V S A G P[H]F N P F K K T[H]G A P T D E
2: T E G D H G F[H]V[H]Q F G D N T Q G C T S A G P[H]F N P L S K K[H]G G P K D E
         80                  90                 100                 110
1: V R[H]V G[D]M G N V K T D E N G V A K G S F K D S L I K L I G P T S V V G R S
2: E R[H]V G[D]L G N V T A D K N G V A I V D I V D P L I S L S G E Y S I I G R T
        120                 130                 140                 150
1: V V I[H]A G Q D D L G K G D T E E S L K T G N A G P R P A C G V I G L T N
2: M V V[H]E K P D D L G R G G N E E S T K T G N A G S R L A C G V I G I A K
```

Fig. 11. Sequence of Cu/Zn superoxide dismutases. (1) Yeast, (2) Bovine erythrocytes. From Johansen *et al.* (1979). The metal binding sites are boxed.

with similar folding patterns. A comparison of the amino-acid sequence of structurally equivalent regions shows that the tripeptide Val–His–Gln at positions 45 to 47 in the bovine enzyme is almost repeated across the dyad as Val–His–Glu at positions 117 to 119. The histidines are Cu ligands. From these and other intramolecular structural comparisons McLachlan proposes that the repeated features found only associated with the active site of the Cu/Zn enzyme might be the result of convergent evolution.

The amino-acid sequences of Mn superoxide dismutase from *E. coli* B (Steinman, 1978) and *Bacillus stearothermophilus* (Brock & Walker, 1980) are known. The *E. coli* enzyme consists of two identical subunits with 205 amino-acid residues whereas the *Bacillus* enzyme has two identical subunits of 203 residues. If one allows two deletions in the *Bacillus* enzyme sequence then 60% of the residues are identical in the two sequences and conserved amino acids are found throughout the subunits. Brock & Walker (1980) suggest that Mn enzymes from the two bacteria may have similar secondary structures but that they are different from the secondary structures predicted for the Cu/Zn enzyme.

Harris, Auffret, Northrop & Walker (1980) determined the amino terminal sequences of superoxide dismutases from seven microorganisms, five containing Fe and two containing Mn (Fig. 12). The partial sequences show that the Fe and Mn enzymes are related to each other in structure but not to the Cu/Zn family of superoxide dismutases. From these sequences together with six others from the same family Harris *et al.* (1980) conclude that there is no clear distinction in sequences between the Mn and Fe superoxide dismutases. However, wide sequence variations between the same metal-containing enzymes from different bacteria

```
 1: A Y Z Z P A L P Y A B B A L Z P H I - X A Z T I G F H Y G K H H A A Y V K T Y X G L V
 2: M H E L P A L P Y E K N A L E P V I - S A E T I E Y H Y G X X H Q T Y V T N L G
 3: A F E L P S L P F D Q D A L E S S K M S A N T L S Y H H G K H H A A Y V K N L N A A I Z G T B M A B
 4: A F E L P P L P Y A H D A L Q P H I - S K E T L E Y H H B K H H N T Y V V N L N N L V P G X T F
 5: S F E L P A L P Y A K D A L A P H I - S A E X I E Y H Y G K
 6: S Y T L P S L P Y A Y D A L E P H F - D K Q T M E I H H T K H H Q T Y V N N A N A A L E S L P E F A
 7: P F E L P A L P Y P Y D A L E P H I - D K E T M N I H H T K H H N T Y V T N L N A A L E G H P D L Q
 8: A F T L P D L P Y A H D A L A A L G M M K E T M E Y H H D I H H K A Y V D N G N K L I A G T
 9: K V T L P D L K W D F G A L E P Y I S G Q I N E L H Y T
10: K H S L P D L P Y D Y G A L E P H I - N A Q I M Q L H X S K
```

Fig. 12. Partial sequences of Fe- and Mn-superoxide dismutases. (1) *Chlorobium thiosulphatophilum* Fe–SOD; (2) *Chromatium vinosum* Fe–SOD; (3) *Spirulina platensis* Fe–SOD; (4) *Pseudomonas ovalis* Fe–SOD; (5) *E. coli* Fe–SOD; (6) *E. coli* Mn–SOD; (7) *Bacillus stearothermophilus* Mn–SOD; (8) *Rhodopseudomonas sphaeroides* Mn–SOD; (9) *S. cerevisiae* Mn–SOD; (10) Human liver mitochondrial Mn–SOD.

were demonstrated. Thus enzymes from anaerobes did not seem to be a particularly related group and were not more closely related to each other than to enzymes from aerobes. Two histidine residues are conserved in all Mn and Fe superoxide dismutases and secondary structure predictions suggest that they are in close proximity in the same α-helix; these histidines may provide ligands for the bound metal. In spite of these similarities in primary and secondary structures of the Mn and Fe enzymes one should expect subtle differences in the metal-binding sites of the two classes of proteins to account for the non-interchangeability of the metal atoms in the reconstitution of active holo-enzymes. Further the Mn and Fe enzymes from the same organism have been immunologically distinguished.

Superoxide dismutase in the study of evolution

Evolutionary schemes based on the phylogenetic distribution of superoxide dismutases have been proposed by Lumsden & Hall (1975), Fridovich (1976), Hall, Lumsden & Tel-Or (1977) and Asada, Kanematsu, Okada & Hayakawa (1980). The cyanide-sensitive Cu/Zn enzyme is present in eukaryotic cytoplasm. The properties of this enzyme have been resistant to evolutionary changes and the enzymes isolated from fungi, plants, birds and mammals are indistinguishable in properties. The Cu/Zn enzyme was found to be absent in all prokaryotic bacteria and algae examined so far except in the luminescent bacterium *Photobacterium leiognathi* (Puget & Michelson, 1974) and in *Paracoccus denitrificans* (Vignais, Henry, Terech & Chabert, 1980). *Photobacterium leiognathi* is a symbiont with pony-fish and gene transfer from the host fish has been speculated to explain the anomalous occurrence of the Cu/Zn enzyme in this organism (Puget, Lavelle & Michelson, 1977). There is no general agreement regarding the occurrence of Cu/Zn superoxide dismutase in the green algae (Chlorophyceae). Thus Asada, Kanematsu & Uchida (1977) surveyed the superoxide dismutase activity in extracts from photosynthetic bacteria, prokaryotic algae (cyanobacteria) and eukaryotic algae (red, green and brown algae, diatoms, *Euglena* and Charophyta) and found that the activity in all these organisms was insensitive both to cyanide and to the antibody against spinach Cu/Zn superoxide dismutase. They concluded that these organisms lack the Cu/Zn enzyme and contain the Fe or Mn enzymes or both. However Henry & Hall (1977) investigated the occurrence and nature of superoxide dismutase in a number of chlorophycean algae and noted the presence of cyanide-sensitive activity in *Chara*, *Nitella* and *Spirogyra*. These authors suggest that the Charales and Conjugales represented by the above species are more closely related to the higher plants than the remainder of the chlorophycean algae and that this would account for the presence of the cyanide-sensitive enzyme in these algae.

The Fe superoxide dismutase was until recently thought to be restricted to prokaryotes and eukaryotic algae. However a recent report of its presence in *Brassica campestris* (Salin & Bridges, 1980) may mean that it also occurs in higher plants. The Mn enzyme occurs in prokaryotes and eukaryotic algae and also in the matrix of mitochondria. The properties of the mitochondrial and prokaryotic Mn enzymes are strikingly similar except that the mitochondrial enzyme contains four subunits and hence has double the molecular weight of the prokaryotic Mn enzyme which contains two identical subunits per molecule. The striking similarities in properties and the high degree of homology in sequences between the Fe and Mn superoxide dismutases (Fig. 12) point to a common ancestor for these two proteins. By the same criterion the Cu/Zn enzyme which bears no sequence homology to the Fe and Mn enzymes would have had an independent origin.

The questions to be answered are: (1) if the sole function of superoxide dismutase in cells is to scavenge the O_2^- formed during oxygen metabolism, what is the role of the enzyme in anaerobic bacteria? (2) Why should there be three different types of enzymes in the cells performing the same type of reaction? To explain the occurrence of superoxide dismutase in strict anaerobes it has been suggested that these cells are occasionally exposed to oxygen and the enzyme is a defence against the potentially catastrophic results of such exposure. Alternatively, the enzyme might be present to protect against O_2^- produced by photolysis. If this is so, the superoxide dismutases probably evolved before there was oxygen in the atmosphere. There were undoubtedly traces of O_2 present in the biosphere during the early days of evolution of life as a result of u.v.-induced photolysis of water (Urey effect) and the effects of ionizing radiation. *Micrococcus radiodurans* contains more superoxide dismutase than most other bacteria (McCord, Keele & Fridovich, 1971). The Fe-enzyme found in many strict anaerobes would have evolved to protect the organisms against this oxygen damage. Even organisms which lack catalase or peroxidase contain superoxide dismutase. However, *Neisseria gonorrhoeae* has no superoxide dismutase but a very high content of catalase.

It should also be remembered that the superoxide dismutase is induced in organisms to cope with environmental changes. Thus in *Escherichia coli* B, the addition of Paraquat caused a pronounced increase in the rate of biosynthesis of Mn superoxide dismutase. Cells with increased content of the Mn enzyme showed increased resistance to oxygen toxicity (Hassan & Fridovich, 1977). Moreover, *E. coli* grown in the absence of oxygen are devoid of the Mn enzyme; exposure of the cells to oxygen induces rapid synthesis of the Mn enzyme. The synthesis of superoxide dismutase could be induced by hyperbaric oxygen to the extent that the resultant strain of bacteria could survive 20 atmospheres of oxygen.

Lumsden & Hall (1975) have suggested a common origin for the Mn dependent oxygen-evolving system of cyanobacterial and plant photosynthesis and the Mn superoxide dismutase. Thus the ancestor of these enzymes in the membrane-bound form would have acted as a catalyst in the photoinduced water-splitting reaction of primitive algae and in soluble form would have acted as a scavenger of superoxide formed during the photosynthetic electron transport. Evidence for this hypothesis may be obtained when the oxygen-evolving enzyme is isolated and characterized.

From an extensive study of the distribution of superoxide dismutase in all types of photosynthetic organisms Asada *et al.* (1977) propose that the Cu/Zn enzyme would have been acquired first by ferns or mosses, land plants appearing in the late Silurian age, 4.2×10^8 years ago. The plants evolving after this period all contain both the cyanide-sensitive and the cyanide-insensitive enzymes.

Osterberg (1974) has proposed that prior to the accumulation of oxygen in the atmosphere by photosynthesis copper occurred in the form of insoluble copper sulphide whereas iron and manganese were available in soluble form for assimilation by organisms. This unavailability of soluble copper may explain the absence of the Cu/Zn superoxide dismutase. The blue copper protein plastocyanin occurs in the photosynthetic electron transfer chains of present-day cyanobacteria and algae, even though their oxygen-evolving type of photosynthesis was presumably evolved at a time when the environment was anaerobic. However in some cyanobacteria and algae plastocyanin can be interchanged with the soluble cytochrome c-553 (Wood, 1978; Bohner, Merkle, Kroneck & Böger, 1980). Thus it is not improbable that primitive algae would have functioned without any copper proteins at all.

The presence of Mn-superoxide dismutase in prokaryotes and in the mitochondrial matrix lends support to the endosymbiotic theory of mitochondrial origin. Amino-terminal sequences of the mitochondrial dismutases are highly homologous to the Mn- and Fe-superoxide dismutases of prokaryotes, but nonhomologous with the sequences of the Cu/Zn dismutase of eukaryotic cytosol (Walker et al., 1977). The recent discovery of Mn and Cu/Zn superoxide dismutases in *Paracoccus denitrificans* (P. M. Vignais, personal communication) also is in agreement with this hypothesis. John & Whatley (1975) have observed that *P. denitrificans* shares many morphological and metabolic features with mitochondria and have proposed that both *P. denitrificans* and mitochondria might have evolved from a common ancestral aerobic bacterium.

DISCUSSION

Primary and tertiary structures of functionally identical proteins from different species allow classification of these species (Dickerson, 1980; Wilson, Carlson & White, 1977). Though protein sequences alone cannot provide enough phylogenetic information, taken together with similar studies on nucleic acids (Stackebrandt & Woese, this volume), morphological characters and fossil records, protein sequences are valuable for determining evolutionary trends. Current evolutionary theory assumes that genes are undergoing mutations, rearrangements and duplications (Cullum & Saedler, this volume) and that these mutational changes are reflected in the protein sequences. Comparison of sequences from species whose times of evolutionary divergence are known from the fossil record and other data allows an estimate of the rate at which mutations are occurring in specific proteins and their corresponding genes. These studies have shown that the rate of change of sequence, as determined by amino-acid replacements, is almost constant for a particular protein and that different proteins evolve at different rates. As sequence evolution is assumed to be mainly divergent, it is possible to construct evolutionary trees from sequences of homologous proteins, utilizing the basic assumption that every living organism, structure or function had ancestors similar to itself, but simpler (Eck & Dayhoff, 1966).

When the first amino-acid sequence of a ferredoxin was determined by Tanaka et al. (1966), Eck & Dayhoff (1966) proposed that this ferredoxin or its prototype may have participated in the metabolism of organisms during an early stage of cellular evolution. This hypothesis was elaborated by Hall, Cammack & Rao (1971) who noted the correlation between the amino-acid composition of clostridial ferredoxins and of meteorites and lunar soils. The observations and assumptions which favour the hypothesis that a ferredoxin-type molecule was functioning in the 'primitive cells' are the following.

1. It is generally believed that the earth's atmosphere was anoxygenic in the early periods of its formation. Organisms which lived in this period should have had an anaerobic type of metabolism as is found in present-day obligate fermentative bacteria. Ferredoxins are essential catalysts in the present-day clostridia; they catalyse oxido-reductions at a potential near to that of the hydrogen electrode.

2. The clostridial ferredoxin molecule contains only 55 amino acids and the latter half of this molecule is homologous with the initial half suggesting gene duplication. The ferredoxin molecule is so small it could be coded by a molecule of the size of transfer RNA.

3. The *Clostridium butyricum* ferredoxin molecule is composed of only 11 different amino acids and a very high proportion of these are the smaller, thermodynamically more stable amino acids. These are the same amino acids which can be synthesized from compounds such as CH_4, NH_3, HCN and HCHO under conditions which simulate the primitive environment. Also, these are the major amino acids detected in the Murray and Murchison meteorites.

4. The active centre of clostridial-type ferredoxins contains an

[4Fe–4S] cluster. Inorganic electron carriers in the form of FeS_2 and $Fe_{0.86}S$ would have been present where the first proteins evolved (Osterberg, 1974). As already noted, polypeptides containing cysteine residues can react spontaneously with these iron sulphides in a reducing atmosphere, forming clusters similar to those present in ferredoxins.

Therefore, unlike the complex prosthetic groups of cytochromes and flavoproteins, iron–sulphur clusters can be produced in a simple non-enzymic process. Many of the synthetic clusters are catalytically as active as the clusters in native ferredoxins. Thus the water-soluble clusters $[Fe_4S_4\ (Ac\ .\ Gly_2\ .\ Cys\ .\ Gly_2\ .\ Cys\ .\ Gly_2\ NH_2)_2]^{2-}$ and $[Fe_4S_4(S\ .\ CH_2\ .\ CH_2\ OH)_4]^{2-}$ can transfer electrons from sodium dithionite to *C. pasteurianum* hydrogenase in a hydrogen-evolving system (Adams *et al.*, 1977). Moreover, an iron–sulphur molybdenum cluster $[Fe_6\ Mo_2\ S_8\ (SCH_2\ CH_2\ OH)_9]^{3-}$ mediated transfer of electrons from illuminated chloroplasts to *C. pasteurianum* hydrogenase or platinum to produce hydrogen (Adams *et al.*, 1980). It is interesting to note in this context that 'Jeevanu' particles prepared by exposing a mixture of mineral salts, transition metal ions and formaldehyde to sunlight (Bahadur, 1975) can also substitute for ferredoxin in biological reactions (Rao *et al.*, 1980). Thus small molecules containing the ferredoxin-type clusters could have acted as electron and hydrogen transfer agents in early cell metabolism.

Evolution of ferredoxins studied by amino-acid sequences

Since ferredoxins are small, coloured, acidic proteins they are comparatively easy to extract and purify. Extensive studies of the primary structures of ferredoxins have been carried out to determine the structure–function and evolutionary relationship of these proteins (Yasunobu & Tanaka, 1973; Rao & Hall, 1977; Matsubara *et al.*, 1980). Ferredoxins from *Clostridia* and other fermentative bacteria with two [4Fe–4S] clusters are highly homologous with 19 out of 56 amino-acid residues invariant – the eight cysteines in particular occupying invariant positions in the protein chain. As already mentioned the three ferredoxins so far sequenced from green sulphur bacteria are slightly longer (61 amino-acid residues) yet contain the same arrangement of cysteine residues in the molecule as is found in the clostridial ferredoxins. The ferredoxin molecule from the purple sulphur bacteria, *Chromatium vinosum*, contains 82 amino-acid residues; however, the cysteine distribution in this protein also is similar to that of clostridial ferredoxins which allows it to have the same type of cluster binding. The mid point potential of this ferredoxin (-490 mV) is more negative than the other eight iron ferredoxins. We unfortunately lack the sequence of any ferredoxins from the *Rhodospirillaceae* which would be valuable in constructing an evolutionary progression of bacterial ferredoxins.

The ferredoxin sequences from blue-green algae and eukaryotic algae are remarkably similar and are closely homologous. The sequences are most useful for taxonomic purposes when used in conjunction with the sequences of other proteins and nucleic acids and with other data. It is noteworthy that there is no clear-cut distinction between the cyanobacterial (blue-green algal) ferredoxin sequences and those of algae and plants, even though in the latter the protein is synthesized on cytoplasmic ribosomes with information from nuclear DNA.

Evolution of the superoxide dismutases

The presence or absence of a particular type of superoxide dismutase can no longer be taken as a rigid criterion for the evolutionary status of an organism. Generally, but not without exceptions, iron and manganese-containing superoxide dismutases are characteristic of prokaryotes. Gram-positive bacteria were most frequently found to contain only the Mn-enzyme while Gram-negative species usually contained both Fe and Mn enzymes. The Cu/Zn enzyme is mainly found in the cytosols of eukaryotes while the predominant enzyme in the mitochondrial matrix is of the manganese type. The similarities in sequences and properties of the bacterial and mitochondrial Mn superoxide dismutase supports the theory that the mitochondria evolved from an endocellular symbiosis between a prokaryote and protoeukaryote.

CONCLUSION

In this article we have discussed the occurrence, properties and interrelationships of two types of metalloproteins, the ferredoxins and superoxide dismutases. Both types of protein contain transition metal ions at the active centre and in both cases the active centre can be removed and the native protein reconstituted by the addition of the respective inorganic elements to the apoprotein. The ferredoxins occur in all types of organisms, evolutionary differences being manifested in the types of iron–sulphur clusters and protein chain length. These two characteristics control the midpoint potentials and reactivity of various ferredoxins. Superoxide dismutases are found in most organisms so far examined, the type of enzyme occurring in an organism depending to a large extent on its evolutionary history.

The encouragement and advice of Professor David Hall in the preparation of this article is gratefully acknowledged.

REFERENCES

ADAMS, M. W. W., RAO, K. K., HALL, D. O., CHRISTOU, G. & GARNER, C. D. (1980). Biological activity of synthetic molybedenum–iron–sulphur, iron–sulphur and iron–selenium analogues of ferredoxin-type centres. *Biochimica Biophysica Acta,* **589**, 1–9.

ADAMS, M. W. W., REEVES, S. G., HALL, D. O., CHRISTOU, G., RIDGE, B. & RYDON, H. N. (1977). Biological activity of synthetic tetranuclear iron–sulphur analogues of the active sites of ferredoxins. *Biochemical and Biophysical Research Communications,* **79**, 1184–91.

ADMAN, E. T. (1979). A comparison of the structures of electron transfer proteins. *Biochimica Biophysica Acta,* **549**, 107–44.

ADAM, E. T., SIEKER, L. C. & JENSEN, L. H. (1973). The structure of a bacterial ferredoxin. *Journal of Biological Chemistry,* **248**, 3987–96.

AMBLER, R. P., DANIEL, M., HERMOSO, J., MEYER, T. E., BARTSCH, R. G. & KAMEN, M. D. (1979). Cytochrome c_2 sequence variation among the recognised species of purple nonsulphur photosynthetic bacteria. *Nature, London,* **278**, 659–60.

ASADA, K., KANEMATSU, S., OKADA, S. & HAYAKAWA, T. (1980). Phylogenetic distribution of superoxide dismutase in organisms and cell organelles. In *Chemical and Biochemical Aspects of Superoxide and Superoxide Dismutase*, ed. J. V. Bannister & H. A. O. Hill, pp. 136–53. Amsterdam: Elsevier/North Holland.

ASADA, K., KANEMATSU, S. & UCHIDA, K. (1977). Superoxide dismutase in photosynthetic organisms: absence of the cuprozinc enzyme in eukaryotic algae. *Archives of Biochemistry Biophysics,* **179**, 243–56.

ASADA, K., YOSHIKAWA, K., TAKAHADI, M., MAEDA, Y. & ENMAYI, K. (1975). Superoxide dismutase from a blue-green alga, *Plectonema boryanum*. *Journal of Biological Chemistry,* **250**, 2801–7.

BAHADUR, K. (1975). Photochemical formation of self-sustaining coacervates. *Zentralblatt für Bakteriologie, Abt. II.* **130**, S.211–18.

BANNISTER, J. V. & HILL, H. A. O., eds. (1980). *Chemical and Biochemical Aspects of Superoxide and Superoxide Dismutase.* Amsterdam: Elsevier/North Holland.

BARTSCH, R. G. (1963). Non-heme iron proteins and *Chromatium* iron protein. In *Bacterial Photosynthesis*, ed. H. Gest, A. San Pietro & L. P. Vernon, pp. 315–26. Yellow Springs, Ohio: Antioch Press.

BOHNER, H., MERKLE, H., KRONECK, P. & BÖGER, P. (1980). High variability of the electron carrier plastocyanin in microalgae. *European Journal of Biochemistry*, **105**, 603–9.

BROCK, C. J. & WALKER, J. E. (1980). Superoxide dismutase from *Bacillus stearothermophilus*. Complete amino-acid sequence of a manganese enzyme. *Biochemistry*, **19**, 2873–82.

BRUSCHI, M. (1979). Amino-acid sequence of *Desulfovibrio gigas* ferredoxin: revisions. *Biochemical and Biophysical Research Communications*, **91**, 623–8.

BUCHANAN, B. B., WOLOSIUK, R. A. & SCHÜRMANN, P. (1979). Thioredoxin and enzyme regulation. *Trends in Biochemical Sciences*, **4**, 93–6.

CAMMACK, R. (1973). 'Super-reduction' of *Chromatium* high-potential iron-sulphur protein in the presence of dimethylsulphoxide. *Biochemical and Biophysical Research Communications*, **54**, 548–54.

CAMMACK, R. (1975). Effects of solvent on the properties of ferredoxins. *Biochemical Society Transactions*, **3**, 482–8.

CAMMACK, R. (1979). Functional aspects of iron–sulphur proteins. In *Metalloproteins: Structure, Function and Clinical Aspects*, ed. U. Weser, pp. 162–84. Stuttgart: Thieme Verlag.

CAMMACK, R., DICKSON, D. P. E. & JOHNSON, C. E. (1977a). Evidence from Mössbauer spectroscopy and magnetic resonance on the active centres of the iron–sulfur proteins. In *Iron–sulphur proteins*, Vol. 3, ed. W. Lovenberg. pp. 283–330. London: Academic Press.

CAMMACK, R., RAO, K. K., HALL, D. O., MOURA, J. J. G., XAVIER, A. V., BRUSCHI, M., LE GALL, J., DEVILLE, A. & GAYDA, J. P. (1977b). Spectroscopic studies of the oxidation-reduction properties of three forms of ferredoxin from *Desulfovibrio gigas*. *Biochimica Biophysica Acta*, **490**, 311–21.

CARTER, C. W. JR. (1977). X-ray analysis of high-potential iron–sulphur proteins and ferredoxins. In *Iron-Sulphur Proteins*. Vol. 3. ed. W. Lovenberg. pp. 158–205. London: Academic Press.

CARTER, C. W. JR., KRAUT, J., FREER, S. T., ALDEN, R. A., SIEKER, L. C., ADMAN, E. & JENSEN, L. H. (1972). A comparison of Fe_4S_4 clusters in high potential iron protein and in ferredoxin. *Proceedings of the National Academy of Sciences, USA*, **69**, 3526–9.

CARTER, K. R., RAWLINGS, J., ORME-JOHNSON, W. A., BECKER, R. R. & EVANS, H. J. (1980). Purification and properties of a ferredoxin from *Rhizobium japonicum* bacteroids. *Journal of Biological Chemistry*, **255**, 4213–23.

DICKERSON, R. E. (1980). Evolution and gene transfer in purple photosynthetic bacteria. *Nature London*, **283**, 210–12.

DICKERSON, R. E., TIMKOVICH, R, & AMASSY, R. J. (1976). The cytochrome fold and the evolution of bacterial energy metabolism. *Journal of Molecular Biology*, **100**, 473–91.

ECK, R. V. & DAYHOFF, M. O. (1966). Evolution of the structure of ferredoxin based on living relics of primitive amino acid sequences. *Science*, **152**, 363–6.

EMPTAGE, M. H., KENT, T. A., HUYNH, B. H., RAWLINGS, J., ORME-JOHNSON, W. H. & MUNCK, E. (1980). On the nature of the iron–sulfur centres in a ferredoxin from *Azotobacter vinelandii*. *Journal of Biological Chemistry*, **255**, 1793–6.

ESTABROOK, W. E., SUZUKI, K., MASON, J. I., BARON, J., TAYLOR, W. E., SIMPSON, E. R., PURVIS, J. & MCCARTHY, J. (1973). Adrenodoxin: an iron–sulfur protein of adrenal cortex mitochondria. In *Iron-sulphur Proteins*. Vol. 1, ed. W. Lovenberg, pp. 193–224. New York: Academic Press.

EVANS, M. C. W., BUCHANAN, B. B. & ARNON, D. A. (1966). A new ferredoxin-dependent carbon reduction cycle in a photosynthetic bacterium. *Proceedings of the National Academy of Sciences, USA,* **55,** 928–34.

EVANS, M. C. W., LORD, A. V. & REEVES, S. G. (1974). The detection and characterization by electron paramagnetic resonance spectroscopy of iron–sulphur proteins and other electron transport components in chromatophores from the purple bacterium *Chromatium. Biochemical Journal,* **138,** 177–83.

FOYER, C. H. & HALL, D. O. (1980). Oxygen metabolism in the active chloroplast. *Trends in Biochemical Sciences,* **5,** 181–91.

FRIDOVICH, I. (1975). Superoxide dismutases. *Annual Review of Biochemistry,* **44,** 147–59.

FRIDOVICH, I. (1976). Oxygen radicals, hydrogen peroxide, and oxygen toxicity. In *Free Radicals in Biology,* ed. W. A. Pryor, pp. 239–77. New York: Academic Press.

FUKUMORI, Y. & YAMANAKA, T. (1979). A high-potential non heme iron protein-linked thiosulfate oxidising enzyme derived from *Chromatium vinosum. Current Microbiology,* **3,** 117–20.

FUKUYAMA, K., HASE, T., MATSUMOTO, S., TSUKIHARA, T., KATSUBE, Y., TANAKA, N., KAKUDO, M., WADA, K. & MATSUBARA, H. (1980). Structure of *S. platensis* [2Fe–2S] ferredoxin and evolution of chloroplast-type ferredoxins. *Nature, London,* **286,** 522–3.

GUNTER, M., RIDGE, B., RYDON, H. N. & SHARPE, R. (1979). Toward a synthesis of *Clostridium butyricum* apoferredoxin: two tetradecapeptides comprising half the sequence (Residues 7–20 and 21–34). *Bioinorganic Chemistry,* **8,** 371–89.

HALL, D. O. (1977). Iron–sulphur proteins and superoxide dismutases in the biology and evolution of electron transport. In *Advances in Chemistry Series,* ed. K. N. Raymond, No. 162. Bioinorganic Chemistry II, pp. 227–49. American Chemical Society.

HALL, D. O., CAMMACK, R. & RAO, K. K. (1971). Role for ferredoxins in the origin of life and biological evolution. *Nature, London,* **233,** 136–8.

HALL, D. O., CAMMACK, R. & RAO, K. K. (1974a). Non-haem iron proteins. In *Iron in Biochemistry and Medicine,* ed. A. Jacobs & M. Worwood, pp. 279–335. London: Academic Press.

HALL, D. O., CAMMACK, R. & RAO, K. K. (1974b). The iron–sulphur proteins: evolution of a ubiquitous protein from model systems to higher organisms. *Origins of Life,* **5,** 363–86.

HALL, D. O., LUMSDEN, J. & TEL-OR, E. (1977). Iron–sulphur proteins and superoxide dismutase in the evolution of photosynthetic bacteria and algae. In *Chemical Evolution of the Early Pre-Cambrian,* ed. C. Ponnamperuma, pp. 191–210. New York: Academic Press.

HALL, D. O., RAO, K. K. & CAMMACK, R. (1975). The iron–sulphur proteins: structure, function and evolution of a ubiquitous group of proteins. *Science Progress,* **62,** 284–317.

HARRIS, J. I., AUFFRET, A. D., NORTHROP, F. D. & WALKER, J. E. (1980). Structural comparisons of superoxide dismutases. *European Journal of Biochemistry,* **106,** 297–303.

HASE, T., MATSUBARA, H., BEN-AMOTZ, A., RAO, K. K. & HALL, D. O. (1980). Purification and sequence determination of two ferredoxins from *Dunaliella salina. Phytochemistry,* **19,** 2065–70.

HASE, T., MATSUBARA, H. & EVANS, M. C. W. (1977). Amino-acid sequence of *Chromatium vinosum* ferredoxin: revisions. *Journal of Biochemistry,* **81,** 1745–9.

HASE, T., WAKABAYASHI, S., MATSUBARA, H., EVANS, M. C. W. & JENNINGS, J. V.

(1978a). Amino acid sequence of a ferredoxin from *Chlorobium thiosulfatophilum* strain Tassajara. *Journal of Biochemistry*, **83**, 1321–5.

HASE, T., WAKABAYASHI, S., MATSUBARA, H., KERSCHER, L., OESTERHELT, D., RAO, K. K. & HALL, D. O. (1978b). Complete amino acid sequence of *Halobacterium halobium* ferredoxin containing N-acetyl lysine residue. *Journal of Biochemistry*, **83**, 1659–90.

HASE, T., WAKABAYSHI, S., MATSUBARA, H., IMAI, T., MATSUMOTO, T. & TOBARI, J. (1979). *Mycobacterium smegmatis* ferredoxin: a unique distribution of cysteine residues constructing iron–sulphur cluster. *FEBS Letters*, **103**, 224–8.

HASE, T., WAKABAYASHI, S., MATSUBARA, H., OHMORI, D. & SUZUKI, K. (1978c). *Pseudomonas ovalis* ferredoxin: similarity to *Azotobacter* and *Chromatium* ferredoxins. *FEBS Letters*, **91**, 315–19.

HASSAN, M. H. & FRIDOVITCH, I. (1977). Regulation of the synthesis of superoxide dismutase in *Escherichia coli*. *Journal of Biological Chemistry*, **252**, 7667–72.

HASSAN, M. H. & FRIDOVITCH, I. (1979). Paraquat and *Escherichia coli*. *Journal of Biological Chemistry*, **254**, 10846–52.

HENRY, L. E. A. & HALL, D. O. (1977). Superoxide dismutases in green algae: An evolutionary survey. *Photosynthetic organelles. Special issue of Plant and Cell Physiology*, 377–82.

HEWITT, J. & MORRIS, J. G. (1975). Superoxide dismutase in some obligately anaerobic bacteria. *FEBS Letters*, **55**, 282–5.

HILL, C. L., RENAUD, J., HOLM, R. H. & MORTENSON, L. E. (1977). Synthetic analogues of the active sites of iron–sulphur proteins. 15. Comparative polarographic potentials of the $[Fe_4S_4(SR)_4]^{2-, 3-}$ *Clostridium pasteurianum* ferredoxin redox couples. *Journal of American Chemical Society*, **99**, 2549–57.

HILL, H. A. O. (1978). The superoxide ion and the toxicity of molecular oxygen. In *New Trends in Bioinorganic Chemistry*. ed. R. J. P. Williams & J. R. R. F. Da Silva, pp. 173–208. London: Academic Press.

HOLM, R. H. (1977). Synthetic approaches to the active sites of iron–sulfur proteins. *Accounts of Chemical Research*, **10**, 427–34.

HUTSON, G. K., ROGERS, L. J., HASLETT, B. G., BOULTER, D. & CAMMACK, R. (1978). Comparative studies on two ferredoxins from the cyanobacterium *Nostoc* strain MAC. *Biochemical Journal*, **172**, 465–77.

HUYNH, B. H., MOURA, J. J. G., MOURA, I., KENT, T. A., LE GALL, J., XAVIER, A. V. & MUNCK, E. (1980). Evidence for a three-iron centre in a ferredoxin from *Desulfovibrio gigas*. *Journal of Biological Chemistry*, **255**, 3242–4.

JOHANSEN, J. T., OVERBALLE-PETERSEN, C., MARTIN, B., HASEMANN, V. & SVENDSEN, I. B. (1979). The complete amino acid sequence of copper, zinc superoxide dismutase from *Saccharomyces cerevisiae*. *Carlsberg Research Communications*, **44**, 201–17.

JOHN, P. & WHATLEY, F. R. (1977). The bioenergetics of *Paracoccus denitrificans*. *Biochimica Biophysica Acta*, **463**, 129–53.

KERSCHER, L. & OESTERHELT, D. (1977). Ferredoxin is the coenzyme of α-ketoacid oxidoreductases in *Halobacterium halobium*. *FEBS Letters*, **83**, 197–201.

KERSCHER, L., OESTERHELT, D., CAMMACK, R. & HALL, D. O. (1976). A new plant-type ferredoxin from halobacteria. *European Journal of Biochemistry*, **71**, 101–7.

KIRBY, T., BLUM, J., KAHANE, I. & FRIDOVITCH, I. (1980). Distinguishing between Mn-containing and Fe-containing superoxide dismutases in crude extracts of cells. *Archives of Biochemistry and Biophysics*, **201**, 551–5.

LANYI, J. K. (1980). Physical chemistry and evolution of salt tolerance in halobacteria. *Origins of Life*, **10**, 161–7.

LUMSDEN, J. & HALL, D. O. (1975). Superoxide dismutase in photosynthetic organisms provides an evolutionary hypothesis. *Nature, London*, **257**, 670–2.

MAGRUM, L. J., LUEHRSEN, K. R. & WOESE, C. R. (1978). Are extreme halophiles actually 'Bacteria'? *Journal of Molecular Evolution*, **11**, 1–8.

MATSUBARA, H., HASE, T., WAKABAYASHI, S. & WADA, K. (1980). Structure and function of chloroplast- and bacterial-type of ferredoxins. In *Evolution of Protein Structure and Function*, ed. D. S. Sigman & M. A. B. Brazier. U.C.L.A. Forum in Medical Science. Vol. 21 (in press). New York: Academic Press.

MCCORD, J. M. & FRIDOVICH, I. (1968). The reduction of cytochrome c by milk xanthine oxidase. *Journal of Biological Chemistry*, **243**, 5743–60.

MCCORD, J. M. & FRIDOVICH, I. (1969). An enzymic function for erythrocuprein *Journal of Biological Chemistry*, **244**, 6049–55.

MCCORD, J. M., KEELE, JR. B. B. & FRIDOVICH, I. (1971). An enzyme-based theory of obligate anaerobiosis: the physiological function of superoxide dismutase. *Proceedings of the National Academy of Sciences, USA*, **68**, 1024–7.

MCLACHLAN, A. D. (1980). Repeated folding pattern in copper-zinc superoxide dismutase. *Nature, London*, **285**, 267–8.

MICHELSON, A. M., MCCORD, J. M. & FRIDOVICH, I. ED. (1977). *Superoxide and Superoxide Dismutases*. London: Academic Press.

MORTENSON, L. E., VALENTINE, R. C. & CARNAHAN, J. E. (1962). An electron transport factor from *Clostridium pasteurianum*. *Biochemical and Biophysical Research Communications*, **7**, 448–54.

MOURA, J. J. G., XAVIER, A. V., HATCHIKIAN, E. C. & LE GALL, J. (1978). Structural control of the redox potentials and of the physiological activity by oligomerization of ferredoxin. *FEBS Letters*, **89**, 177–9.

NORROD, P. & MORSE, S. A. (1979). Absence of superoxide dismutase in some strains of *Neisseria gonorrhoeae*. *Biochemical and Biophysical Research Communications*, **90**, 1287–94.

OHNISHI, T., BLUM, H., SATO, S., NAKAZAWA, K., HON-NAMI, K. & OSHIMA, T. (1980). A stable *Thermus thermophilus* iron–sulfur protein containing only one binuclear and one tetrameric cluster. *Journal of Biological Chemistry*, **255**, 345–8.

ORME–JOHNSON, W. H. (1972). Tryptic cleavage of *Clostridium acidi-urici* apoferredoxin and reconstitution of the separated fragments. *Biochemical Society Transactions*, **1**, 30–31.

ORME-JOHNSON, W. H. & BEINERT, H. (1969). On the formation of superoxide anion radical during the reaction of reduced iron–sulfur proteins with oxygen. *Biochemical and Biophysical Research Communications*, **36**, 905.

ORME-JOHNSON, W. H. & HOLM, R. H. (1978). Identification of iron–sulfur clusters in proteins. *Methods in Enzymology*, **53**, 268–76.

ORME–JOHNSON, W. H. & SANDS, R. H. (1973). Probing iron–sulfur proteins with EPR and ENDOR spectroscopy. In *Iron–sulfur Proteins* Vol. 2, ed. W. Lovenberg, pp. 195–238. New York: Academic Press.

OSTERBERG, R. (1974). Origins of metal ions in biology. *Nature, London*, **249**, 382–3.

PALMER, G. (1973). Current insights into the active centre of spinach ferredoxin and other iron–sulfur proteins. In *Iron–sulfur Proteins*. Vol. 2, ed. W. Lovenberg, pp. 285–325. London: Academic Press.

PALMER, G. (1975). Iron–sulfur proteins. In *The Enzymes*. Vol. 12, part B, ed. P. D. Boyer, pp. 2–56.

PASTERNAK, R. F. & HALLIWELL, B. (1979). Superoxide dismutase activities of an iron porphyrin and other iron complexes. *Journal of the American Chemical Society*, **101**, 1026–31.

PROBST, I., MOURA, J. J. G., MOURA, I., BRUSCHI, M. & LE GALL, J. (1980). Isolation and characterisation of a rubredoxin and an [8Fe–8S] ferredoxin from *Desulfuromonas acetoxydans*. *Biochimica Biophysica Acta,* **502,** 38–44.

PUGET, K., LAVELLE, F. & MICHELSON, A. M. (1977). Superoxide dismutase from procaryote and eukaryote bioluminescent organisms. In *Superoxide and Superoxide Dismutases,* ed. J. M. Michelson, J. McCord & I. Fridovich, pp. 139–150. London: Academic Press.

PUGET, K. & MICHELSON, A. M. (1974). Isolation of a new copper-containing superoxide dismutase: bacteriocuprin. *Biochemical and Biophysical Research Communications,* **58,** 830–8.

QUE, L., ANGLIN, J. R., BOBRICK, M. A., DAVISON, A. & HOLM, R. H. (1974). Synthetic analogs of the active sites of iron–sulfur proteins. IX. Formation and some electronic reactivity properties of Fe_4S_4 Glycyl-L-cysteinyl-glycyl oligopeptide complexes obtained by ligand substitution reactions. *Journal of the American Chemical Society,* **96,** 6042–8.

RAO, K. K., ADAMS, M. W. W., MORRIS, P., HALL, D. O., RANGANAYAKI, S. & BAHADUR, K. (1980). Biophotolysis of water for H_2 production via natural and artificial catalytic systems. In *Proceedings of the International Symposium on Biological Applications of Solar Energy.* ed. A. Gnanam, S. Krishnaswamy & J. S. Kahn. pp. 199–204. Madras: Macmillan, India.

RAO, K. K. & HALL, D. O. (1977). Chemistry and evolution of ferredoxins and hydrogenases. In *The Evolution of Metalloenzymes, Metalloproteins and Related Materials.* ed. G. J. Leigh, pp. 39–64. London: Symposium Press.

RAO, K. K., HALL, D. O. & CAMMACK R. (1981). The photosynthetic apparatus. In *Biochemical Evolution,* ed. H. Gutfreund. Cambridge University Press.

RICHARDSON, D. C. (1977). The three-dimensional structure of Cu/Zn superoxide dismutase. In *Superoxide and Superoxide Dismutases,* ed. A. M. Michelson, J. M. McCord & I. Fridovich. pp. 217–23. London: Academic Press.

SALIN, M. L. & BRIDGES, S. M. (1980). Isolation and characterization of an iron-containing superoxide dismutase from a eucaryote, *Brassica campestris*. *Archives of Biochemistry and Biophysics,* **201,** 369–74.

SCHERINGS, G., HAAKER, H. & VEEGER, C. (1977). Regulation of nitrogen fixation by Fe–S protein II in *Azotobacter vinelandii*. *European Journal of Biochemistry,* **77,** 621–30.

SCHNEIDER, K. & SCHLEGEL, H. (1981). Production of superoxide radicals by soluble hydrogenase from *Alcaligenes eutrophus* H16. *Biochemical Journal,* **193,** 99–107.

STEINMAN, H. M. (1978). The amino-acid sequence of mangano superoxide dismutase from *Escherichia coli* B. *Journal of Biological Chemistry,* **253,** 8708–20.

STEINMAN, H. M., NAIK, V. R., ABERNETHY, J. L. & HILL, R. L. (1974). Bovine erythrocyte superoxide dismutase. *Journal of Biological Chemistry,* **249,** 7326–38.

STELLWAGEN, E. (1978). Haem exposure as the determinant of oxidation-reduction potential in haem proteins. *Nature, London,* **275,** 73–4.

STEPHENS, P. J., THOMSON, A. J., DUNN, J. B. R., KEIDERLING, T. A., RAWLINGS, J., RAO, K. K. & HALL, D. O. (1978). Circular dichroism and magnetic circular dichroism of iron–sulphur proteins. *Biochemistry,* **17,** 4770–8.

STOUT, C. D., GHOSH, D., PATTABHI, V. & ROBBINS, A. H. (1980). Iron–sulfur clusters in *Azotobacter* ferredoxin at 2.5Å resolution. *Journal of Biological Chemistry,* **255,** 1797–800.

SWEENEY, W. V., BEARDEN, A. J. & RABINOWITZ, J. C. (1974). The electron paramagnetic resonance of oxidised *Clostridial* ferredoxins. *Biochemical and Biophysical Research Communications,* **59,** 188–94.

SWEENEY, W. V. & RABINOWITZ, J. C. (1980). Proteins containing 4Fe–4S clusters: an overview. *Annual Reviews of Biochemistry,* **49,** 139–61.

TANAKA, M., HANIU, M., YASUNOBU, K. T., EVANS, M. C. W. & RAO, K. K. (1975). The amino-acid sequence of ferredoxin 2 from *Chlorobium limicola. Biochemistry,* **14,** 1938–43.

TANAKA, M., NAKASHIMA, T., BENSON, A., MOWER, H. & YASUNOBU, K. T. (1966). The amino acid sequence of *Clostridium pasteurianum* ferredoxin. *Biochemistry,* **5,** 1666–80.

TEDRO, S. M., MEYER, T. E. & KAMEN, M. D. (1979). Primary structure of a high potential, four-iron–sulfur ferredoxin from the photosynthetic bacterium *Rhodospirillum tenue. Journal of Biological Chemistry,* **254,** 1495–500.

TEWES, F. J. & THAUER, R. K. (1979). Purification and properties of ferredoxin from *Ruminococcus albus. FEMS Microbiology Letters,* **6,** 375–7.

THOMPSON, C. L., JOHNSON, C. E., DICKSON, D. P. E., CAMMACK, R., HALL, D. O., WESER, U. & RAO, K. K. (1974). Mössbauer effect in the eight-iron ferredoxin from *Clostridium pasteurianum.* Evidence for the state of the iron atoms. *Biochemical Journal,* **139,** 97–103.

VIGNAIS, P. M., HENRY, M. F., TERECH, A. & CHABERT, J. (1980). Production of superoxide anion and superoxide dismutase in *Paracoccus denitrificans.* In *Chemical and Biochemical Aspects of Superoxide and Superoxide Dismutase,* ed. J. V. Bannister & H. A. O. Hill. pp. 154–9. Amsterdam: Elsevier/North Holland.

WALKER, J. E., AUFFRET, A. D., BROCK, C. J. & STEINMANN, H. M. (1977). Structural comparisons of superoxide dismutases. In *Chemical and Biochemical Aspects of Superoxide and Superoxide Dismutase,* ed. J. V. Bannister & H. A. O. Hill, pp. 212–22. Amsterdam: Elsevier/North Holland.

WALTER, M. R., BUICK, R. & DUNLOP, J. S. R. (1980). Stromatolites 3400–3500 M yr old from the North Pole area, Western Australia. *Nature, London,* **284,** 443–5.

WERBER, M. M. & MEVARECH, M. (1978). Induction of a dissimilatory reduction pathway of nitrate in *Halobacterium* of the Dead Sea. *Archives of Biochemistry and Biophysics,* **186,** 60–5.

WILSON, A. C., CARLSON, S. S. & WHITE, T. J. (1977). Biochemical evolution. *Annual Reviews of Biochemistry,* **46,** 573–639.

WOESE, C. R., GIBSON, J. & FOX, G. E. (1980). Do genealogical patterns in purple photosynthetic bacteria reflect interspecific gene transfer? *Nature, London,* **282,** 212–14.

WOOD, P. M. (1978). Interchangeable copper and iron proteins in algal photosynthesis. Studies on plastocyanin and cytochrome *c*-552 in *Chlamydomonas. European Journal of Biochemistry,* **87,** 9–19.

YASUNOBU, K. T. & TANAKA, M. (1973). The types, distribution in nature, structure-function and evolutionary data of the iron–sulfur proteins. In *Iron–sulfur Proteins,* ed. W. Lovenberg, Vol. 2, pp. 27–130. New York: Academic Press.

YOCH, D. C. & CARITHERS, R. P. (1979). Bacterial iron–sulfur proteins. *Microbiological Reviews,* **43,** 384–421.

YOCH, D. C., CARITHERS, R. P. & ARNON, D. I. (1977). Isolation and characterisation of bound iron–sulfur proteins from bacterial photosynthetic membranes. I. Ferredoxins III and IV from *Rhodospirillum rubrum* chromatophores. *Journal of Biological Chemistry,* **252,** 7553–60.

ZUBIETA, J. & DALTON, H. (1973). Ferredoxin from *Veillonella alcalescens. Biochimica Biophysica Acta,* **322,** 133–40.

THE EVOLUTION OF METABOLIC PATHWAYS

HANS KREBS

Metabolic Research Laboratory, Nuffield Department of Clinical Medicine, Radcliffe Infirmary, Oxford OX2 6HE, UK

INTRODUCTION

No attempt is made in this chapter to survey the huge literature on the evolution of metabolic pathways, as this subject has been covered adequately in recent years by other authors. Instead I will discuss the evolution of metabolic pathways on the basis of a somewhat novel approach which, I think, has not been sufficiently explored in the past and which is fruitful. In dealing with the question of why certain pathways have evolved, my starting point is one of the basic principles of evolution by natural selection. This states that in a competitive environment, the chances of survival are greatest if optimal use is made of resources; or, as Orgel & Crick (1980), have recently formulated it – replacing survival of the individual by survival of the genotype: 'The theory of natural selection, in its more general formulation, deals with the competition between replicating entities. It shows that, in such a competition, the more efficient replicators increase in number at the expense of their less efficient competitors. After a sufficient time, only the most efficient replicators survive'.

As the number of pathways is very large I have to select a few examples which illustrate the merits of my approach.

THE EVOLUTION OF THE CITRIC ACID CYCLE

The evolutionary axiom defined in the introduction leads to the question: 'Has the citric acid cycle advantages over other pathways of acetate oxidation?' It is the function of this cycle to provide energy in a form that can be utilized by living cells – ATP. Hence the most advantageous mechanism of acetate oxidation is that which gives the highest yield of ATP. In evaluating the potential yield of ATP it has to be borne in mind that the bulk of the synthesis of ATP is not directly connected with the degradation of foodstuffs, but with the transport of hydrogen atoms or electrons from NADH to molecular oxygen. Thus it is a prerequisite of ATP synthesis that the primary process in degradation is dehydrogenation.

The question has been considered whether a direct oxidation of acetate via glycollate, glyoxylate, formaldehyde and formate can be an alternative to the citric acid cycle (Krebs, 1947). In the past no answer could be given to this question. Professor Jack Baldwin (personal communication) recently suggested to me a satisfactory answer. He pointed out that acetate, either free or as acetyl-coenzyme A, cannot be dehydrogenated and therefore the first step cannot donate electrons to the mitochondrial electron transport chain. Acetate can be oxidized only by molecular oxygen through monooxygenase ('hydroxylase' or 'mixed function oxidase'). This enzyme system, which is located not in the mitochondria, but in the microsomal fraction, catalyses the insertion of one oxygen atom of molecular oxygen into the organic substrate while the second oxygen atom is reduced to water. This reduction requires a second substrate to donate the electrons. In most cases the second substrate is $NADPH_2$. The general equation of an oxygenase reaction is therefore

$$R.CH_3 + O_2 + NADPH_2 \rightarrow R.CH_2OH + H_2O + NADP$$

where the CH_3 of $R.CH_3$ stands for any methyl group which, because of the nature of R, cannot be dehydrogenated. R can be $-H$, or $-COOH$, or, as in the case of toluene or a xylene, a benzene ring.

Not only is an oxygenation of acetate wasteful because it cannot be coupled with ATP synthesis, but in addition, the consumption of one $NADPH_2$ for the reduction of the second oxygen atom of O_2 robs the cell of the potential synthesis of 3 ATP molecules. The quantitative aspects of the waste are as follows: when acetate is oxidized through the citric acid cycle there is a net gain of 10 pyrophosphate bonds of ATP. Twelve pyrophosphate bonds are formed by the reactions of the cycle but two are used for the conversion of acetate into acetyl coenzyme A. Were acetate 'directly' degraded via glycollate then 6 ATP would be prevented from being formed by the oxygenase reaction (with its need for $NADPH_2$). If all three stages of dehydrogenation from glycollate to carbon dioxide were brought about by NAD- or NADP-linked dehydrogenases, then no more than 6 ATP could be formed, because 1 $NADH_2$ is needed in the oxygenase reaction. This is a maximum value. The only glycollate oxidizing enzyme (EC 1.1.3.1.) known is a flavoprotein and therefore can at most yield 2 ATP.

The quantitative aspects are slightly different when acetate, in the form of acetyl CoA, is formed by the degradation of fatty acids or carbohydrate, but the principle remains the same in that the cyclic pathway is more efficient than a 'direct' degradation of acetyl CoA.

When acetyl CoA is metabolized by the citric acid cycle 12 ATP are formed during one turn of the cycle. When long-chain fatty acids are the precursor of acetyl CoA, two pyrophosphate bonds are needed to form acyl-CoA. Acetyl CoA then arises by β-oxidation without any further expenditure of ATP. Hence, for a C_{18} saturated fatty acid, 2/9 pyrophosphate bonds are spent on average before acetate enters the cycle and the ATP yield of the cycle is therefore $12 - 0.22 = 11.78$ pyrophosphate bonds for one turn of the cycle.

The concept that a 'direct' oxidation of acetate is inefficient not only explains why such a pathway cannot compete with another pathway; it also offers an explanation of why the alternative pathway must be cyclic. The principle is this: when a metabolic sequence of reactions cannot occur efficiently by a direct route but requires a primary attachment of a substance to another low molecular weight molecule (such as acetate to oxaloacetate), then the 'other low molecular weight molecule' must be regenerated when the metabolic utilization of the substrate (such as acetate) has been completed. This is necessary: if it were not so, the 'other low molecular weight molecule' would have to be produced afresh. This would be wasteful, if not impossible; if the 'other low molecular weight molecule' were not regenerated it would yield a by-product in stoichiometric proportions, which would be nonsensical.

OPTIMAL USE OF RESOURCES

The prediction that a cyclic pathway of acetate degradation is more efficient than the linear direct route of acetate oxidation does not postulate the specific details of the citric acid cycle. That acetate happens to react as the thioester of coenzyme A and condenses with oxaloacetate, may be explained by the well documented assumption that the pathways leading to citrate and thence to α-oxoglutarate evolved long before oxygen appeared in the atmosphere, in connection with the synthesis of glutamate. Coenzyme A is assumed to have arisen on the surface of the earth very early, possibly before 'life', together with the chemically closely related nucleic acids and other coenzymes (Oparin, 1957; Handler, 1961; King, 1980).

Thus in the evolution of the citric acid cycle use was made of mechanisms already existing in connection with other functions. It is indeed a general principle of evolution that multiple use is made of given resources. Amino acids have many other functions apart from being building blocks of proteins. For example glycine is a precursor of porphyrins and glutathione, serine of ethanolamine and choline and hence of phospholipids, glutamine of purines, aspartate of pyrimidines. These multiple uses represent specific cases of the principle stated earlier concerning optimal use of resources.

ADVANTAGE OF ADDITIONAL CONTROL POINTS

Apart from being more efficient in making optimal use of the energy released by the combustion of acetate, the citric acid cycle has the advantage over a more simple linear reaction sequence because it offers additional possibilities for metabolic control.

The cycle is regulated by several factors, one being the availability of oxaloacetate, the first step of the cycle (Bowman, 1966; Garland, 1968; Randle, England & Denton, 1970; Williamson & Cooper, 1980). As oxaloacetate reacts in the manner of a catalyst, a cyclic organization has the advantage that regeneration of the catalyst becomes faster when the rate of the cycle becomes faster. Thus oxaloacetate is not only a catalyst of the ordinary type promoting a chemical process, it also provides control of a feedback type. Furthermore, the concentration of oxaloacetate can be adjusted by the redox state of the NAD-couple (Taegtmeyer, personal communication) and by the tendency of aspartate amino transferase to establish an equilibrium state of its substrates (Bowman, 1966; Randle *et al.*, 1970).

OTHER METABOLIC CYCLES

Some metabolic cycles are precisely analogous to the citric acid cycle in that the substrate, before it can react, has to combine with another substance of low molecular weight and this substance must then be regenerated. The following are examples.

Glyoxylate Cycle. This cycle, a central mechanism in the microbial synthesis of cell constituents from acetate, brings about the synthesis of succinate from 2 molecules of acetyl CoA;

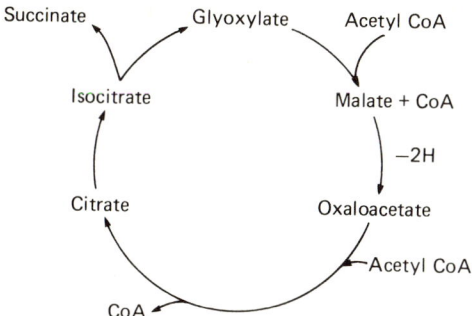

The overall result of the cycle is

$$2 \text{ acetyl CoA} - 2H \rightarrow \text{succinate} + \text{CoA}$$

It was suspected long ago that a direct reductive condensation of acetate to form succinate according to the scheme

$$2 \text{ acetate} - 2H \rightarrow \text{succinate}$$

might play a role in biological oxidation, but all attempts to demonstrate this reaction have failed (Thunberg, 1920; Knoop, 1923; Wieland, 1925). Later, after the discovery of acetyl CoA, an analogous condensation of 2 molecules of acetyl CoA seemed unlikely because it would necessitate the simultaneous attachment of 2 molecules of acetyl CoA plus a hydrogen-accepting coenzyme to one enzyme protein, and the large size of acetyl CoA (molecular weight 811) argues against the possibility of the required spatial arrangement on the enzyme molecule.

As the diagram of the glyoxylate cycle indicates, the two molecules of acetyl CoA combine in two stages with low molecular weight molecules, the first being glyoxylate and the second oxaloacetate. Glyoxylate may be looked upon as the primary 'low molecular weight substance' required for the build-up of the succinate from acetate. When succinate has been formed, glyoxylate is regenerated. The succinate via known pathways serves as a precursor of carbohydrates and amino acids.

Chemical knowledge does not provide feasible alternative pathways which would achieve the synthesis of cell constituents from acetate. Evolution has led to what would seem a perfect mechanism. The glyoxylate cycle is another example of 'multiple use of resources' in that the stages between malate and isocitrate in the glyoxylate cycle are identical with stages in the citric acid cycle. The glyoxylate cycle is indeed a modification of the citric acid cycle. By evolving no more than two new enzymes – malate synthase (EC 4.1.3.2) and isocitrate lyase (EC 4.1.3.1) – an energy releasing mechanism has been transformed into a component of a biosynthetic system.

Photosynthesis

There was at one time extensive discussion and testing of the possibility that in photosynthesis light promotes a direct reduction of carbon dioxide to formaldehyde.

$$CO_2 + H_2O \xrightarrow{light} HCOH + O_2$$

This hypothesia, however, had to be discarded (Florkin, 1979). The carbon of formaldehyde has the same redox level as glucose carbon and the idea that glucose was formed from formaldehyde by some kind of polymerization was tempting. From the work of Calvin's school (Quayle *et al.*, 1954) it is known that carbon dioxide is not reduced directly but combines with another low molecular weight substance, 1,5-ribulose bisphosphate. This splits into two molecules of 3-phosphoglycerate, and the reductive step of photosynthesis is the conversion of 3-phosphoglycerate to triosephosphate. When six molecules of ribulose bisphosphate have reacted with carbon dioxide, twelve molecules of 3-phosphoglycerate are formed, and the reduction of these gives six molecules of triosephosphate. Two of these form one glucose molecule and the remaining ten triosephosphates give six ribulose bisphosphate molecules, thus completing the cycle.

Pentose phosphate cycle

Somewhat different in detail, but the same in principle, is the optimal efficiency achieved by the pentose phosphate cycle. It is the function of the pentose phosphate cycle to generate $NADPH_2$ for reductive biosyntheses in the cytosol, above all for the synthesis of fat from carbohydrate. $NADPH_2$ is formed by glucose-6-phosphate dehydrogenase plus 6-phospho-gluconate dehydrogenase. Thus one molecule of glucose-6-phosphate forms two molecules of $NADPH_2$, with ribulose-5-phosphate as a by-product. A complex series of reactions then reconverts this by-product to glucose-6-phosphate in such a way that six molecules of glucose-6-phosphate entering the cycle are converted into five molecules of glucose-6-phosphate while one undergoes degradation to 6 CO_2 and 12 $NADPH_2$. Thus NADP acts in place of oxygen as hydrogen acceptor in this degradation of glucose. The free energy changes in this pathway are slight, hence the pentose phosphate cycle disposes effectively and economically of an unwanted by-product.

These considerations refer to situations where $NADPH_2$ is needed in excess of pentose phosphate, as is the case when fat is synthesized from carbohydrate. When pentose phosphate only is needed, it is formed through the non-oxidative section of the cycle, i.e. through the actions of hexosephosphate isomerase, transketolase, transaldolase, xylulose epimerase and pentose phosphate isomerase.

DEFINITION OF METABOLIC CYCLES

The preceding considerations apply to 'metabolic cycles' which are defined as processes in which an overall chemical change is brought about by continuous cyclic reaction sequences. The citric acid cycle is a prototype: the overall change is the combustion of acetate, achieved by the cyclic formation of di- and tricarboxylic acids. Other types of cyclic process exist where no overall changes occur. An example is the storage cycle of glycogen: in times of plenty, glucose is converted into glycogen, and in times of need the glycogen is reconverted into glucose. As the synthesis of glycogen and degradation proceed by different pathways, glycogen metabolism as a whole is a cyclic process, but the cycle is not a continuous one, the two phases taking place at different times. Another type of cycle is the transport cycle. Thus, haemoglobin transports oxygen from the lungs to the peripheral tissues, the γ-glutamyl cycle of Meister transports amino acids into cells, the methionine cycle transfers methyl groups from methyltetrahydrofolate to various acceptors, and the methylation of homocysteine and the methylation of various substrates occur at different times, different sites and involve different pathways. There are numerous other types of cycle – all distinct from metabolic cycles as defined above.

LINEAR REACTION SEQUENCES

Cycling sequences may be advantageous when a metabolite of low molecular weight has to be attached to another low molecular weight compound in order to make it more reactive. Metabolic cycles do not, however, necessarily have an advantage over non-cyclic (i.e. linear reaction) sequences. The anaerobic degradations of carbohydrates such as the lactic acid and alcoholic fermentations are linear metabolic processes and the question of why they have evolved is also relevant to these non-cyclic processes. Why are phosphorylated intermediates involved in fermentation? Although some of the phosphorylated sugars were discovered during the first two decades of this century, the significance of their formation could not be appreciated at that time, and most early investigators thought that a primary step in fermentation must be a fission of glucose into two 3-carbon products. They therefore tested the behaviour, especially in yeast cells, of 3-carbon compounds and their expectations were encouraged by the finding of Neuberg (1913) that methylglyoxal readily forms lactic acid, and that pyruvate forms carbon dioxide and acetyldehyde which in turn is reduced to ethanol. Indeed Neuberg made the statement 'methylglyoxal is the essence of glycolysis'.

The main intermediary stages of the glycolytic pathway were clarified during the 1920s and 1930s. It remained however a mystery why so many of the intermediate compounds were phosphorylated. ATP was discovered during the study of the pathway of glycolysis but its function was interpreted as acting as a coenzyme in several steps – such as the hexokinase and phosphofructokinase reactions. Only when it was realized that ATP is not an ordinary cofactor but the energy source closest to the contraction process in muscle did it become clear that the formation of phosphorylated intermediates is an essential feature of obtaining utilizable energy from the anaerobic breakdown of carbohydrate. Thus the pathway of glycolysis evolved because it is an efficient way of utilizing the energy released in the anaerobic degradation (Kalckar, 1969).

PROPIONIC ACID FERMENTATION

When lactate is fermented anaerobically by Propionibacterium, the balance of the substrate changes is approximately

3 Lactate + ADP + Pi = 2 Propionate + Acetate + CO_2 + ATP

The formation of propionate (Wood & Stjernholm, 1962) is not brought about by a simple reduction of lactate but by a complex 'detour' involving the following steps:

Lactate → pyruvate → oxaloacetate → malate → fumarate → succinate → succinyl-CoA → methylmalonyl-CoA → propionyl CoA → propionate.

The question of why this pathway evolved and why there is no direct reduction of lactate to propionate via acrylate can be satisfactorily answered by consideration of general principles of organic chemistry.

A conversion of lactate into propionate was originally suspected to proceed in two stages, the first being the elimination of water and the formation of acrylate, followed by the reduction of the latter. This would be analogous to the conversion of malate into succinate. However an elimination of water from lactic acid is impossible because of the general characteristics of the structure of lactate (March, 1977). A 'detour' is therefore essential if bacteria are to derive utilizable energy anaerobically from lactate. It is a remarkable and essential feature of the 'detour' (Wood & Stjernholm, 1962; Wood, 1972) that it does not involve the expenditure of ATP. The carboxylation of pyruvate is brought about by methylmalonyl CoA: pyruvate carboxyl transferase (EC 2.1.3.1) which catalyses the following reaction.

Methylmalonyl CoA + Pyruvate = Propionyl CoA + Oxaloacetate
Succinyl CoA is formed from propionyl CoA by a CoA transferase
   Propionyl CoA + Succinate → Succinyl CoA + Propionate

Only one molecule of ATP is formed for three molecules of lactate fermented. It arises from acetyl-CoA via acetylphosphate. This implies that the organism cannot afford to spend ATP on the reduction of lactate to propionate. Considering the impossibility of a direct reduction via acrylate, one may say that evolution has perfected a pathway that is economically ideal. It enables *Propionibacterium shermanii* to grow anaerobically with lactate as a source of energy, as it does in the manufacture of Emmental and Gruyère cheeses.

PATHWAY OF LONG-CHAIN FATTY ACID METABOLISM

Saturated fatty acids are rather stable compounds at neutral pH and low temperatures. They become reactive when they are attached to the sulphur atom of coenzyme A (Lynen, 1973). The acyl-CoA bond affects the reactivity of the hydrogen atoms at the α-and β-carbon of the chain, and under the influence of an acyl-CoA dehydrogenase these hydrogen atoms are transferred to FAD which results in the formation of trans Δ^2-enoyl CoA. The next step is the addition of water to the double bond between carbons 2 and 3. This addition of water, according to modern knowledge of organic reaction mechanisms (March, 1977) necessarily places the hydroxyl group of water at the β-carbon; it cannot be placed at the α-carbon. Thus the evolution of the pathway of β-oxidation of long-chain fatty acids was not entirely a matter of chance but conditioned by the limited scope of the chemical reactions that can take place under conditions compatible with life. β-Oxidation of course is also a highly efficient mechanism of degradation because it breaks up the fatty acid chain into a most versatile product, acetyl-coenzyme A. It would be difficult to conceive an alternative pathway of degradation which would be equally efficient in respect of the yield of utilizable energy, or as starting material for biosyntheses.

The activating influence of the $-$CO.S.CoA group on the fatty acid chain extends to the immediately adjacent carbon atoms only. Dehydrogenation at other positions of the chain therefore requires other mechanisms. Indeed desaturation at the carbon atoms 9 and 10 is known to be brought about by monooxygenase in animal tissues as well as in microorganisms. Yeast cells do not grow anaerobically in the absence of unsaturated fatty acids, but traces of oxygen are sufficient to enable the cells to grow. The need for oxygen is explained by the fact that desaturation is achieved by monooxygenase (Bloomfield & Bloch, 1960; Light, Lennarz & Bloch, 1962) which catalyses the following reaction.

Stearyl CoA + NADPH$_2$ + O$_2$ $\rightarrow$ Oleyl CoA + NADP + 2H$_2$O

The obstacles to dehydrogenation by NAD or FAD are thermodynamic. The monooxygenase system has a more positive redox potential and is the only one of the given redox systems that can attack the hydrogen atom far away from the carboxyl group. For the same reason the ω-oxidation of long-chain fatty acids is catalysed by a monooxygenase. Clearly the monooxygenases, together with dioxygenases (which catalyse the reaction RH$_2$ + O$_2$ $\rightarrow$ R(OH)$_2$) have evolved because they are powerful oxidizing agents, which fulfil many important functions. Apart from those already mentioned, these functions include the initial stages of the breakdown of tryptophan and phenylalanine, the synthesis of steroids and the detoxication of foreign substances such as drugs.

SURVIVAL OF MICROORGANISMS

Making optimal use of resources in a competitive environment is not as vital to the survival of microorganisms as it is to the survival of higher organisms because starvation in microorganisms does not necessarily lead to death but usually to cessation of growth and reproduction. Many organisms form spores when nutrients become scarce. Others can survive in a dried state or at low temperatures when their metabolic activities will drop to near zero. Some microorganisms are inefficient in their use of available resources, in that they do not oxidize nutrients to completion. Acetic acid bacteria oxidize ethanol only to acetic acid, glycerol to dihydroxyacetone and, when there is an excess of substrate, glucose to gluconic acid. Under some conditions *Aspergillus* does not dissimilate glucose completely but forms citric acid as an end-product. The phycomycete *Rhizopus* accumulates fumarate. Microorganisms often can afford such waste, higher organisms cannot. Some organisms degrade carbohydrate anaerobically by a modification of the standard pathway in which phosphogluconate is an intermediate – the Entner–Doudoroff pathway – although this pathway yields only one ATP per molecule of glucose compared with two produced by the standard Embden–Meyerhof pathway. The Entner–Doudoroff pathway is limited to a few groups of microorganisms and does not occur in higher organisms which need greater efficiency.

Lower efficiency shows itself also in other ways. Regulatory mechanisms of glycolysis are less developed at the stage of phosphofructokinase in the cellular slime mould *Dictyostelium* (Baumann & Wright, 1968). In many microorganisms the efficiency of oxidative phosphorylation is less than optimal, sometimes with the P/O ratio only one or two instead of three (Linnane & Haslam, 1970).

Some microorganisms have a competitive advantage in their high capacity for adaptation. They have evolved a potential for synthesizing enzymes enabling them to utilize substances which cannot serve as a source of energy in higher organisms (Gray & Postgate, 1976).

PATHWAYS AND ENZYME FAMILIES

The recent elucidation of the secondary, tertiary and quaternary structures of the enzymes of glycolysis has revealed close structural relationships between the various enzymes. As Blake (1974) has remarked, 'One of the most fascinating questions about the formation of life and its immediate development is how the enzymes needed to sustain basic life processes were derived. It seems reasonable that the relatively large number of enzymes required by even a primitive bacterium were derived from a much smaller number of basic, relatively non-specific enzymes – probably by gene duplication and subsequent divergence leading to a diversification of function and narrowing of specificity. On this basis it would be expected that the large number of present-day enzymes could be grouped into a much smaller number of families.' He points out that the structural relationship indicating such a family '... was first seen clearly when the structures of liver alcohol dehydrogenase and glyceraldehyde 3-phosphate dehydrogenase were compared with the homologous lactate and malate dehydrogenase molecules. The polypeptides chains of all four dehydrogenases, composed of between 327 and 374 amino-acid residues, are folded into two distinct units or domains. One of these domains is responsible for binding the common NAD cofactor and the other domain carries the binding side for specific substrate of each enzyme and its catalytic site.' See also Blake (1977). Analogous family relations have recently come to light for peptide hormones (Blundell & Humbel, 1980) and for proteases (Hartley, 1979).

Biologists have for centuries been accustomed to the concept of the existence of more or less closely related species and genera, linked as are families by common ancestry. Now it is clear that there are also families at the biochemical level, as is indeed expected from the concept of evolution.

CONCLUDING COMMENTS

The main theme of this article is the idea that it is productive to ask why evolution has led to specific metabolic pathways. The examples chosen for discussion – the citric acid cycle and other metabolic cycles, glycolysis and alcoholic fermentation, propionic fermentation, fatty acid metabolism – make it clear that these pathways are optimally effective for life by making maximum use of the given resources. The laws of chemistry – the properties of atoms and molecules – impose limits on possible pathways. They predict why certain reactions can take place and why others cannot. Confronted with these, the chemist would find it very difficult, if not impossible, to better the efficiency which evolution has achieved.

The question of *how* evolution proceeds is answered by the concept of random mutation and natural selection. The question of *why* biochemical evolution has proceeded as it has, has not been given sufficient attention in the past. Biochemists exploring metabolic pathways have usually aimed at establishing the reactions which make up the pathway, and at examining the mechanism of action of the enzymes operating the pathways (Walsh, 1979). But they have not concerned themselves much with the question of why pathways have arisen. There is now a great deal of information which makes it possible to comment fruitfully on this question and to formulate new working hypotheses for experimental work.

REFERENCES

BAUMANN, P. & WRIGHT, B. E. (1968). The phosphofructokinase of *Dictostelium discoideum*. *Biochemistry*, **7**, 3653–61.

BLAKE, C. C. F. (1974). Evolution of nucleotide-binding proteins. *Nature, London*, **250**, 284–5.

BLAKE, C. C. F. (1977). The nucleotide-binding fold revisited. *Nature, London*, **267**, 482–3.

BLOOMFIELD, D. K. & BLOCH, K. (1960). The formation of Δ^9-unsaturated fatty acids. *Journal of Biological Chemistry*, **235**, 337–45.

BLUNDELL, T. L. & HUMBEL, R. E. (1980). Hormone families: pancreatic hormones and homologous growth factors. *Nature, London*, **287**, 781–7.

BOWMAN, R. H. (1966). Effects of diabetes, fatty acids and ketone bodies on tricarboxylic acid cycle metabolism in the perfused rat heart. *Journal of Biological Chemistry*, **241**, 3041–8.

FLORKIN, M. (1979). The photosynthetic cycle of carbon reduction. In *A History of Biochemistry; Comprehensive Biochemistry*, vol. **33A**, ed. M. Florkin & E. H. Stotz, pp. 81–108. Amsterdam: Elsevier.

GARLAND, P. B. (1968). Control of citrate syntesis in mitochondria. In *The Metabolic Roles of Citrate*, ed. T. W. Goodwin, pp. 41–60. New York: Academic Press.

GRAY, T. R. & POSTGATE, J. R. ed. (1976). *The Survival of Vegetative Microbes*. Symposia of the Society for General Microbiology, **26**, London: Cambridge University Press.

HANDLER, P., (1961). Evolution of the coenzymes. In *Proceedings of the 5th International Congress of Biochemistry*, ed. A. I. Oparin, pp. 149–57. New York: Macmillan.

HARTLEY, B. S. (1979). Evolution of enzyme structure. *Proceedings of the Royal Society* B, **205**, 443–52.
KALCKAR, H. M. (1969). *Biological Phosphorylations.* Englewood Cliffs, New Jersey: Prentice-Hall.
KING, G. A. M. (1980). Evolution of coenzymes. *BioSystems*, **13**, 23–45.
KNOOP, F. (1923). Wie werden unsere Hauptnährstoffe in Organen verbrannt und wechselseitig ineinander übergeführt? *Klinische Wochenschrift.* **2**, 60–63.
KREBS, H. A. (1947). Cyclic processes in living matter. *Enzymologia*, **12**, 88–100.
LIGHT, R. J., LENNARZ, W. J. & BLOCH, K. (1962). The metabolism of hydroxystearic acid in yeast. *Journal of Biological Chemistry*, **237**, 1793–800.
LINNANE, A. W. & HASLAM, J. M. (1970). The biogenesis of yeast mitochondria. In *Current Topics in Cellular Regulation*, **2**, 101–72. Academic Press: London.
LYNEN, F. (1973). The pathway from 'activated acetic acid' to the terpenes and fatty acids. In *Nobel Lectures: Physiology and Medicine* (1963–1970), pp. 103–38. New York: Elsevier.
MARCH, J. (1977). *Advanced Organic Chemistry; Reaction Mechanisms and Structure*, 2nd edn. New York, London: McGraw-Hill.
NEUBERG, C. (1913). *Die Gärungsvorgänge und der Zuckerumsatz der Zelle.* Jena: Gustav Fischer.
OPARIN, A. I. (1957). *The Origin of Life on the Earth*, 3rd edn, translated by Ann Synge. London: Oliver & Boyd.
ORGEL, L. E. & CRICK, F. H. C. (1980). Selfish DNA: the ultimate parasite. *Nature, London*, **284**, 604–7.
QUAYLE, J. R., FULLER, R. C., BENSON, A. A. & CALVIN, M. (1954). Enzymatic carboxylation of ribulose diphosphate. *Journal of the American Chemical Society*, **76**, 3610.
RANDLE, P. J., ENGLAND, P. J. & DENTON, R. M. (1970). Control of the tricarboxylate cycle and its interactions with glycolysis during acetate utilisation in rat heart. *Biochemical Journal*, **117**, 677–95.
THUNBERG, T. (1920). Zur Kenntnis des intermediären Stoffwechsels und der dabei wirksamen Enzyme. *Skandinavisches Archiv für Physiologie*, **40**, 1–91.
WALSH, C. (1979). *Enzymatic Reaction Mechanisms.* San Francisco: W. H. Freeman.
WIELAND, H. (1925). Mechanismus der Oxydation und Reduktion in der lebenden Substanz. In *Handbuch der Biochemie*, ed. C. Oppenheimer, **2**, 252–272. Jena: Gustav Fischer.
WILLIAMSON, J. R. & COOPER, R. H. (1980). Regulation of the citric acid cycle in mammalian systems. *FEBS Letters*, **117**, K73–K83.
WOOD, H. G. (1972). My life and carbon dioxide. In *Miami Winter Symposium.* pp. 1–54. New York, **3**, ed. J. F. Woessner Jr. & F. Huijing. Academic Press.
WOOD, H. G. & STJERNHOLM, R. L. (1962). Assimilation of carbon dioxide by heterotrophic organisms. In *The Bacteria*, vol. II, ed. I. C. Gunsalus & R. Y. Stanier, pp. 41–117. New York: Academic Press.

THE EVOLUTION OF METABOLIC REGULATION

SIMON BAUMBERG

Department of Genetics, University of Leeds, Leeds, LS2 9JT, UK

INTRODUCTION

As so often when trying to deal with an evolutionary topic – and particularly if it concerns microorganisms – one is tempted merely to catalogue isolated studies and then link these with a more-or-less plausible thread of speculation. I am going to proceed basically in this way, but with an approach that may not be entirely conventional. Faced with the fascinating variety of metabolic controls in the microbial world, most of which look so obviously adaptive, we are inclined to adopt what Gould & Lewontin (1979), in an amusing and polemical article, describe as 'the Panglossian paradigm'. I cannot do better than quote their summary.

'An adaptationist programme has dominated evolutionary thought in England and the United States during the past 40 years. It is based on faith in the power of natural selection as an optimizing agent. It proceeds by breaking an organism into unitary "traits" and proposing an adaptive story for each considered separately. Tradeoffs among competing selective demands exert the only brake upon perfection; non-optimality is thereby rendered as a result of adaptation as well. We ... attempt to reassert a competing notion ... that organisms must be analysed as integrated wholes, with Baupläne so constrained by phyletic heritage, pathways of development and general architecture that the constraints themselves become more interesting and more important in delimiting pathways of change than the selective force that may mediate change when it occurs. We fault the adaptationist programme for its failure to distinguish current utility from reasons for origin ...; for its unwillingness to consider alternatives to adaptive stories; for its reliance upon plausibility alone as a criterion for accepting speculative tales; and for its failure to consider adequately such competing themes as random fixation of alleles, production of non-adaptive structures by developmental correlation with selected features ..., the separability of adaptation and selection, multiple adaptive peaks, and current utility as an epiphenomenon of non-adaptive structures.'

It will be interesting to try to apply these ideas in the course of this review.

Discussion of the evolution of a character must cover two aspects: the historical pattern of descent of the character; and the causes of this pattern, among which selection of advantageous phenotypes will be important but not exclusive. My approach will be to describe diverse systems, then speculate first on their present or past selective advantage (no doubt thereby falling into the Panglossian trap), and second on possible evolutionary pathways. I am not going to deal with evolution in the laboratory, reviewed recently by Clarke (1978, 1980). I will also take for granted the possibility of genetic transfer between strains and species, whether involving plasmids (see e.g. Bennett & Richmond, 1978), transposable elements (Calos & Miller, 1980), or other processes and entities. The reader is also referred to previous reviews relating to topics covered here: Hartley (1974, 1979), Ornston & Parke (1977), and Clarke (1974, 1979).

Metabolic regulation can act at the level of preformed enzyme or other protein, or of protein synthesis. The following account is divided into four sections: the kinds of control observed at the level of preformed enzyme and their selective advantages; the evolutionary origins of controls at the level of preformed enzyme; the kinds of control observed at the level of protein synthesis and their selective advantages; and the evolutionary origins of controls at the level of protein synthesis.

CONTROLS AT THE LEVEL OF PREFORMED ENZYME AND THEIR SELECTIVE ADVANTAGES

The obvious mode of control to discuss is via changes in enzyme activity. However, I will take the view that spatial organization of enzyme activities is also part of metabolic regulation.

Controls utilizing changes in enzyme activity

The most familiar process here is *modulation*, whereby the activity of a protein is altered by the binding of a (usually low molecular weight) *effector*. The inhibited or non-activated form of a modulatable enzyme usually shows some residual activity. Modulatable proteins are generally homooligomeric (although sometimes made up of more than one kind of polypeptide chain); this oligomeric structure is thought to be implicated in another feature, the sigmoidal response to effector concentration (Monod, Changeux & Jacob, 1963; Monod, Wyman & Changeux, 1965; Atkinson, 1970; Koshland, 1970; Wyman, 1972).

Perhaps the most familiar form of modulation is *feedback inhibition*, where a product of a later step in a pathway inhibits an earlier enzyme; other forms are *energy-linked controls* and *precursor activation*.

Feedback inhibition is characteristic of biosynthetic pathways, including those which also act catabolically (*amphibolic* pathways: Davis, 1961; Sanwal, 1970; Sanwal, Kapoor & Duckworth, 1971). The inhibitor is often the end-product of the pathway, and the inhibited enzyme the first enzyme of the sequence specifically leading to that end-product: for instance, in all organisms studied, tryptophan inhibits anthranilate synthase (Fig. 1) (Crawford, 1975). Another example is citrate synthase in the Gram-negative facultative anaerobes, *Escherichia coli* and its relatives (Fig. 2). During anaerobic growth, these organisms generate energy by fermentation; under these conditions, there is no α-oxoglutarate dehydrogenase activity, so that the citric acid cycle is reduced to two branches ending in succinyl-coenzyme A (CoA) and α-oxoglutarate respectively. In these organisms, citrate synthase, now the first enzyme in its branch, is inhibited by α-oxoglutarate (Weitzman & Danson, 1976; Weitzman, 1980). In amphibolic pathways, the inhibitor is often the substrate at the next branch point: for instance, phospho*enol*pyruvate (PEP) inhibits *E.coli* phosphofructokinase, the inhibition being reversed by the substrate fructose-6-phosphate (Blangy, Buc & Monod, 1968).

Branched, purely biosynthetic pathways show an interesting diversity of patterns, well exemplified by those for the aromatic amino-acid family (reviewed in Umbarger, 1978; Clarke, 1979). The first enzyme of the common pathway to chorismate (see Fig. 1), is 3-deoxy-D-*arabino*-heptulosonate 7-phosphate (DAHP) synthase. From chorismate, one branch leads to tryptophan; another step, catalysed by chorismate mutase, yields prephenate, from which diverge the paths to phenylalanine and tyrosine. The first steps in the major routes from prephenate to the last two are catalysed by prephenate dehydratase and prephenate dehydrogenase respectively. In *E.coli* and *Salmonella typhimurium* (and also in *Neurospora crassa*) there are three DAHP synthases: one, DAHP synthase (phe), inhibited by phenylalanine, a second, DAHP synthase (tyr),

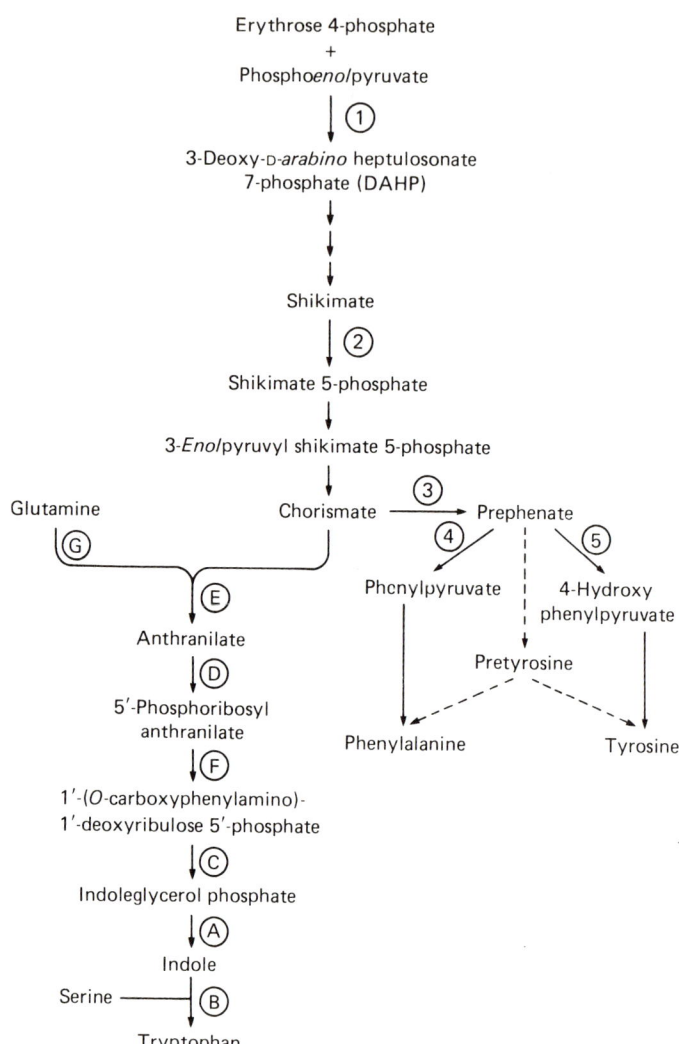

Fig. 1. Pathways of aromatic amino acid biosynthesis, featuring metabolites and enzymes mentioned in the text. Enzymes: 1, DAHP synthase; 2, shikimate kinase; 3, chorismate mutase; 4, prephenate dehydratase; 5, prephenate dehydrogenase; A, tryptophan synthase α component; B, tryptophan synthase β component; C, indoleglycerol phosphate synthase; D, anthranilate phosphoribosyltransferase; E, anthranilate synthase catalytic component; F, phosphoribosylanthranilate isomerase; G, anthranilate synthase glutamine-binding component. Dashed lines refer to steps not present in all micro-organisms mentioned.

inhibited by tyrosine, and the third, DAHP synthase (trp), inhibited by tryptophan. There are two isoenzymes of chorismate mutase, both parts of bifunctional polypeptides: one (chorismate mutase P) combines this activity with prephenate dehydratase, the other

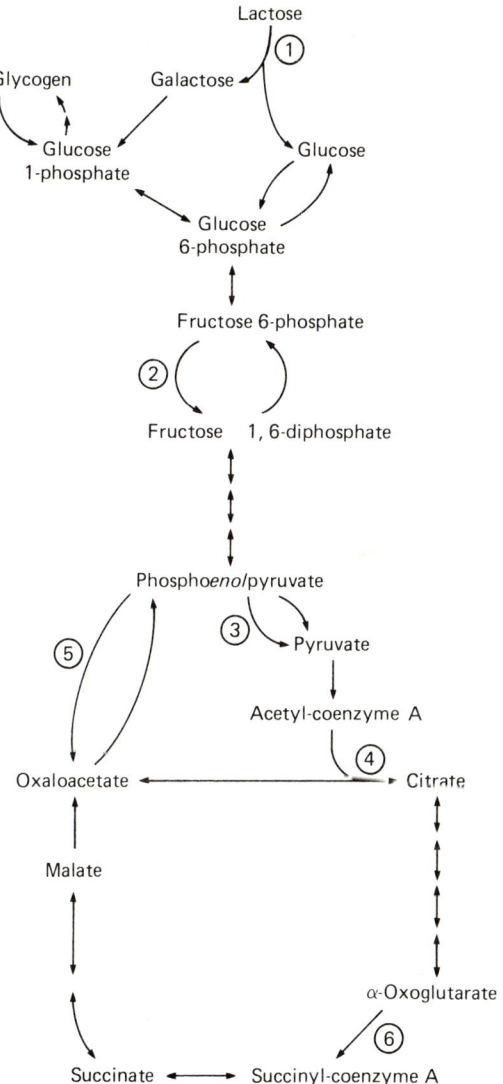

Fig. 2. Amphibolic pathways of sugar catabolism and neogenesis, featuring metabolites and enzymes mentioned in the text. Enzymes: 1, β-galactosidase; 2, phosphofructokinase; 3, pyruvate kinase; 4, citrate synthase; 5, α-oxoglutarate dehydrogenase; 6, phospho*enol*pyruvate carboxylase. In this figure alone, single- and double-headed arrows are used to distinguish thermodynamically irreversible and reversible steps respectively.

(chorismate mutase T) with prephenate dehydrogenase. Both activities of chorismate mutase P/prephenate dehydratase are inhibited by phenylalanine; prephenate dehydrogenase, but not chorismate mutase T, is inhibited by tyrosine.

In the pseudomonads, there is only a single DAHP synthase. In *Pseudomonas aeruginosa* this is (approximately cumulatively) inhibited by phenylpyruvate, tryptophan and tyrosine, the last-named being the most potent inhibitor (Jensen, Calhoun & Stenmark, 1973). Balance between the pathways may be achieved through the nature of the controls at the later stages, termed by Jensen *et al.* (1973) the *channel-shuttle* model. Chorismate mutase and prephenate dehydrogenase activities are carried on a single polypeptide; phenylalanine inhibits both activities, the latter much more severely than the former. Chorismate mutase activity is also strongly substrate-inhibited by chorismate; prephenate dehydratase activity is activated by tyrosine. Prephenate dehydrogenase, on the other hand, appears to be uncontrolled (Calhoun, Pierson & Jensen, 1973; Patel, Pierson & Jensen, 1977). Suppose that tyrosine were in excess; the prephenate produced by chorismate mutase action would be converted preferentially to phenylalanine, because of the activation of prephenate dehydratase and the *channelling* (discussed below) of freshly formed prephenate to the adjacent prephenate dehydrogenase part of the enzyme molecule. If phenylalanine were in excess, prephenate, formed by the only partially inhibited chorismate mutase, would flow towards tyrosine because of the inhibition of the dehydratase. Hence an imbalance in either tyrosine or phenylalanine would correct itself.

Finally, in the bacilli there is, as with the pseudomonads, only a single DAHP synthase. However, its control is radically different. In *Bacillus subtilis*, this activity is carried on a single polypeptide, together with chorismate mutase activity, and this protein aggregates with another, which in the combination shows shikimate kinase activity. The DAHP synthase and shikimate kinase activities are inhibited by chorismate and/or prephenate, apparently via the binding of the latter to the mutase active site; the chorismate mutase activity is inhibited by prephenate (Huang *et al.*, 1975). Since, as will be seen shortly, these two compounds may accumulate in response to excess of, or imbalance in, the amino-acid end-products, this pattern of control has been termed *sequential feedback inhibition* (Rebello & Jensen, 1970). Prephenate dehydrogenase is inhibited by tyrosine; prephenate dehydratase by either phenylalanine or tryptophan, the inhibition by tryptophan being antagonized by tyrosine. The net result of these controls, together with the universal inhibition of anthranilate synthase by tryptophan mentioned earlier, is that an excess of any of the amino acids would push intermediates towards

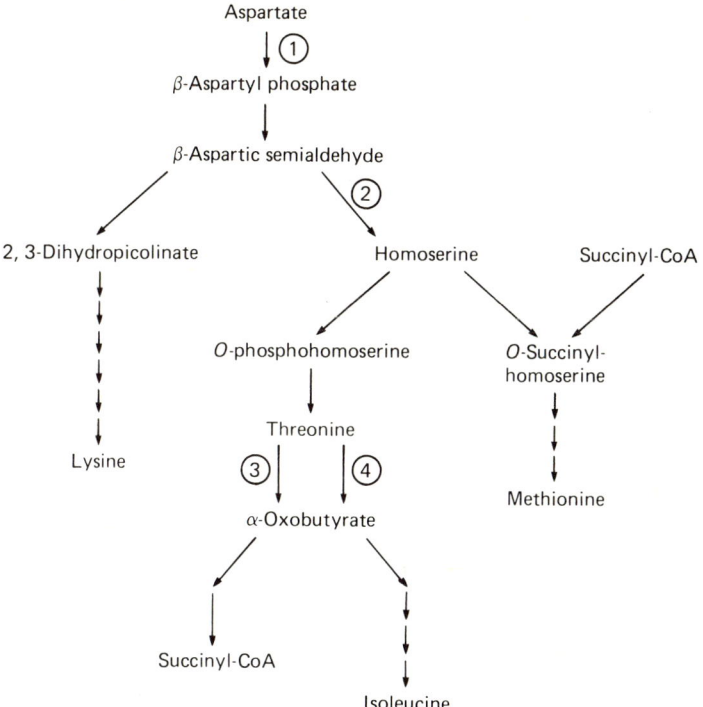

Fig. 3. Pathways of biosynthesis of the aspartate family of amino acids in *E. coli*, featuring metabolites and enzymes mentioned in the text. Enzymes: 1, aspartokinase; 2, homoserine dehydrogenase; 3, (catabolic) threonine dehydratase; 4, (biosynthetic) threonine dehydratase.

the others, while an excess of all would cut down flow into the common pathway by inhibition of DAHP synthase.

Another branched pathway system to which I will also refer later is that for synthesis of the aspartate family of amino acids (Fig. 3), in particular the common pathway as far as homoserine (reviewed by Umbarger, 1978). In *E.coli* and *S. typhimurium* there are three aspartokinases and two homoserine dehydrogenases. Aspartokinase I and homoserine dehydrogenase I are carried on a single bifunctional polypeptide chain, and both activities are inhibited by threonine, though to different extents. Aspartokinase II and homoserine dehydrogenase II are also carried on a single bifunctional polypeptide chain: neither activity is inhibited by any amino acid, although the synthesis of this polypeptide is repressed by methionine. Repression will be discussed in more detail later. Aspartokinase III is inhibited by lysine, and also by hexose monophosphates, AMP and (weakly) by various hydrophobic amino acids unrelated to lysine. As with DAHP synthases, this isoenzyme pattern seems to be peculiar to the enterobacteria. In the pseudomonads, *Rhodopseudomonas capsulata*, *Bacillus polymyxa* and *B. licheniformis*, there are single aspartokinases showing various patterns of inhibition involving threonine and lysine.

Energy-linked controls are characteristic of catabolic (including amphibolic) pathways, and feature modulation by 'signal' metabolites, such as purine nucleotides, phosphate and pyrophosphate, whose level indicates the energy state of the cell. Examples from purely catabolic peripheral pathways (usually involving the first enzyme of the sequence) include the biodegradative threonine dehydratase of *E.coli*, which is activated by adenosine 5'-monophosphate (AMP) (Umbarger & Brown, 1957; Shizuta & Hayaishi, 1976) and the histidase of *Ps. aeruginosa*, whose inhibition by pyrophosphate is relieved by AMP or guanosine 5'-diphosphate (GDP) (Lessie & Neidhardt, 1967). Examples among amphibolic enzymes are numerous (Sanwal, 1970); e.g. phosphofructokinase is activated by adenosine 5'-diphosphate (ADP) or GDP (Blangy *et al.*, 1968). A related form of control may be regulation by reduced nicotinamide adenine dinucleotide (NADH). A very interesting pattern is shown by citrate synthase (Weitzman & Danson, 1976; Weitzman, 1980). These may be divided (with few exceptions) into the following categories:

(*a*) 'Large' (Molecular weight *ca.* 250 000); inhibition by NADH antagonized by AMP; inhibition by α-oxoglutarate: in facultatively anaerobic Gram-negative bacteria.

(*b*) 'Large'; inhibition by NADH not antagonized by AMP; no inhibition by α-oxoglutarate: in strictly aerobic Gram-negative bacteria.

(*c*) 'Large'; no inhibition by NADH; inhibition by α-oxoglutarate or succinyl-CoA: in cyanobacteria.

(*d*) 'Small' (Molecular weight *ca.* 100 000); inhibition by ATP: in Gram-positive bacteria, eukaryotes.

The difference between 'large' and 'small' enzymes probably relates to quaternary structure, the former being tetramers and the latter dimers of similarly sized subunits.

As with patterns of control in the aromatic amino-acid pathways, the conservation of regulatory type within taxonomic group is striking. Indeed, citrate synthase properties have been used as an aid to taxonomy (Weitzman & Jones, 1975).

Precursor or 'feedforward' activation, the converse of feedback inhibition, seems to be unique to amphibolic pathways (Sanwal, 1970; Sanwal, Kapoor & Duckworth, 1971). Examples in *E.coli* are the activation by fructose-1,6-diphosphate of pyruvate kinase and PEP carboxylase.

Susceptibility of an enzyme to modulation can be affected by covalent modification, as illustrated by *E.coli* glutamine synthetase. This enzyme is a homododecamer, and can be enzymically adenylylated at one specific tyrosine residue per subunit, this process also being subject to complex controls (Stadtman & Ginsburg, 1974). The adenylylated and non-adenylylated forms differ in their activity and modulation response to several effectors. It is suggested that they play different roles in nitrogen metabolic control of synthesis of a variety of enzymes, as will be discussed below (Magasanik *et al.*, 1974; Magasanik, 1976, 1980).

Enzyme inhibition may also be produced by the binding of another protein rather than a small molecule effector. The best-known example of this is the inactivation of the *Saccharomyces cerevisiae* ornithine carbamoyltransferase by arginase. The former is substrate inhibited by ornithine, and the inhibition seems to involve enhancement of this inhibition upon interaction with arginase to which arginine has bound (Penninckx, Simon & Wiame, 1974; Penninckx, 1975). A similar interaction may occur in *B. subtilis* (Issaly & Issaly, 1974).

To end this section, a type of control should be mentioned which appears to be less important in microorganisms than in Metazoa, namely controlled protein inactivation, in the sense of irreversible loss *in vivo* of catalytic activity (see Switzer, 1977, for review). In the microbial eukaryotes, it is often noted that on changing carbon or nitrogen source, unneeded preformed enzymes become inactivated. In prokaryotes, this mechanism often operates on entry into the stationary phase or, for spore-forming bacteria, at the onset of sporulation (Maurizi & Switzer, 1980). An interesting case is *E.coli* aspartokinase III which becomes adenylylated in late stationary growth phase, concomitant with (and possibly leading to) loss of activity (Niles & Westhead, 1973 *a, b*).

The role of spatial organization of enzyme activities

This takes two forms: juxtaposition of enzyme activities, and compartmentation.

Juxtaposition of enzyme activities may be achieved in two ways:

by aggregation of separate polypeptides, and by carrying the two activities on the same polypeptide (it may be noted that in the latter case the activities invariably lie on clearly delimited and separate segments of the chain). We have already seen examples of each in those systems considered above, and from the large number of further instances available I will describe some in the tryptophan pathway. As seen from Fig. 1, the five steps actually invoke seven distinct activities. The conversion of chorismate to anthranilate requires not only anthranilate synthase (function E), but also the function G that enables glutamine to be used rather than NH_4^+; and the final reaction involves the conversion of indoleglycerol phosphate to indole (mediated by function A), followed by the condensation of the latter with serine to yield tryptophan (function B). In *E.coli*, the G function is carried on the same polypeptide as the D function (anthranilate phosphoribosyltransferase); this bifunctional chain aggregates with that bearing E function. The F (phosphoribosylanthranilate isomerase) and C (indoleglycerol phosphate synthase) activities are carried on a single polypeptide. Finally, A and B functions are carried on separate polypeptide chains which aggregate to give the tryptophan synthase complex. These patterns alter as we go to other organisms. In *Proteus* and *Serratia* species, G and D functions are on separate polypeptides; G still aggregates with E, but D is now separate. In *Ps.aeruginosa, Ps.putida* and *B.subtilis*, all the functions are now carried by separate polypeptides (see review by Crawford, 1975). In *Neurospora crassa*, D function is on its own; E (and G?) are on one polypeptide, F and C on another, these two polypeptide chains aggregating; A and B are carried on a single chain (Hulett & DeMoss, 1975; Matchett & DeMoss, 1975).

From the above, a generalization suggests itself, namely that juxtaposition of enzyme activities may occur in two situations which sometimes overlap: when the activities catalyse successive steps in a pathway, and when one activity presumably provides a feedback inhibition site for the other (the only possible exception among the small number of cases cited is that of the tryptophan pathway E(G?)–F–C complex in *N.crassa*, since the D activity which would complete the succession of steps is apparently missing and there is no evidence that the F and C activities are feedback inhibited). The latter situation has been proved to obtain in some cases. Proteolytic cleavage of the *B.subtilis* DAHP synthase–chorismate mutase–shikimate kinase complex yields a fragment with only a non-feedback-inhibitable DAHP synthase (Huang, Montoya & Nester,

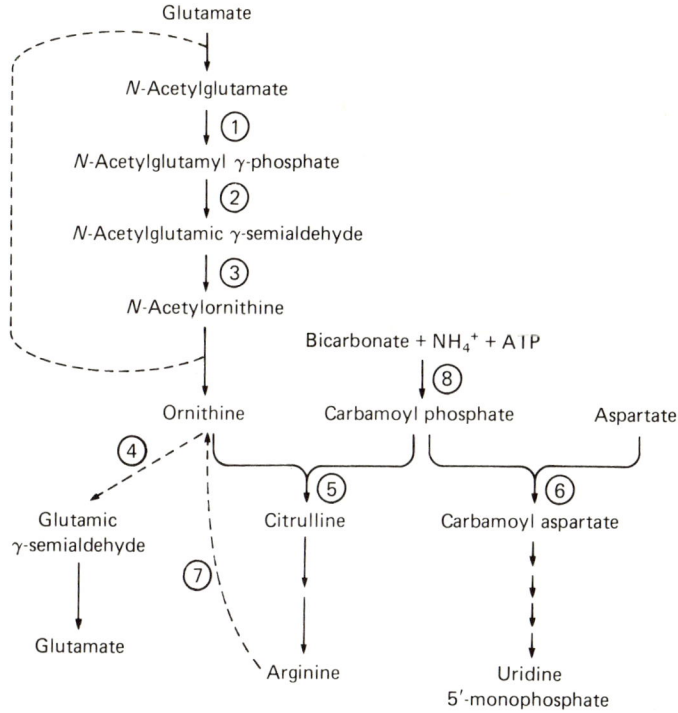

Fig. 4. Pathways of arginine and pyrimidine biosynthesis, featuring metabolites and enzymes mentioned in the text. Enzymes: 1, N-acetyl-glutamate kinase; 2, N-acetylglutamic γ-semialdehyde dehydrogenase; 3, N-acetylornithine γ-aminotransferase; 4, ornithine aminotransferase; 5, ornithine carbamoyltransferase; 6, aspartate carbamoyltransferase; 7, arginase; 8, carbamoyl phosphate synthetase. Dashed lines refer to steps not present in all micro-organisms mentioned.

1975). Similarly, limited proteolysis of the *E.coli* aspartokinase I–homoserine dehydrogenase I yields a carboxy-terminal-derived homodimer carrying a non-feedback-inhibitable homoserine dehydrogenase activity (Cohen & Dautry-Varsat, 1980).

True compartmentation is probably only found in the microbial eukaroytes. The best example in this context is that of arginine and pyrimidine synthesis in *N.crassa* (Fig. 4). These two pathways are linked through carbamoyl phosphate, the substrate of the aspartate carbamoyltransferase (pyrimidine pathway) and ornithine carbamoyltransferase (arginine pathway) (reviewed by Makoff & Radford, 1978). Generally in bacteria there is one carbamoylphosphate synthetase, whose activity is regulated in various ways by both pyrimidine derivatives and arginine (in *E.coli*, for instance, this enzyme is inhibited by uridine 5'-monophosphate (UMP) and activated by ornithine: aspartate carbamoyltransferase is also feedback-inhibited by various pyrimidine nucleotides, cytidine 5'-triphosphate (CTP) being the most potent). In fungi, there are usually two carbamoylphosphate synthetases. One of these is carried on the same polypeptide as aspartate carbamoyltransferase, and is inhibited by pyrimidine nucleotides. The other is involved in the arginine pathway, since, for example, in *S.cerevisiae* it is inhibited by arginine. In *N.crassa,* the two carbamoylphosphate synthetases are located in different cell compartments. The 'pyrimidine-specific' activity is located in the nucleus. The 'arginine-specific' activity is found in the mitochondrion together with ornithine carbamoyltransferase, which catalyses the condensation of carbamoyl phosphate with ornithine to yield citrulline, and all the enzymes leading from glutamate to ornithine except acetylglutamate kinase; this enzyme, which is feedback-inhibited by arginine, is situated in the cytosol. The enzymes catalysing the conversion of citrulline to arginine, and arginase, which breaks down arginine to ornithine and urea, are cytosolic. Most of the endogenously produced ornithine and arginine is sequestered into vesicles, and only when arginine is provided exogenously do these compounds begin to move into the cytosol (whereupon they are catabolized) (Cybis & Davis, 1975; Karlin, Bowman & Davis, 1976; Weiss & Davis, 1977).

Possible selective advantages of controls involving enzyme activity or spatial location

For every observation described above, a plausible *a priori* case can be made for the control phenotype concerned being of selective advantage. Indeed, the argument is usually self-evident. Feedback inhibition cuts down flux through a pathway whose end-product is in good supply. Energy-linked controls enhance flux through energy-yielding pathways in conditions of energy dearth and diminish it in conditions of energy sufficiency. Precursor activation mobilizes an activity late in a pathway in readiness for an increased supply of substrate which might otherwise transiently accumulate. It is noteworthy that those enzymes that show modulation are just the ones whose changes in activity will be most effective: they are usually thermodynamically irreversible (else mass action could counteract the effects of modulation), and they come at branch points, i.e. a modulatable enzyme catalyses one of two or more possible transformations of its substrate. In branched pathways, complex patterns of inhibition and activation prevent flow into the different branches from getting out of balance. Finally, the juxtaposition of sequential enzyme activities, whether via aggregation of different chains or multifunctional polypeptides, permits 'metabolic channelling' of an intermediate that is thereby prevented from diffusing freely. This can ensure that an intermediate with alternative metabolic fates is channelled into a particular pathway – this might be the explanation, for example, for the juxtaposition of pyrimidine nucleotide-inhibitable carbamoyl phosphate synthetase and aspartate carbamoyltransferase referred to above. When this effect is absent, channelling may reduce the transient flux needed to start up the pathway since no pool of intermediate needs to accumulate, or it may prevent wasteful breakdown of substances unstable *in vivo*.

Leaving aside the plausibility of such arguments, what evidence is there that they are correct? If enzyme modulation is so economically important to the cell, variants that have lost this function through mutation should be at a selective disadvantage to the parent type. As far as I know, there is only one study in which such a result has been reported: that of Baich & Johnson (1968) with a mutant of *E.coli* W that had lost feedback control of proline biosynthesis. Against this, there is the purely anecdotal point, mentioned by Clarke (1979), that 'decontrolled' mutants generally show no sign of instability during culture maintenance. The universality of modulation in particular pathways also favours the *a priori* arguments: for instance, feedback inhibition in some form or another occurs in most amino-acid biosynthetic pathways.

It is when we try to account for the remarkable variety in control patterns for a single enzyme or pathway that doubts creep in. The point I wish to argue is that although a particular control pattern may play a satisfactory rôle in the life of the organism at present, it may not be in any useful sense 'optimal'. A given feature may be historical, meaning both that it may have been advantageous in an evolutionary progenitor under very different conditions, and that it may have resulted from a contingent event whose result has become fixed by accident. The common observation of constancy of control pattern within a group of related organisms which have diverged to fill widely different ecological niches, suggests that the pattern may have been present in a successful ancestor whose progeny diversified by adaptive radiation: it may then have been easier to superimpose modifications on the original control pattern where necessary, than abandon it in favour of some potentially more nearly ideal one.

Nevertheless, although control patterns are broadly conserved, there are differences between related species or even between different strains of the same species – for instance, the predominant aspartokinase varies from one *E. coli* strain to another (Cohen, Stanier & Le Bras, 1969). This variation is likely to be considerable, but is obscured by the understandable tendency for most work to be done on one or a small number of strains of a given species. Its origin and rôle is obscure, in common with those of most kinds of intraspecies variation in microorganisms except for those properties where it is variation *per se* that is advantageous: surface antigens, restriction-modification specificity, heterokaryon incompatibility and heterothallism. Are particular kinds of regulatory pattern adaptations to particular niches? Or could the major determining events be fixation of control alleles by genetic drift, or 'hitch-hiking' – e.g. by accompanying an allele which enhances fitness in some direct way such as increasing virulence – followed by further genetic changes to optimize the phenotype making use of the presence of the additional, control allele?

Other characteristics of an organism may constrain the kinds of adaptation possible in given selective conditions. In the systems described above, it is intriguing to note that particular groups of organisms have a predilection for particular regulatory patterns. For example, in the common pathways of both the aromatic and aspartate families of amino acids, the coliforms possess isoenzymes (DAHP synthase, chorismate mutase, aspartokinase-homoserine dehydrogenase), each specifically controlled by one of the endproducts. This device seems curiously wasteful: why encode more enzymes than necessary, and, if the pseudomonads and bacilli can manage with only one of each of these enzymes, why not the coliforms? One is driven to consider implausible ideas such as that *E. coli* enzymes do not easily evolve the more complex patterns of control that would allow one enzyme to suffice – it may be remembered, though, that Hartley (1974) was similarly driven by the finding that, whereas the *Ps. aeruginosa* amidase readily mutated to altered substrate specificity, certain *E. coli* and *Klebsiella aerogenes* enzymes appeared not to do so, to suggest that *Ps. aeruginosa* enzymes may have evolved to be potentially flexible in substrate specificity.

Another preference for particular patterns is shown by *S. cere-*

visiae and *N. crassa*, where many activities which in the bacteria are isolated or carried on separate polypeptide chains in aggregates, are now carried on bi- or multi-functional polypeptides. The extreme examples of this tendency are the *N. crassa* polypeptide which carries five consecutive activities of the common aromatic amino-acid pathway, mediating conversion of DAHP to 3-*eno*lpyruvyl shikimate 5-phosphate (Gaertner & Cole, 1976; Lumsden & Coggins, 1977; see also Welch & Gaertner, 1976, 1980, for an account of the 'catalytic facilitation' produced by interaction of this with other enzymes in the pathways) and the fatty acid synthetase complex of *S. cerevisiae*, with three activities on each of two chains (Tauro *et al.*, 1974; Knobling *et al.*, 1975). The only interpretation here seems to be that of Bonner, DeMoss & Mills (1965), who suggested in effect that multifunctionality is the most highly evolved form of enzyme juxtaposition and would therefore be expected to be more frequent in the microbial eukaryotes.

A similar problem applies to citrate synthase. Inhibition by α-oxoglutarate or succinyl-CoA is a form of feedback inhibition; that by ATP is an energy-linked control. That by NADH can be rationalized as a form of the latter (Sanwal, 1970; Sanwal, Kapoor & Duckworth, 1971), but why it should be restricted to Gram-negative bacteria is obscure. Sanwal (1970) points out that eukaryotes, with their energy-yielding biochemical machinery compartmentalized in mitochondria, might need fewer and different controls of activity of the relevant enzymes; but it is hard to see why this should also be true of Gram-positive bacteria. AMP antagonism of NADH inhibition, which is characteristic of strictly aerobic, and absent in facultatively anaerobic, Gram-negative bacteria, is understandable: the aerobic organisms depend on the citric acid cycle for energy production, so that inhibition of its operation must be reversible by a 'low-energy' signal such as AMP, while the latter organisms can generate energy by fermentation alone, so that it is key glycolytic enzymes which need to respond to such signals (Weitzman & Danson, 1976; Weitzman, 1980). Danson *et al.* (1979) have reported the isolation of an *E. coli* mutant in which the enzyme is insensitive to both NADH and α-oxoglutarate. The mutant enzyme is now of the 'small' type and is inhibited by ATP. This remarkable result suggests that the enterobacterial enzyme retains the possibility of adopting eukaryotic or Gram-positive characteristics, and more importantly highlights a theme also illustrated in the discussion below on control at the level of protein synthesis:

namely, that highly conserved (and therefore presumably fundamental) differences in regulatory phenotype may result from only slight changes at the genetic level.

ORIGINS OF CONTROLS AT THE LEVEL OF PREFORMED ENZYME

I am only going to tackle here the evolutionary origins of modulatable proteins and juxtaposed activities. For the former, we note that effectors bind *allosterically,* i.e. at sites distinct from substrate binding sites. Another useful term will be *domain,* a three-dimensional region of a protein concerned with a particular function, often the binding of a specific ligand. As mentioned previously, modulatable proteins are homooligomers, although the identical units (protomers) may themselves be composite. For instance, the *E.coli* aspartate carbamoyltransferase is a dodecamer made up of six catalytic subunits and six regulatory subunits which bear the binding sites for the feedback inhibitor pyrimidine nucleotides. In most cases, however, the catalytic and regulatory sites are on the same polypeptide.

How can we imagine evolution of modulatable proteins, enzyme aggregates and multifunctional polypeptides to start? The most reasonable way is to postulate a 'proto-peptide', relatively unspecific and conformationally 'floppy', so that binding of a ligand to one part of the molecule may perturb the binding site for another ligand in another part; weak activatory or inhibitory effects that can be strengthened by selection may thus be expected (Weber, 1975). The proto-peptide is encoded by a 'proto-gene'. Scattered proto-genes may come together in clusters while remaining separate, or may fuse to yield genes specifying bifunctional polypeptides. Conversely, a proto-gene may suffer duplication, to give either two scattered genes or (if the duplication is tandem) two contiguous or fused genes: since there is redundant genetic information, one copy could diverge and acquire altered or new functions, for which reason gene duplication has been postulated as a potent evolutionary mechanism (Ohno, 1970; Koch, 1972; Hartley, 1974). So the regulatory domain of a modulatable protein, whether carried on the same polypeptide or not, could derive from a proto-peptide unrelated to the catalytic proto-peptide but possessing, say, a binding site for the appropriate effector; or it could derive from the catalytic proto-peptide itself. If two juxtaposed enzyme activities, whether part of the same polypeptide or not, did not derive from the same proto-peptide, one or both could have evolved *ab initio* from proto-peptides with incipient activities of the same kind, or they could have been 'recruited' (seen by Jensen, 1976, as a likely route to new functions), following duplication and divergence of proto-genes under selection for a different, though perhaps related, activity. Finally, we can imagine the proto-peptides to be 'sticky', in that they readily bind weakly to one another. This again gives a route for selection to work, if this incipient aggregation confers an advantage, such as association of catalytic and regulatory subunits (as in *E.coli* aspartate carbamoyltransferase), or oligomerization of a modulatable enzyme to sharpen the sigmoidal response to effector, or aggregation leading to channelling of an intermediate. The initial weak binding must be imagined to involve hydrophilic surface residues, evolving to stronger binding with more hydrophobic residues. From these postulates, the required evolutionary models are self-evident.

Which is the more likely of the above two suggestions for the direction of evolution leading to two (or more) domains (e.g. for effector binding and potentially modulatable activity, or sequential enzyme activities) on a single polypeptide chain: centrifugal or centripetal? Is it more likely to involve partial proto-gene duplication to yield, after divergence of one region, an allosteric or bifunctional polypeptide; or is the recruitment of a regulatory domain, or the fusion of two initially scattered proto-genes to give eventually a bifunctional chain, more probable? Of course, each may operate in different systems (or indeed in the same system at different times). The two hypotheses have different corollaries: the centrifugal one, that, when derived from a single polypeptide, the regulatory and catalytic domains of an allosteric protein, or the two domains of a bifunctional enzyme, might show some similarity of amino-acid sequence or tertiary structure to each other; the centripetal one, that, certainly for the allosteric protein, and possibly for the bifunctional enzyme, for the reasons noted above, a recruited domain should resemble an apparently unrelated protein or domain thereof.

Before going further, some additional experimental information concerning the clustering of genes of related function in bacteria is needed (see Riley & Anilionis, 1978). This phenomenon, virtually absent in eukaryotes (Fincham, Day & Radford, 1979), is always found to varying degrees in the different bacterial groups, being more pronounced in the enterobacteria and *B. subtilis* than in the pseudomonads. By and large, organization in contiguous clusters is most pronounced for genes of peripheral catabolic sequences, and is more variable for genes of biosynthetic pathways. In some cases, genes of related function, even when not completely clustered, tend to lie closer together than expected on a purely random basis (e.g. in *E.coli* the *pro* and *arg* genes: Bachmann & Low, 1980; see also Riley & Anilionis, 1978). This was described (Leidigh & Wheelis, 1973; Wheelis, 1975) in one particularly spectacular case as 'supra-operonic organisation', for genes of catabolism of a variety of

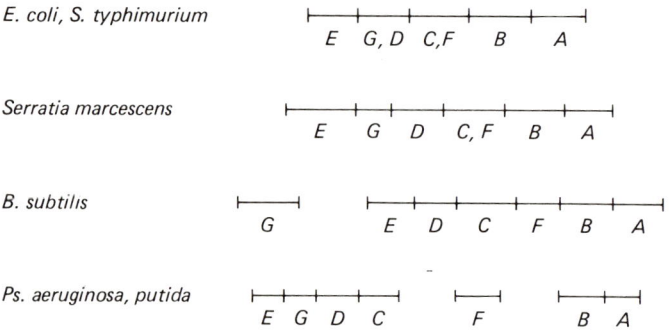

Fig. 5. Organization of genes of tryptophan biosynthesis in various bacteria. Regulatory genes known (*E. coli, S. typhimurium*) or presumed (*B. subtilis, Ps. aeruginosa* and *putida*) to encode repressors are unlinked to any of the structural genes. In *Ps. aeruginosa*, this regulatory gene controls the *trpEGDC* cluster only. In *B. subtilis*, the *trp* G gene is linked to one or more genes of *p*-aminobenzoate biosynthesis.

aromatic compounds by *Ps. putida*; the same term would seem to be appropriate for the genes of aromatic amino-acid biosynthesis in *B. subtilis* (Hoch & Nester, 1973; Nester & Montoya, 1976*a,b*). Fig. 5 illustrates clustering of genes of the tryptophan terminal sequence in various bacteria.

There are two hypotheses to explain clustering that do not relate to the functional expression of clustered genes (others that do will be discussed below). Horowitz (1945, 1965) put forward the classic centrifugal model of retrograde evolution of pathways. He suggested that under 'primeval soup' conditions (Orgel, 1973), the exhaustion of an amino-acid led to evolution of an activity (I will use the previous proto-gene–proto-peptide terminology for convenience), that could make it from the most suitable precursor (say P1).

When that in turn ran out, the process had to be repeated; since the original proto-peptide had P1 as substrate and the new one would need P1 as product (hence a P1 binding site should be on both), the new one could arise following tandem duplication of the original one, and would hence initially be contiguous. This process might be repeated many times, generating a gene cluster: an attractive feature is that each enzyme, down to the first in the sequence, could maintain an 'ancestral memory' of the end-product binding site of the original proto-peptide, which could evolve when advantageous into an allosteric site for feedback inhibition. This hypothesis suggests that some resemblance might persist between different enzymes in a given pathway.

An alternative explanation for clustering consistent with a centripetal direction of evolution (Clowes, 1960) is that it would be advantageous because only fragments (often small) of chromosomal DNA are transferred between prokaryotic cells; and transfer of isolated genes of a pathway, as opposed to the complete set, is unlikely to confer anything useful on a recipient.

What evidence is there for or against any of these evolutionary processes? Very little. For one allosteric protein, bovine glutamate dehydrogenase, it has been claimed (Engel, 1973) that the regulatory site could have evolved by internal duplication of a substrate binding site. For the aspartokinase-homoserine dehydrogenase system of *E.coli* K12, an extremely interesting recent finding is that all the isoenzymes, namely aspartokinase I-homoserine dehydrogenase I, aspartokinase II-homoserine dehydrogenase II and aspartokinase III, show immunological cross-reactivity, only detectable with one denatured enzyme as antigen, and antibody raised against another denatured enzyme; when native enzymes were used either as antigen in the test or to raise the antibody no cross-reaction could be detected (Zakin *et al.*, 1978; Cohen & Dautry-Varsat, 1980; Mouhli *et al.*, in press). The data suggest that the different aspartokinase and homoserine dehydrogenase domains contain regions that have evolved from common ancestors, presumably following genetic duplication. The results of sequencing studies on aspartokinase II-homoserine dehydrogenase II (the gene encoding aspartokinase I-homoserine dehydrogenase I having been sequenced: Katinka *et al.*, 1980) now in progress (G. N. Cohen, personal communication) will be of great interest. Finally, one instance has been claimed of homology between two enzymes catalysing sequential steps of a pathway (Ornston & Yeh, 1979). The enzymes are the muconolactone isomerases of *Ps. putida* and *Acinetobacter calcoaceticus,* and the β-oxoadipate *enol*-lactone hydrolase II of the latter organism: the pathway is for catabolism of aromatic compounds to β-oxoadipate and thence to citric-acid cycle intermediates (Ornston & Parke, 1977; Clarke, 1979). The enzymes are inducible, but details of the inducibility vary: in *Ps. putida,* there is a single *enol*-lactone hydrolase induced by its product β-oxoadipate, while in *A. calcoaceticus* there are two hydrolases, I and II, induced by different precursors. Ornston & Yeh (1979) show that, when the *A. calcoaceticus* isomerase amino-terminal sequence is compared with the first 29 amino-terminal residues of the same organism's hydrolase II, the best match gives only five identical amino acids, but if the same comparison is made with the *Ps. putida* isomerase, the best match produces ten identities. This suggests a common origin of the sequences compared, which could be consistent with the Horowitz model: however, the authors put forward a different hypothesis, namely 'that genetic rearrangements introduced the homologous regions of the isomerase and the hydrolase after the catalytic sites of the enzymes had been established'.

Can we deduce anything about the direction of evolution from the genetic organization of the tryptophan (*trp*) biosynthetic genes in different bacteria illustrated in Fig. 5? Although formally both the centrifugal and centripetal hypotheses could account for the observations, it seems to be generally agreed (see e.g. Crawford, 1975) that the centripetal version is easier to accept; for instance, that the location of the *trp G* gene in *B. subtilis* reflects a similar isolation of its homologues before their incorporation with *trp E, D* and *C* in the other organisms (assuming that they are indeed homologous, for which Crawford quotes some evidence), or that the bifunctional G–D and C–F polypeptides of the coliforms represent gene fusion.

CONTROLS AT THE LEVEL OF ENZYME SYNTHESIS AND THEIR SELECTIVE ADVANTAGES

To start with, some conceptual and terminological disentangling is necessary. Selection will usually operate through the phenotype, while the evolutionary history of the complex systems for regulation of enzyme synthesis needs to be considered at the genotypic level. I am therefore going to consider in this section the variety of control phenotypes, and postpone to the next section discussion of genetic mechanisms. It is not immediately apparent where phenotype ends and genetic mechanism begins: I shall take the former to include (i) the changes in rates of enzyme synthesis attendant on changes (primarily chemical) in environment, and (ii) the identity of the signal compound or metabolite responsible for those changes. Genetic organization of regulated genes, which it might be argued is also directly subject to selection, will be considered together with mechanisms, in the next section.

Current terminology does not make it easy to preserve the distinction between the phenotypic levels. The commonest observation is that when some substance is added to a culture growing on a given medium, rates of synthesis of certain enzymes may increase (*induction*) or decrease (*repression*). The substance that causes induction may be called the inducer, although it is often not the signal metabolite mechanistically responsible (e.g. in the case of product induction, see below): the latter, when identified, assumes the title of inducer, and some circumlocution has to be devised for the former. However, the substance that causes repression cannot be called the repressor, since this term has been pre-empted for a particular kind of mechanism; we end up having to use the clumsy *repressing metabolite,* though, when this is known to work by interaction with a protein repressor, it can be described as a *co-repressor*. The increase in rate of enzyme synthesis accompanying removal of repressing metabolite from the medium is termed *derepression*. Another semantic problem sometimes arises when the molecular mechanism of a control phenotype is determined: either induction or repression may work via *activation* (see next section), in which case the signal metabolite should perhaps be referred to as the *co-activator*.

Control of enzyme synthesis has been most thoroughly studied in biosynthetic and peripheral catabolic pathways. In biosynthetic pathways, by far the commonest phenotype is end-product repression; this is found, for instance, in virtually every amino-acid pathway in *E.coli* K12. It should be noted that partial or even full repression may result from an endogenously produced pool of repressing metabolite, even when the latter is not present in the medium. Although the end-product may itself be the repressing metabolite (as it appears to be, for instance, in the *E.coli* tryptophan (*trp*) system: Squires *et al.*, 1973; Squires, Lee & Yanofsky, 1975), this is not invariably the case. For instance, in *Ps. aeruginosa* and *Ps. putida* (Crawford, 1975) the E, G, D and C polypeptides are repressed by tryptophan, the F polypeptide is synthesized *constitutively*, i.e. at a constant rate apparently unaffected by pathway-related metabolites, and the A and B components of tryptophan synthase are induced by that enzyme's substrate, indoleglycerol phosphate (it will be noted from Fig. 5 that genes in the same cluster are under the same control).

Although repression of biosynthetic enzymes is very common, it is, by contrast with feedback inhibition, far from universal. Further, its commonness or rarity is a characteristic of different groups of organisms. In the pseudomonads it is often relatively inapparent: for instance, all the enzymes of pyrimidine biosynthesis in *Ps. aeruginosa* are constitutive (Isaac & Holloway, 1968). Of the eight enzymes of arginine biosynthesis (Fig. 4), only ornithine carbamoyltransferase is appreciably repressed, *N*-acetylglutamic semialdehyde dehydrogenase is slightly repressed (Isaac & Holloway, 1972) and one enzyme, *N*-acetylornithine δ-aminotransferase, is induced, probably because it also functions catabolically as ornithine aminotransferase (Voellmy & Leisinger, 1975). In *Chromobacter violaceum*, even the tryptophan enzymes are entirely constitutive (Wegman & Crawford, 1968). In microbial eukaryotes repression of biosynthetic enzymes is less pronounced than in bacteria, both in the proportion of regulated enzymes and the extent of repression (Metzenberg, 1972; Fincham, Day & Radford, 1979).

Another interesting point is that repressibility may vary from different isolates of a given species. For example, the arginine (*arg*) biosynthetic enzymes of *E.coli* are repressed in the K12 and W strains, but are constitutive or slightly inducible in the B strain (Jacoby & Gorini, 1967, 1969; Karlström & Gorini, 1969). Of 20 recent clinical isolates of *E.coli*, five showed repression ratios of less than 2, i.e. were more comparable to B than to K12 and W (S. J. Collinson & S. Baumberg, unpublished results). Again, any such general tendency would be obscured by the inclination of workers to concentrate on a very small number of strains.

In branched pathways, control of synthesis, like control of activity, of the enzymes of the common sequences is effected in complex ways by end-products (and sometimes intermediates), acting sometimes in concert and sometimes individually. We may refer back to the aromatic amino-acid pathways described earlier (see reviews by Umbarger, 1978, and Clarke, 1979). In *E.coli*, DAHP synthases (phe), (tyr) and (trp) are repressed respectively by phenylalanine + tryptophan, tyrosine and tryptophan; chorismate mutase T/prephenate dehydrogenase is repressed by tyrosine. In *Ps. aeruginosa,* the single DAHP synthase is constitutive. In *B. subtilis,* the aggregating DAHP synthase/chorismate mutase and shikimate kinase are repressed by tyrosine (this repression being enhanced by phenylalanine), which also represses the terminal enzymes of its synthetic pathway.

In catabolic pathways, there are usually two kinds of control, responsive respectively to presence/absence of substrate and levels of metabolites or energy within the cell. Induction in the presence of substrate is universal; however, it is very often found that the substrate has to be metabolized by the first enzyme of the pathway (which must be present, together with any inducible uptake system, at a sufficient basal level under non-inducing conditions) to give the true inducer (product induction). The latter may be the first pathway intermediate: e.g. the inducer of the histidine degradative (*hut*) enzymes is urocanate in *S. typhimurium, Klebsiella aerogenes* and *Ps. aeruginosa,* although it is histidine in *B. subtilis* (Lessie & Neidhardt, 1967; Magasanik *et al.,* 1974; Magasanik, 1976, 1980). Or the inducer may be a side-product: e.g. the inducer of the *E.coli* lactose (*lac*) system in the presence of lactose is *allo*lactose (Jobe & Bourgeois, 1972). In the long pathways of aromatic degradation in the pseudomonads and *Acinetobacter,* it is often found that enzymes are induced in sequential blocks, the product of one block causing induction of the next block, and so on. For instance, in *Ps. putida* the enzymes that convert D-mandelate to benzoate are induced by D- or L-mandelate or phenylglyoxylate (the latter two being the products of the first and second enzymes of the pathway respectively); benzoate induces the enzyme that converts it to catechol; the next set of enzymes, degrading catechol to β-oxoadipate *enol*-lactone, are induced by *cis,cis*-muconate; and the enzymes for the final steps of the pathway are induced by β-oxoadipate (Clarke & Ornston, 1975; Ornston & Parke, 1977; Clarke, 1979). This pattern is termed *sequential induction.* Substrate-related induction of catabolic systems is widespread in the microbial world, being equally common in those groups (pseudomonads, fungi) in which end-product repression is relatively infrequent.

Controls providing sensitivity to intracellular levels of metabolites or energy within the cell are termed (carbon) *catabolite repression* and *nitrogen metabolite regulation* or *repression* according to whether the response is to availability of carbon/energy or nitrogen. For instance, in *E.coli,* inducible sugar catabolic systems show catabolite repression by glucose, whereas in *Ps. aeruginosa* glucose catabolic enzymes are repressed by citrate (Hamlin, Ng & Dawes, 1967). The signal metabolite for carbon catabolite repression in the enterobacteria (and possibly also in *Ps. aeruginosa:* Smyth & Clarke, 1975) is cyclic 3', 5'-AMP(cAMP) (Pastan & Adhya, 1976), whose level correlates inversely with growth rate. In *K. aerogenes,* the enzymes of histidine catabolism show catabolite repression by glucose under conditions of nitrogen sufficiency; under nitrogen starvation, carbon catabolite repression is relieved due to active intervention of a control system (whose mechanism will be discussed below) sensitive to the nitrogen status of the cell. It appears in this case that there is not so much a signal metabolite as an enzyme, glutamine synthase, whose state of adenylylation acts as the signal (Magasanik *et al.,* 1974; Magasanik, 1976, 1980).

A similar relief of histidase from carbon catabolite repression (in this case caused by succinate) in nitrogen-limited conditions has been reported by Potts & Clarke (1976) in *Ps. aeruginosa.* In fungi, NH_4^+ often represses enzymes that catabolize nitrogenous compounds (and the permeases for uptake of the latter) (see Fincham, Day & Radford, 1979).

There are also a number of metabolic controls (most studied in bacteria) relating to growth rate and the overall rates of RNA and protein synthesis. Many of these affect macromolecular components of the machinery for gene expression, such as ribosomal RNA and proteins and transfer RNA (tRNA), whose rates of synthesis increase with growth rate. The signal metabolite (in *E.coli* at any rate) appears to be 5'-diphosphoguanosine 3'-diphosphate (ppGpp) (see review by Nierlich, 1978). Other controls related to growth rate, about which little is known, affect constitutive or inducible/repressible systems, such as the *E.coli* tryptophan system (*metabolic regulation*: Rose & Yanofsky, 1972; Platt, 1980).

Finally, it may be noted that control is necessarily implied, even for constitutive enzymes, by a set level of synthesis, and, for different enzymes in a pathway, by their relative levels. It might seem *a priori* reasonable that in an inducible or repressible system, the different proteins should be formed in a fixed ratio: however, such strict *co-ordinate control* is far from universal, as will be seen later.

Possible selective advantages of metabolic controls at the level of protein synthesis

Just as for controls at the level of enzyme activity, the reasons for the evolution of regulation at the level of synthesis seem obvious. The former allow economy in production of metabolites (or energy) by rapidly adapting flux through a pathway to need, while the latter allows economy in protein synthesis by more slowly adapting levels of enzymes (and other proteins such as permeases) to need. End-product repression, induction, carbon catabolite repression and nitrogen metabolite repression, growth-related control of ribosomal components and tRNA – what sensible queries can be raised about their existence? The same general caveats can be issued as for controls at the level of activity, noted above. My more specific comments will be directed at end-product repression – and its frequent absence. If repression is such a boon to a microorganism, how is it that, for many enzymes or systems, pseudomonads and microbial eukaryotes (and *C. violaceum,* for the *trp* system) can do so well without it? It will be argued (see e.g. Koch, 1974) that *E.coli*'s ecological niche subjects it to frequent fluctuations in nutrient level, so that end-product repression is highly advantageous; whereas the organisms in which repression is less frequent are adapted to a less nutritionally variable environment. This is plausible and may be true, but there is little independent evidence to support it. Precise information on the nutritional (and indeed most other) characteristics of the microbial environment is usually hard to come by. One class of organisms which has been claimed to be incapable of the usual range of controls of enzyme synthesis is the cyanobacteria (Carr, 1973), with the suggestion that this was connected with their autotrophic character. This contention has been challenged by Doolittle (1979) on what seem reasonable grounds.

But the variations in regulatory patterns between different groups are less surprising than those between different strains of the same or closely related species. To go back to the differences in repressibility of arginine enzymes in different strains of *E.coli:* this must mean either that the differences are of selective value and that these organisms are adapted to different niches, or that in their recent environments selection for repressibility has been relaxed, so that the control phenotypes have diverged by drift. If the latter is true, under what circumstances – and when – were the complex mechanisms of repression selected?

As with controls at the level of enzyme activity, few competition experiments have been carried out with isogenic strains differing only in control phenotype. Zamenhof & Eichhorn (1967) found that a *B. subtilis* mutant derepressed for the tryptophan biosynthetic enzymes was outgrown by the wild-type strain. But once again, there is the observation that regulatory mutants do not seem disadvantaged under normal stock maintenance conditions.

ORIGINS OF CONTROLS AT THE LEVEL OF ENZYME SYNTHESIS

I said above that speculation on the evolutionary history of patterns of control at the level of protein synthesis must be based on their genetic mechanisms. What are these? We have sufficiently detailed knowledge of the metabolic systems only for *E. coli* and *S. typhimurium*, and I will limit discussion to these.

It is useful to use the concept of a *control device*; every system generally has two or more of these, each acting relatively independently of the others. The best documented are: repressor–operator interaction; activator–DNA binding site interaction; attenuation (modulation of transcription termination efficiency); and RNA polymerase subunit addition, modification or replacement. The last-named has not been shown for certain to be involved in any purely metabolic system (Scaife, 1973; but see Greenblatt *et al.*, 1980, an intriguing paper, possibly with important implications) and so will not be referred to further here.

Repressor–operator interaction (it is hard to avoid calling this 'repression', in spite of the obvious confusion) has two main components: an oligomeric repressor protein and an operator DNA region to which repressor binds. The operator overlaps a promoter, at which RNA polymerase must bind to initiate transcription of one or more adjacent genes; binding of repressor to operator blocks access of RNA polymerase to the promoter and thus reduces frequency of transcription initiation, to an extent that depends on the fraction of time the operator remains free, leaving the promoter open. Repressors possess a DNA-binding domain at the amino-terminal end (Beyreuther, 1980; Miller, 1980; Weber & Geisler, 1980), and beyond this a domain that binds an effector whose binding affects the affinity of repressor for its operator: repressors may thus be regarded as allosteric proteins in a general sense (see

Barkley & Bourgeois, 1980). In inducible systems such as *E.coli lac*, binding of effector – the inducer – antagonizes operator binding; in repressible systems, such as *E.coli trp*, binding of effector – the co-repressor – enhances it (Platt, 1980).

It may be noted that whether a system is inducible or repressible depends only on whether binding of effector to repressor results in the operator-binding conformation of the repressor being less or more stable respectively than the operator-non-binding conformation. It is therefore not surprising that small mutational alterations in a repressor can cause a change from one control phenotype to another, or between either of these phenotypes and an equally balanced, constitutive level of expression. *lacI* (repressor gene) mutations are known that render the *lac* operon repressible by its normal inducers (Barkley & Bourgeois, 1980; Miller, 1980). The *arg* control phenotype of *E.coli* B mentioned above can alter to that of the K12 strain by a single point mutation (Jacoby & Gorini, 1969). Such observations indicate that, as for controls at the level of enzyme activity, drastic phenotypic differences may result from small genetic changes.

In regard to differences between related organisms, comparison of the *trp* systems of *E. coli* K12 and of *Shigella dysenteriae* 16 has proved very interesting. This *Shigella* strain is naturally auxotrophic for several amino acids, including tryptophan; the tryptophan auxotrophy is due to two mutations within *trpE* (Manson & Yanofsky, 1976a). Miozzari & Yanofsky (1978) have shown that in the 65-base pair DNA segment preceding the transcription startpoint which includes the promoter and operator, there are only two base-pair differences in sequence between *E. coli* K12 and *Sh. dysenteriae* 16. Nevertheless, transcription from the *Sh. dysenteriae* promoter is maximally only a tenth that in *E. coli*, and is much less repressible. It was shown that only one of these mutations, lying in the region of overlap between promoter and operator, is responsible for these effects. These results are clearly of great general interest in view of the frequency with which natural bacterial isolates are encountered that are partly or completely auxotrophic, often for nutrients which closely related organisms can manufacture for themselves. In most such cases, the genetic and enzymic bases for the auxotrophy are unknown, although it is often found that the strains can revert to prototrophy (e.g. tryptophan-requiring *Salmonella typhosa:* Fildes, Gladstone & Knight, 1933) suggesting that the block is due to one or more revertible mutations in an otherwise functional gene. It has been suggested (Miozzari & Yanofsky, 1978; Crawford & Stauffer, 1980) that this state of affairs is advantageous to *Sh. dysenteriae* 16 in an environment usually adequately supplied with tryptophan: the strain produces low semi-constitutive levels of the *trp* enzymes, while the pathway can be made functional by mutation if needed. There seems an undesirably teleological flavour to this argument, which may disappear as such studies are extended to other strains of *Shigella* and other enterobacteria auxotrophic for tryptophan. If an ancestral tryptophan prototroph, after long-continued growth in the presence of tryptophan loses unused *trp* functions by random mutations, these could presumably be deletions or frameshift mutations as well as the base-pair substitutions found in *Sh. dysenteriae* 16. It seems unlikely that base changes could be favoured because of the likely need to revert. Any *trpE* mutation (if not strongly polar) would still permit the strain to utilize small amounts of the tryptophan precursors, anthranilate and/or indole, which can be taken up from the gut.

Activator–DNA binding site interaction (activation) also has two components, an oligomeric activator protein and the DNA site to which it binds. The latter is adjacent to or part of the promoter for one or more contiguous genes. Binding of activator to its DNA site in some way enhances transcription initiation at the adjoining promoter; the favoured idea has been that activator binding in some way stimulates 'melting' (partial or incipient denaturation) of the promoter, but recent results with the *E. coli lac* (Reznikoff & Abelson, 1980) and galactose (*gal*) operons (Adhya & Miller, 1979; DiLauro *et al.*, 1979; de Crombrugghe & Pastan, 1980) suggest that the DNA-bound activator may interact directly with RNA polymerase. Like repressors, activators are allosteric; usually, binding of effector (*co-activator* would be a useful term) is needed for adoption of the DNA-binding conformation.

The best-known examples are the catabolite activator protein (or catabolite repressor protein) (CAP or CRP), which, with the effector cAMP, is responsible for the catabolite repression response in the coliforms (Pastan & Adhya, 1976; Reznikoff & Abelson, 1980; for suggestions that catabolite repression may involve more than interaction of the CAP–cAMP complex with its binding site next to the promoter, see Ullmann, Joseph & Danchin, 1979; Wanner, Kodaira & Neidhardt, 1978); and the *araC* protein, a repressor in the absence of its effector L-arabinose, and an activator in its presence (Lee, 1980). Another system is that responsible for the coliform nitrogen metabolite repression response. It has been proposed (Magasanik, 1976, 1980) that under nitrogen starvation conditions, glutamine synthetase is largely non-adenylylated, in which form it acts as an activator of nitrogen-yielding systems such as *hut* and also of its own synthesis – the latter is reasonable since in combination with glutamine:α-oxoglutarate aminotransferase (GOGAT) it furnishes a route of NH_4^+ assimilation of higher affinity than glutamate dehydrogenase, which functions under conditions of NH_4^+ sufficiency. This stimulation of its own synthesis makes glutamine synthetase an example of *autogenous regulation* (see below). It has, however, been suggested that this model may be over-simplified (Pahel & Tyler, 1979; Kustu *et at.*, 1979; Leonardo & Goldberg. 1980).

Attenuation differs from the other two devices in influencing, not the frequency of initiation of transcription of the controlled genes, but the frequency with which transcription, once initiated, reaches the regulated gene(s): the region of transcribed DNA between the transcription and translation startpoints (the *leader sequence*) contains an *attenuator* or *terminator* site at which a proportion of transcripts terminate. This proportion can be controlled. In certain enterobacterial amino-acid biosynthetic pathways (the only metabolic systems at present known to show attenuation), a complicated attenuation control operates with the following characteristics (Platt, 1980; Crawford & Stauffer, 1980):

(*a*) The leader messenger RNA (mRNA) contains multiple overlapping self-complementary regions (four in *E. coli trp*), each capable in theory of forming a unique *stem-loop* structure (the corresponding DNA region possesses a two-fold rotational axis of symmetry). Attenuation occurs just beyond the end of the last of these regions.

(*b*) Early in the leader mRNA there is a potential ribosome-binding site, followed by an AUG triplet at which translation could begin and one or more in-phase polypeptide chain termination triplets further on, at which it could stop. In *E. coli trp,* this hypothetically translatable region spans parts of the first two self-complementary regions.

(*c*) Within this hypothetically translatable region (whose putative product is called the *leader peptide*) are several codons, contiguous or close together, for the amino acid whose pathway is being regulated or for related amino acids known to affect expression of the cluster (e.g. two UGG's in the case of *E. coli trp*).

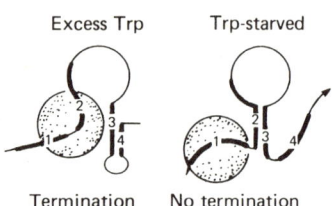

Fig. 6. Model for attenuation in the *E. coli trp* operon. Under conditions of excess tryptophan the ribosome (the shaded circle) translating the newly transcribed leader mRNA will synthesize the complete leader peptide. During this synthesis the ribosome will mask regions 1 and 2 of the mRNA and prevent the formation of stem and loop 1–2 or 2–3. Stem and loop 3–4 will be free to form and signal the RNA polymerase molecule (not shown) transcribing the leader sequence to terminate transcription. Under conditions of tryptophan starvation, charged tryptophanyl-tRNA will be limiting and the ribosome will stall at the adjacent tryptophan codons in the leader peptide coding region. Because only region 1 is masked, stem and loop 2–3 will be free to form as regions 2 and 3 are synthesized. Formation of stem and loop 2–3 will exclude the formation of stem and loop 3–4, which is required as the signal for transcription termination. Therefore, RNA polymerase will continue transcription into the structural genes. (Figure and legend adapted from Oxender *et al.*, 1979).

(*d*) The proportion of transcripts that terminate at the attenuator is influenced by mutations or nutritional conditions affecting the proportion of charged to uncharged aminoacyl-tRNA for the amino acid concerned; the more charged tRNA there is, the more attenuation occurs.

Yanofsky and his colleagues have suggested a model for *trp* (Fig. 6: Oxender, Zurawski & Yanofsky, 1979; Zurawski & Yanofsky, 1980; see also Crawford & Stauffer, 1980) in which attenuation occurs most efficiently when the attenuator mRNA is adjacent to a stem-loop structure. The probability of this structure forming can be predicted to correlate with conditions favouring attenuation. The effect of attenuation is not unlike repression, and it is not clear why some systems (such as *trp*) should possess both. Other enterobacterial systems, such as *his* (Johnston *et al.*, 1980), may possess only attenuation.

How might repressors and activators have evolved? Since they are allosteric proteins, the explanations for their derivation may be the same as those considered earlier for modulatable enzymes. Effectors for repressors either are, or resemble, ligands for enzymes of the regulated pathway: inducers will be substrates or products of the first enzyme, co-repressors products of the last. It is therefore tempting to suppose that an effector-binding domain might have originated from the ligand-binding domain of such an enzyme. However, amino-acid or DNA sequences have so far failed to show any of the homologies that might be expected. For instance, the *lac* repressor shows no homology with β-galactosidase (Beyreuther *et al.*, 1973). It may be that three-dimensional structures will show similarities not reflected in sequence relationships. The DNA-binding domains could have originated from those domains in enzymes involved in (or from proto-peptides evolving towards) replication, repair, or restriction/modification.

It is now emerging that many repressors repress their own synthesis. This is true for the *S. typhimurium hut* repressor, whose gene (*hutC*) is part of an operon together with other genes for the enzymes in this pathway (Magasanik, 1980), and for the *E. coli trp* and *arg* repressor genes (*trpR* and *argR*) which are unlinked to the corresponding structural genes (Zurawski *et al.*, 1981, and C. Yanofsky, personal communication; T. Eckhardt, personal communication). Perhaps polypeptides can most readily develop DNA-binding ability for an adjacent sequence in the first instance.

How might the DNA regions that specifically bind regulatory proteins – operators and activator-binding sites – have evolved? It is hard to see how they could derive other than from sequences recognised (possibly at first fortuitously) by a polypeptide already possessing a domain for binding an appropriate effector. Their partial twofold rotational symmetry (Dickson *et al.*, 1975; Pribnow, 1979; Barkley & Bourgeois, 1980; Reznikoff & Abelson, 1980) is capable of various explanations: the most likely is, perhaps, as a response (involving duplication and limited divergence) to symmetry in the binding protein, which in turn presumably reflects oligomerizability developed (as for any allosteric protein) to sharpen a threshold response to effector.

Attenuation, alone among control devices, requires no components additional to the cell's transcription–translation machinery. In the systems described, it could have evolved through the development and extension of a leader mRNA gradually accumulating triplets which provided the features of the model depicted in Fig. 5, the essential element being the codons for the regulatory amino acid(s).

Possible reasons for observed mixtures of control devices

In most of the well-studied coliform systems, more than one control device is found. For instance, *E. coli trp* shows repression and attenuation; *lac* and *ara*, repression and activation (*lac* only by the

CAP–cAMP system, *ara* by the *araC* protein-L-arabinose system also); *K. aerogenes hut,* repression and activation by both the CAP–cAMP and glutamine synthetase systems. A coherent attempt at explaining these observations has been made by Savageau. His theoretical exploration (1976) indicates, firstly, that a 'repressor-controlled mechanism is selected for regulation in an environment in which there is low demand for expression of the operon, whereas an activator-controlled mechanism is selected for regulation in an environment in which there is high demand for expression of the operon'. His second point refers to autogenous regulation, envisaged (Cove, 1974; Goldberger, 1974) as a form of control in which an enzyme regulates its own synthesis: he suggests that, on the basis of five criteria of functional effectiveness of inducible catabolic systems, 'repressor-controlled sytems with autogenous regulation are superior to the corresponding system with classical [i.e. non-autogenous] regulation according to all five criteria for functional effectiveness. Just the opposite is true for activator-controlled systems: according to all these same criteria, the classically regulated system is superior to the corresponding autogenously regulated systems' (Savageau, 1979). He has adduced evidence that both the first and second deductions (1974*a, b,* 1976, 1979) are borne out by the mechanisms of known systems.

This interesting treatment provides a comprehensive framework for the bewildering variety of regulatory mechanisms, and is at least partially in accord with the data. It will no doubt become further testable as knowledge accumulates both about the details of control systems, especially in organisms other than the coliforms, and about microbial nutritional ecology. A difficulty may lie in assimilating attenuation into Savageau's treatment. It will also be important to elucidate the rôle, in the *his* and *ilva* systems, of the first enzymes, which appear to play some part, possibly peripheral, in regulation, thus enabling these systems to be classed as showing autogenous control (Calhoun & Hatfield, 1975). The description of the *trp* and *arg* systems as classical, rather than autogenous, may be affected by the discovery (alluded to above) of the self-repression of *trpR* and *argR.*

Factors affecting gene organization

It has already been mentioned that, in bacteria, the genes encoding enzymes (and permeases) of a given metabolic system may be clustered to varying extents for different organisms and different systems in the same organisms.

Such a cluster is often referred to as an 'operon', though the term operon should strictly be confined to one or several genes whose transcription is under the control of a unique promoter/operator region. However, the term operon is often applied to a cluster of genes transcribed at the same time under any kind of control. In some instances, systems traditionally thought of as 'operons', e.g. the *S. typhimurium* histidine (*his*) cluster, have not been linked to a defined operator. Scattered genes, or those in different clusters, are in some instances regulated by the same devices, in which event they are said to constitute a *regulon* (Maas & Clark, 1964). For instance, the *trp* regulon in *E.coli* consists of the *trpEDCBA* cluster and *aroH*, the gene for DAHP synthase (trp), and *trpR*, all of which are under the control of the *trpR*-encoded repressor (Zurawski *et al.*, 1981). In the pseudomonads, *trpEGDC*, *trpF*, and *trpAB* constitute separate regulons. Clustered genes may be transcribed on a single mRNA molecule, as is probably the case for the *S. typhimurium his* genes, but their organization may be more complex. For instance, *E. coli trpEDCBA* and *argCBH* both have weak internal non-repressible promoters, so that the repressed levels of *trpCBA* and *argH* polypeptides are higher than those of *trpED* and *argCB* (Crawford, 1975; Elseviers *et al.*, 1972; Cunin *et al.*, 1975). Many clusters are transcribed on more than one mRNA: *E. coli araCBAD* forms the transcription units, *araBAD* and *araC* (Lee, 1980); *S. typhimurium hutIGCUH* is transcribed in two groups, *hutIGC* and *hutUH* (Magasanik, 1980). *E. coli argECBH* (Elseviers *et al.*, 1972; Bretscher & Baumberg, 1976; Boyen *et al.*, 1978; Charlier *et al.*, 1979) and *bioABFCD* (Otsuka & Abelson, 1979; Campbell, 1979) both show *divergent transcription* from a single regulatory region, lying between *argE* and *C* in the first case, and *bioA* and *B* in the second.

Two possible explanations were mentioned earlier for gene clustering in bacteria. A third, put forward initially by Jacob & Monod (1961*a*, *b*), is simply that it enables genes of related function to be co-ordinately controlled by transcription on a single mRNA. Although this often happens, especially, it would appear, in catabolic systems, the above exceptions suggest that its importance is not overriding. It has been proposed that the deviation from strict co-ordination resulting from the more complex types of organization is of selective value, but it is not clear that this is always plausible. Once again, one is tempted to ask: might they be survivals of past selective conditions, contingent historical accidents, or the adventitious results of constraints of which we are ignorant? An interesting model for the origin of divergently transcribed clusters has been proposed by Charlier *et al.* (1979).

Final comments on evolutionary problems of controls at the transcriptional level

One can think up plausible forms of control at the level of transcription that are not known to be used by microorganisms, and ask if there are manifest reasons why these should not have evolved.

First, there is translational control. This has been proposed from time to time for particular systems, for instance in the arginine biosynthetic enzymes in *E. coli* (Vogel & Vogel, 1974), although it seems unlikely to play a significant rôle there (Cunin *et al.*, 1975). The form of translational control suggested by Vogel and Vogel, in which mRNA half-life is regulated by a repression-style mechanism, seems *a priori* highly advantageous in retaining the flexibility of response of labile mRNA while mitigating its apparent wastefulness. One might think also that modulation of exonucleolytic attack by RNAses could be easily achieved in theory if transcripts were self-complementary at their ends, yielding double-stranded 'hairpins' to which specific regulatory proteins could bind as easily as to double-stranded DNA.

This hypothetical device resembles actual controls in being adaptive rather than clonal: i.e., adaptation is achieved by a phenotypic response in all members of a population rather than by the selection of genetic variants (not necessarily just mutants). But the latter adaptation occurs in microorganisms in a variety of contexts. The carriage of information that is advantageous only occasionally – antibiotic resistance, bacteriocin production, ability to utilize unusual nutrients, pathogenicity – on plasmids present only in subpopulations within a species (see Broda, 1979) may be viewed in this way. Perhaps more obviously relevant are the 'switch' mechanisms, involving inversion or translocation of DNA, by which genetic variants are constantly generated in certain systems. These have now been shown to include phase variation in *Salmonella* (Zieg *et al.*, 1980), host range variations via the invertible G loop in coliphage Mu (van de Putte, Cramer & Giphart–Gassler, 1980), and mating type switch in *S. cerevisiae* (Blair & Herskowitz, 1979;

Kushner, Blair & Herskowitz, 1979). Instead of maintaining elaborate environment-sensing mechanisms, why could not microorganisms have developed mechanisms to switch metabolic pathways on or off at random in a small proportion of the population, the minority type being selectable by an appropriate environment? An example of such a mechanism could be the lysine non-utilizing strain of *Ps. aeruginosa* described by Rahman & Clarke (1980), in which the phenotype is due to a regulatory allele that can readily mutate to permit utilization. It is tempting also to suggest that the 'mutable' lactose fermentation phenotype shown by many enterobacterial systems such as *Shigella sonnei* (Bergey, 1974) may be an example of just this mechanism. Indeed, this phenotype is found in the so-called *mutabile* strains of *E. coli*, which were used in some of the earliest experiments demonstrating the mutational nature of bacterial variation (see Jacob & Wollman, 1961) but have apparently since disappeared from bacterial genetics. Similar phenomena may occur in other carbon source utilization systems in *E. coli* (Woodward & Charles, 1980; Hill & Charles, 1980).

In conclusion, I regret that lack of space prevents discussion of a number of topics:

(*a*) Comparison of homologous sequences of regulatory and other components (both DNA and protein) in different organisms, in particular the data for enterobacterial *trp* regions from Yanofsky's laboratory (reviewed in Crawford & Stauffer, 1980).

(*b*) Functional divergence, or lack of it (Manson & Yanofsky, 1976*b*) inferred from studies on the interactions between regulatory components from different organisms (see e.g. Baumberg *et al.*, 1980).

(*c*) Regulatory systems of gene expression in yeasts and fungi (reviewed in Fincham, Day & Radford, 1979; Arst, 1981).

Most of the ideas put forward in this review that are not in the references cited were suggested in discussions with departmental colleagues, Professor P. H. Clarke, Dr C. Scazzochio, or Dr W. K. Maas. I thank Drs G. N. Cohen, T. Eckhardt and C. Yanofsky for communication of unpublished results.

REFERENCES

ADHYA, S. & MILLER, W. (1979). Modulation of the two promoters of the galactose operon of *Escherichia coli*. *Nature, London*, **279**, 492–4.

ARST, H. N. (1981). Control of genetic expression of fungi. In *Genetics as a tool in*

Microbiology. Symposium of the Society for General Microbiology, **31**, 131–60. Cambridge University Press.

ATKINSON, D. E. (1970). Enzymes as control elements in metabolic regulation. In *The Enzymes*, 3rd edn, vol. I, ed. P. D. Boyer, pp. 461–89. New York, London: Academic Press.

BACHMANN, B. J. & LOW, K. B. (1980). Linkage map of *Escherichia coli* K–12, edition 6. *Microbiological Reviews*, **44**, 1–56.

BAICH, A. & JOHNSON, M. (1968). Evolutionary advantage of a control of a biosynthetic pathway. *Nature, London*, **218**, 464–5.

BARKLEY, M. D. & BOURGEOIS, S. (1980). Repressor recognition of operator and effectors. In *The Operon*, 2nd edn, ed. J. H. Miller & W. S. Reznikoff, pp. 177–220. Cold Spring Harbor, N.Y.: Cold Spring Harbor Laboratory.

BAUMBERG, S., CORNELIS, G., PANAGIOTAKOPOULOS, M. & ROBERTS, M. (1980). Expression of the lactose transposon Tn951 in *Escherichia coli*, *Proteus* and *Pseudomonas*. *Journal of General Microbiology*, **119**, 257–62.

BENNETT, P. M. & RICHMOND, M. H. (1978). Plasmids and their possible influence on bacterial evolution. In *The Bacteria*, vol. VI: *Bacterial Diversity*, ed. L. N. Ornston & J. R. Sokatch, pp. 1–69. New York, San Francisco, London: Academic Press.

BERGEY (1974). Bergey's Manual of Determinative Bacteriology. 1st Edn. (1923); 2nd Edn (1926); 3rd Edn (1930); 4th Edn (1934); 5th Edn (1939); 6th Edn (1948). The Williams and Wilkins Co., Baltimore, 7th Edn (1957) Ballière, Tindall & Cox, London. 8th Edn (1974) The Williams and Wilkins Co., Baltimore.

BEYREUTHER, K. (1980). Chemical structure and functional organization of *lac* repressor from *Escherichia coli*. In *The Operon*, 2nd edn, ed. J. H. Miller & W. S. Reznikoff, pp. 123–54. Cold Spring Harbor, N.Y.: Cold Spring Harbor Laboratory.

BEYREUTHER, K., ADLER, K., GEISLER, N. & KLEMM, A. (1973). The amino-acid sequence of the *lac* repressor. *Proceedings of the National Academy of Sciences, USA*, **70**, 2576–80.

BLANGY, D., BUC, H. & MONOD, J. (1968). Kinetics of the allosteric interactions of phosphofructokinase from *Escherichia coli*. *Journal of Molecular Biology*, **31**, 13–35.

BONNER, D. M., DEMOSS, J. A. & MILLS, S. E. (1965). The evolution of an enzyme. In *Evolving Genes and Proteins*, ed. V. Bryson & H. J. Vogel, pp. 305–18. New York: Academic Press.

BOYEN, A., CHARLIER, D., CRABEEL, M., CUNIN, R., PALCHAUDHURI, S. & GLANSDORFF, N. (1978). Studies on the control region of the bipolar *argECBH* operon of *Escherichia coli*. I. Effect of regulatory mutations and IS2 insertions. *Molecular and General Genetics*, **161**, 185–96.

BRETSCHER, A. P. & BAUMBERG, S. (1976). Divergent transcription of the *argECBH* cluster of *Escherichia coli* K-12. Mutations which alter the control of enzyme synthesis. *Journal of Molecular Biology*, **102**, 205–20.

BRODA, P. (1979). *Plasmids*. Oxford, San Francisco: W. H. Freeman & Company.

CALHOUN, D. H. & HATFIELD, G. W. (1975). Autoregulation of gene expression. *Annual Review of Microbiology*, **29**, 275–99.

CALHOUN, D. H., PIERSON, D. L. & JENSEN, R. A. (1973). Channel-shuttle mechanism for the regulation of phenylalanine and tyrosine synthesis at a metabolic branch point in *Pseudomonas aeruginosa*. *Journal of Bacteriology*, **113**, 241–51.

CALOS, M. P. & MILLER, J. H. (1980). Transposable elements. *Cell*, **20**, 579–95.

CAMPBELL, A. M. (1979). Structure of complex operons. In *Biological Regulation and Development; vol. 1, Gene Expression,* ed. R. F. Goldberger, pp. 19–56. New York, London: Plenum Press.

CARR, N. G. (1973). Metabolic control and autotrophic physiology. In *The Biology of Blue-Green Algae,* ed. N. G. Carr & B. A. Whitton, pp. 39–65. Oxford: Blackwell Scientific Publications.

CHARLIER, D., CRABEEL, M., CUNIN, R. & GLANSDORFF, N. (1979). Tandem and inverted repeats of arginine genes in *Escherichia coli:* structural and evolutionary considerations. *Molecular and General Genetics,* **174,** 75–88.

CHARLIER, D., CRABEEL, M., PALCHAUDHURI, A., CUNIN, R. & GLANSDORFF, N. (1978). Heteroduplex analysis of insertion mutations in the bipolar *argECBH* operon of *Escherichia coli. Molecular and General Genetics,* **161,** 175–94.

CLARKE, P. H. (1974). The evolution of enzymes for the utilisation of novel substrates. In *Evolution in the Microbial World. Symposium of the Society for General Microbiology,* **24,** 183–217. Cambridge University Press.

CLARKE, P. H. (1978). Experiments in microbial evolution. In *The Bacteria, vol. VI: Bacterial Diversity,* ed. L. N. Ornston & J. R. Sokatch, pp. 137–218. New York, San Francisco, London: Academic Press.

CLARKE, P. H. (1979). Regulation of enzyme synthesis in the bacteria: a comparative and evolutionary study. In *Biological Regulation and Development, vol. 1: Gene Expression,* ed. R. F. Goldberger, pp. 109–70. New York, London: Plenum Press.

CLARKE, P. H. (1980) Experiments in microbial evolution: new enzymes, new metabolic activities. *Proceedings of the Royal Society of London* B, **207,** 385–404.

CLARKE, P. H. & ORNSTON, L. N. (1975). Metabolic pathways and regulation. In *Genetics and Biochemistry of Pseudomonas,* ed. P. H. Clarke & M. H. Richmond, pp. 191–340. London, New York, Sydney, Toronto: John Wiley & Sons.

CLOWES, R. C. (1960). Fine genetic structure as revealed by transduction. In *Microbial Genetics. Symposium of the Society for General Microbiology,* **10,** 92–114. Cambridge University Press.

COHEN, G. N. & DAUTRY-VARSAT, A. (1980). The aspartokinases-homoserine dehydrogenases of *Escherichia coli.* In *Multifunctional Proteins,* ed. H. Bisswanger & E. Schmincke-Ott, pp. 49–121. New York, London, Sydney & Toronto: John Wiley & Sons.

COHEN, G. N., STANIER, R. Y. & LE BRAS, G. (1969). Regulation of the biosynthesis of amino acids of the aspartate family in coliform bacteria and pseudomonads. *Journal of Bacteriology,* **99,** 791–801.

COVE, D. J. (1974). Evolutionary significance of autogenous regulation. *Nature, London,* **251,** 256.

CRAWFORD, I. P. (1975). Gene rearrangements in the evolution of the tryptophan synthetic pathway. *Bacteriological Reviews,* **39,** 87–120.

CRAWFORD, I. P. & STAUFFER, G. V. (1980). Regulation of tryptophan biosynthesis. *Annual Review of Biochemistry,* **49,** 163–95.

CUNIN, R., BOYEN, A., POUWELS, P., GLANSDORFF, N. & CRABEEL, M. (1975). Parameters of gene expression in the bipolar *arg ECBH* operon of *Escherichia coli* K-12. The question of translational control. *Molecular and General Genetics,* **140,** 51–60.

CYBIS, J. & DAVIS, R. H. (1975). Organization and control in the arginine biosynthetic pathway of *Neurospora. Journal of Bacteriology,* **123,** 196–202.

DANSON, M. J., HARFORD, S. & WEITZMAN, P. D. J. (1979). Studies on a mutant form of *Escherichia coli* citrate synthase desensitised to allosteric effectors. *European Journal of Biochemistry,* **101,** 515–21.

DAVIS, B. D. (1961). The teleonomic significance of biosynthetic control mechanisms. *Cold Spring Harbor Symposia on Quantitative Biology,* **26,** 1–10.
DE CROMBRUGGHE, B. & PASTAN, I. (1980). Cyclic AMP, the cyclic AMP receptor protein, and their dual control of the galactose operon. In *The Operon,* 2nd edn, ed. J. H. Miller & W. S. Reznikoff, pp. 303–324. Cold Spring Harbor, N. Y.: Cold Spring Harbor Laboratory.
DICKSON, R. C., ABELSON, J., BARNES, W. M. & REZNIKOFF, W. S. (1975). Genetic regulation: the *lac* control region. *Science,* **187,** 27–35.
DILAURO, R., TANIGUCHI, T., MUSSO, R. & DE CROMBRUGGHE, B. (1979). Unusual location and function of the operator in the *Escherichia coli* galactose operon. *Nature, London,* **279,** 494–500.
DOOLITTLE, W. F. (1979). The cyanobacterial genome, its expression, and the control of that expression. *Advances in Microbial Physiology,* **20,** 1–102.
ELSEVIERS, D., CUNIN, R., GLANSDORFF, N., BAUMBERG, S. & ASCHROFT, E. (1972). Control regions within the *argECBH* gene cluster of *Escherichia coli* K-12. *Molecular and General Genetics,* **117,** 349–66.
ENGEL, P. S. (1973). Evolution of enzyme regulator sites: evidence for partial gene duplication from amino-acid sequence of bovine glutamate dehydrogenase. *Nature, London,* **241,** 118–20.
FILDES, P., GLADSTONE, G. P. & KNIGHT, B. C. J. G. (1933). The nitrogen and vitamin requirements of *B. typhosus*. *British Journal of Experimental Pathology,* **14,** 189–96.
FINCHAM, J. R. S., DAY, P. R., & RADFORD, A. (1979). *Fungal genetics,* 4th edn. Oxford, London, Edinburgh, Melbourne: Blackwell Scientific Publications.
GAERTNER, F. & COLE, K. (1976). The protease problem in *Neurospora*. Structural modification of the *arom* multienzyme system during its extraction and isolation. *Archives of Biochemistry and Biophysics,* **177,** 566–73.
GOLDBERGER, R. F. (1974). Autogenous regulation of gene expression. *Science,* **183,** 810–16.
GOULD, S. J. & LEWONTIN, R. C. (1979). The spandrels of San Marco and the Panglossian paradigm: a critique of the adaptionist programme. *Proceedings of the Royal Society of London* B, **205,** 581–98.
GREENBLATT, J., LI, J., ADHYA, S., FRIEDMAN, D. I., BARON, L. S., REDFIELD, B., KUNG, H.-F. & WEISSBACH, H. (1980). L factor that is required for β-galactosidase synthesis is the *nusA* gene product involved in transcription termination. *Proceedings of the National Academy of Sciences, USA,* **77,** 1991–4.
HAMLIN, B. T., NG, F. M.-W. & DAWES, E. A. (1967). Regulation of enzymes of glucose metabolism in *Pseudomonas aeruginosa* by citrate. In *Microbial Physiology and Continuous Culture,* ed E. O. Powell, C. G. T. Evans, R. E. Strange & D. W. Tempest, pp. 211–19. London: Her Majesty's Stationery Office.
HARTLEY, B. S. (1974). Enzyme families. In *Evolution in the Microbial World. Symposium of the Society for General Microbiology,* **24,** 151–82. Cambridge University Press.
HARTLEY, B. S. (1979). Evolution of enzyme structure. *Proceedings of the Royal Society of London* B, **205,** 443–52.
HILL, S. & CHARLES, H. P. (1980). Genetical analysis of natural variation in sucrose utilisation in *Escherichia coli*. *The Society for General Microbiology Quarterly,* **7,** 82.
HOCH, J. A. & NESTER, E. W. (1973). Gene-enzyme relationships of aromatic acid biosynthesis in *Bacillus subtilis*. *Journal of Bacteriology,* **116,** 59–66.
HOROWITZ, N. H. (1945). On the evolution of biochemical syntheses. *Proceedings of the National Academy of Sciences, USA,* **31,** 153–7.

Horowitz, N. H. (1965). The evolution of biochemical syntheses – retrospect and prospect. In *Evolving Genes and Proteins*, ed. V. Bryson & H. J. Vogel, pp. 15–23. New York: Academic Press.

Huang, L., Montoya, A. & Nester, E. (1975). Purification and characterization of shikimate kinase enzyme activity in *Bacillus subtilis*. *Journal of Biological Chemistry*, **250**, 7675–81.

Hulett, F. M. & DeMoss, J. A. (1975). Subunit structure of anthranilate synthetase from *Neurospora crassa*. *Journal of Biological Chemistry*, **250**, 6648–52.

Isaac, J. H. & Holloway, B. W. (1968). Control of pyrimidine biosynthesis in *Pseudomonas aeruginosa*. *Journal of Bacteriology*, **96**, 1732–41.

Isaac, J. H. & Holloway, B. W. (1972). Control of arginine synthesis in *Pseudomonas aeruginosa*. *Journal of General Microbiology*, **73**, 427–38.

Issaly, I. M. & Issaly, A. S. (1974). Control of ornithine carbamoyltransferase activity by arginase in *Bacillus subtilis*. *European Journal of Biochemistry*, **49**, 485–95.

Jacob, F. & Monod, J. (1961a). Genetic regulatory mechanisms in the synthesis of proteins. *Journal of Molecular Biology*, **3**, 318–56.

Jacob, F. & Monod, J. (1961b). On the regulation of gene activity. *Cold Spring Harbor Symposia on Quantitative Biology*, **26**, 193–211.

Jacob, F. & Wollman, E. L. (1961). *Sexuality and the genetics of bacteria*. New York, London: Academic Press.

Jacoby, G. A. & Gorini, L. (1967). Genetics of control of the arginine pathway in *Escherichia coli* B and K. *Journal of Molecular Biology*, **24**, 41–50.

Jacoby, G. A. & Gorini, L. (1969). A unitary account of the repression mechanism of arginine biosynthesis in *Escherichia coli*. I. The genetic evidence. *Journal of Molecular Biology*, **39**, 73–87.

Jensen, R. A. (1976). Enzyme recruitment in evolution of new function. *Annual Review of Microbiology*, **30**, 409–25.

Jensen, R. A., Calhoun, D. H. & Stenmark, S. L. (1973). Allosteric inhibition of 3-deoxy-D-arabino-heptulosonate-7-phosphate synthetase by tyrosine, tryptophan and phenylpyruvate in *Pseudomonas aeruginosa*. *Biochemica et Eiophysica Acta*, **293**, 256–68.

Jobe, A. & Bourgeois, S. (1972). *lac* repressor-operator interaction. VI. The natural inducer of the *lac* operon. *Journal of Molecular Biology*, **69**, 397–408.

Johnston, H. M., Barnes, W. M., Chumley, F. G., Bossi, L. & Roth, J. R. (1980). Model for regulation of the histidine operon of *Salmonella*. *Proceedings of the National Academy of Sciences, USA*, **77**, 508–12.

Karlin, J. N., Bowman, B. J. & Davis, R. H. (1976). Compartmental behaviour of ornithine in *Neurospora crassa*. *Journal of Biological Chemistry*, **251**, 3948–55.

Karlström, O. & Gorini, L. (1969). A unitary account of the repression mechanism of arginine biosynthesis in *Escherichia coli*. II. Application to the physiological evidence. *Journal of Molecular Biology*, **39**, 89–94.

Katinka, M., Cossart, P., Sibilli, L., Saint Girons, I., Chalvignac, M. A., Lebras, G., Cohen, G. N. & Yaniv, M. (1980). Nucleotide sequence of the *thrA* gene of *Escherichia coli*. *Proceedings of the National Academy of Sciences, USA*, **77**, 5730–3.

Knobling, A., Schiffmann, D., Sickinger, H. D. & Schweitzer, E. (1975). Malonyl and palmityl transferase-less mutants of the yeast fatty-acid-synthetase complex. *European Journal of Biochemistry*, **56**, 359–67.

Koch, A. L. (1972). Enzyme volution: I. The importance of untranslatable intermediates. *Genetics*, **72**, 297–316.

Koch, A. L. (1971). The adaptive responses of *Escherichia coli* to a feast and famine existence. *Advances in Microbiological Physiology*, **6**, 147–217.

Koshland, D. E. Jr., (1970). The molecular basis for enzyme regulation. In *The Enzymes*, 3rd edn, vol. I, ed. P. D. Boyer, pp. 341–96. New York, London: Academic Press.

Kushner, P. J., Blair, L. C. & Herskowitz, I. (1979). Control of yeast cell types by mobile genes: a test. *Proceedings of the National Academy of Sciences, USA*, **76**, 5264–8.

Kustu, K., Burton, D., Garcia, E., McCarter, L. & McFarland, N. (1979). Nitrogen control in Salmonella: regulation by the *glnR* and *glnF* gene products. *Proceedings of the National Academy of Sciences, USA*, **76**, 4576–80.

Lee, N. (1980). Molecular aspects of *ara* regulation. In *The Operon*, 2nd edn; ed. J. H. Miller & W. S. Reznikoff, pp. 389–409. Cold Spring Harbor, N.Y.: Cold Spring Harbor Laboratory.

Leidigh, B. J. & Wheelis, M. L. (1973). The clustering on the *Pseudomonas putida* chromosome of genes specifying dissimilatory functions. *Journal of Molecular Evolution*, **2**, 235–42.

Leonardo, J. M. & Goldberg, R. B. (1980). Regulation of nitrogen metabolism in glutamine auxotrophs of *Klebsiella pneumoniae*. *Journal of Bacteriology*, **142**, 99–110.

Lessie, T. G. & Neidhardt, F. C. (1967). Formation and operation of the histidine-degrading pathway in *Pseudomonas aeruginosa*. *Journal of Bacteriology*, **93**, 1800–10.

Lumsden, J. & Coggins, J. R. (1977). The substructure of the *arom* multienzyme complex of *Neurospora crassa*. A possible pentafunctional polypeptide chain. *Biochemical Journal*, **161**, 599–607.

Maas, W. K. & Clark, A. J. (1964). Studies on the mechanism of repression of arginine biosynthesis in *Escherchia coli*. II. Dominance of repressibility in diploids. *Journal of Molecular Biology*, **8**, 365–70.

Magasanik, B. (1976). Classical and postclassical modes of regulation of the syntheses of degradative bacterial enzymes. *Progress in Nucleic Acids Research and Molecular Biology*, **17**, 99–115.

Magasanik, B. (1980). Regulation in the *hut* system. In *The Operon*, 2nd edn, ed. J. H. Miller & W. S. Reznikoff, pp. 373–87. Cold Spring Harbor, N.Y.: Cold Spring Harbor Laboratory.

Magasanik, B., Prival, N. H., Brenchley, J. E., Tyler, B. M., DeLeo, A. B., Streicher, S. L., Bender, R. A. & Paris, C. G. (1974). Glutamine synthetase as a regulator of enzyme synthesis. *Current Topics in Cellular Regulation*, **8**, 119–38.

Makoff, A. J. & Radford, A. (1978). Genetics and biochemistry of carbamoyl phosphate biosynthesis, and its utilization in the pyrimidine biosynthetic pathway. *Microbiological Reviews*, **42**, 307–28.

Manson, M. D. & Yanofsky, C. (1976a). Naturally occurring sites within the *Shigella dysenteriae* tryptophan operon severely limit tryptophan biosynthesis. *Journal of Bacteriology*, **126**, 668–78.

Manson, M. D. & Yanofsky, C. (1976b). Tryptophan operon regulation in interspecific hybrids of enteric bacteria. *Journal of Bacteriology*, **126**, 679–89.

Matchett, W. H. & DeMoss, J. A. (1975). The subunit structure of the tryptophan synthase from *Neurospora crassa*. *Journal of Biological Chemistry*, **250**, 2941–6.

Maurizi, M. R. & Switzer, R. L. (1980). Proteolysis in bacterial sporulation. *Current Topics in Cellular Regulation*, **16**, 164–224.

Metzenberg, R. L. (1972). Genetic regulatory systems in Neurospora. *Annual Review of Genetics*, **6**, 111–32.

MILLER, J. H. (1980). The *lacI* gene: its role in *lac* operon control and its use as a genetic system. In *The Operon*, 2nd edn, ed. J. H. Miller & W. S. Reznikoff, pp. 31–88. Cold Spring Haroor, N.Y.: Cold Spring Harbor Laboratory.

MIOZZARI, G. & YANOFSKY, C. (1978). Naturally occurring promoter down mutation: nucleotide sequence of the *trp* promoter/operator/leader region of *Shigella dysenteriae* 16. *Proceedings of the National Academy of Sciences, USA*, **75**, 5580–84.

MONOD, J., CHANGEUX, J.-P. & JACOB, F. (1963). Allosteric proteins and cellular control systems. *Journal of Molecular Biology*, **6**, 306–29.

MONOD, J., WYMAN, J. & CHANGEUX, J.-P. (1965). On the nature of allosteric transitions: a plausible model. *Journal of Molecular Biology*, **12**, 88–118.

MOUHLI, H., ZAKIN, M. M., RICHAUD, C. & COHEN, G. N. (1980). Detection of the homology among the aspartokinase I-homoserine dehydrogenase I and the aspartokinase III from *E. coli* K12 by immunochemical cross-reactivity between denatured species. *Biochemistry International*, in press.

NESTER, E. W. & MONTOYA, A. (1967*a*). An enzyme common to histidine and aromatic amino acid biosynthesis in *Bacillus subtilis. Journal of Bacteriology*, **126**, 699–705.

NESTER, E. W. & MONTOYA, A. (1976*b*). Involvement of a histidine locus in tyrosine and phenylalanine synthesis in *Bacillus subtilis*. In *Microbiology 1976*, ed. D. Schlessinger, pp. 141–4. Washington: American Society for Microbiology.

NIERLICH, D. P. (1978). Regulation of bacterial growth, RNA and protein synthesis. *Annual Review of Microbiology*, **32**, 393–432.

NILES, E. G. & WESTHEAD, E. W. (1973*a*). The variable subunit structure of lysine-sensitive aspartylkinase from *Escherichia coli* TIR-8. *Biochemistry*, **12**, 1715–22.

NILES, E. G. & WESTHEAD, E. W. (1973*b*). *In vitro* adenylation of lysine-sensitive aspartylkinase from *Escherichia coli* TIR-8. *Biochemistry*, **12**, 1723–9.

OHNO, S. (1970). *Evolution by gene duplication*. London: Allen & Unwin.

ORGEL, L. E. (1973). *The origins of life: molecules and natural selection*. London: Chapman and Hall.

ORNSTON, L. N. & PARKE, D. (1977). The evolution of induction mechanisms in bacteria: insights derived from the study of the β-ketoadipate pathway. *Current Topics in Cellular Regulation*, **12**, 209–62.

ORNSTON, L. N. & YEH, W. K. (1979). Origins of metabolic diversity: evolutionary divergence by sequence repetition. *Proceedings of the National Academy of Sciences, USA*, **76**, 3996–4000.

OTSUKA, A. & ABELSON, J. (1979). The regulatory region of the biotin operon in *Escherichia coli. Nature, London*, **276**, 689–94.

OXENDER, D. L., ZURAWSKI, G. & YANOFSKY, C. (1979). Attenuation in the *Escherichia coli* tryptophan operon: role of RNA secondary structure involving the tryptophan codon region. *Proceedings of the National Academy of Sciences, USA*, **76**, 5524–8.

PAHEL, G. & TYLER, B. M. (1979). A new regulatory gene for glutamine synthetase in *Escherichia coli. Proceedings of the National Academy of Sciences, USA*, **76**, 4544–8.

PASTAN, I. & ADHYA, S. (1976). Cyclic adenosine 5'-monophosphate in *Escherichia coli. Bacteriological Reviews*, **40**, 527–51.

PATEL, N., PIERSON, D. L. & JENSEN, R. A. (1977). Dual enzymatic routes to L-tyrosine and L-phenylalanine via pretyrosine in *Pseudomonas aeruginosa. Journal of Biological Chemistry*, **252**, 5839–46.

PENNINCKX, M. (1975). Interaction between arginase and L-ornithine carbamoyltransferase in *Saccharomyces cerevisiae*. The regulatory sites of arginase. *European Journal of Biochemistry*, **58**, 533–8.

PENNINCKX, M., SIMON, J. P. & WIAME, J. M. (1974). Interaction between arginase and L-ornithine carbamoyltransferase in *Saccharomyces cerevisiae*. Purification of *S. cerevisiae* enzymes and evidence that these enzymes as well as ratliver arginase are trimers. *European Journal of Biochemistry*, **49**, 429–42.

PLATT, T. (1980). Regulation of gene expression in the tryptophan operon of *Escherichia coli*. In *The Operon*, 2nd edn, ed. J. H. Miller & W. S. Reznikoff, pp. 263–302. Cold Spring Harbor, N.Y.: Cold Spring Harbor Laboratory.

POTTS, J. R. & CLARKE, P. H. (1976). The effect of nitrogen limitation on catabolite repression of amidase, histidase and urocanase in *Pseudomonas aeruginosa*. *Journal of General Microbiology*, **93**, 377–87.

PRIBNOW, D. (1979). Genetic control signals in DNA. In *Biological Regulation and Development, vol. 1: Gene Expression*, ed. R. F. Goldberger, pp. 219–78. New York, London: Plenum Press.

RAHMAN, M. & CLARKE, P. H. (1980). Genes and enzymes of lysine catabolism in *Pseudomonas aeruginosa*. *Journal of General Microbiology*, **116**, 357–69.

REBELLO, J. L. & JENSEN, R. A. (1970). Metabolic interlock: the multi-metabolite control of prephenate dehydratase activity in *Bacillus subtilis*. *Journal of Biological Chemistry*, **245**, 3738–44.

REZNIKOFF, W. S. & ABELSON, J. N. (1980). The *lac* promoter. In *The Operon*, 2nd edn, eds. J. H. Miller & W. S. Reznikoff, pp. 221–43. Cold Spring Harbor, N.Y.: Cold Spring Harbor Laboratory.

RILEY, M. & ANILIONIS, A. (1978). Evolution of the bacterial genome. *Annual Review of Microbiology*, **32**, 519–60.

ROSE, J. K. & YANOFSKY, C. (1972). Metabolic regulation of the tryptophan operon of *Escherichia coli:* repressor independent regulation of transcription initiation frequency. *Journal of Molecular Biology*, **69**, 103–18.

SANWAL, B. D. (1970). Allosteric controls of amphibolic pathways in bacteria. *Bacteriological Reviews*, **34**, 20–39.

SANWAL, B. D., KAPOOR, M. & DUCKWORTH, H. W. (1971). The regulation of branched and converging pathways. *Current Topics in Cellular Regulation*, **3**, 1–115.

SAVAGEAU, M. A. (1974*a*). Genetic regulatory mechanisms and the ecological niche of *Escherichia coli*. *Proceedings of the National Academy of Sciences, USA*, **71**, 2453–5.

SAVAGEAU, M. A. (1974*b*). Comparison of classicial and autogenous systems of regulation in inducible operons. *Nature, London*, **252**, 546–9.

SAVAGEAU, M. A. (1976). *Biochemical Systems Analysis: a Study of Function and Design in Molecular Biology*. Reading, Massachusetts: Addison-Wesley.

SAVAGEAU, M. A. (1979). Autogenous and classical regulation of gene expression: a general theory and experimental evidence. In *Biological Regulation and Development, vol. 1: Gene Expression*, ed. R. F. Goldberger, pp. 57–108. New York, London: Plenum Press.

SCAIFE, J. G. (1973). Control of transcription in bacteria. *British Medical Bulletin*, **29**, 214–19.

SHIZUTA, Y. & HAYAISHI, O. (1976). Regulation of biodegradative threonine deaminase. *Current Topics in Cellular Regulation*, **11**, 99–146.

SMYTH, P. F. & CLARKE, P. H. (1975). Catabolite repression of *Pseudomonas aeruginosa* amidase: the effect of carbon source on amidase synthesis. *Journal of General Microbiology*, **90**, 81–90.

SQUIRES, C. L., LEE, F. D. & YANOFSKY, C. (1975). Interaction of the *trp* repressor and RNA polymerase with the *trp* operon. *Journal of Molecular Biology*, **92**, 93–111.

SQUIRES, C. L., ROSE, J. K., YANOFSKY, C., YANG, H.-L. & ZUBAY, G. (1973). Tryptophanyl-tRNA and tryptophanyl-tRNA synthetase are not required for *in vitro* repression of the tryptophan operator. *Nature New Biology, London*, **245**, 131–3.

STADTMAN, E. R. & GINSBURG, A. (1974). The glutamine synthetase of *Escherichia coli*: structure and control. In *The Enzymes*, 3rd edn, vol. X, ed. P. D. Boyer, pp. 755–807. New York, London: Academic Press.

STRATHERN, J. N., BLAIR, L. C. & HERSKOWITZ, I. (1979). Healing of *mat* mutations and control of mating type interconversion of the mating type locus in *Saccharomyces cerevisiae*. *Proceedings of the National Academy of Sciences, USA*, **76**, 3425–9.

SWITZER, R. L. (1977). The inactivation of microbial enzymes *in vivo*. *Annual Review of Microbiology*, **31**, 135–57.

TAURO, P., HOLZNER, U., CASTORPH, H., HILL, F. & SCHWEITZER, E. (1974). Gnetic analysis of non-complementing fatty acid synthetase mutants in *Saccharomyces cerevisiae*. *Molecular and General Genetics*, **129**, 131–48.

ULLMAN, A., JOSEPH, E. & DANCHIN, A. (1979). Cyclic AMP as a modulator of polarity in polycistronic transcriptional units. *Proceedings of the National Academy of Sciences, USA*, **76**, 3194–7.

UMBARGER, H. E. (1978). Amino acid biosynthesis and its regulation. *Annual Review of Biochemistry*, **47**, 533–606.

UMBARGER, H. E. & BROWN, B. (1957). Threonine deamination in *Escherichia coli*. II. Evidence for two L-threonine deaminases. *Journal of Bacteriology*, **73**, 105–12.

VAN DE PUTTE, P., CRAMER, S. & GIPHART-GASSLER, M. (1980). Invertible DNA determines host specificity of bacteriophage Mu. *Nature, London*, **286**, 218–22.

VOELLMY, R. & LEISINGER, T. (1975). Dual role for N-acetylornithine-5-aminotransferase from *Pseudomonas aeruginosa* in arginine biosynthesis and arginine catabolism. *Journal of Bacteriology*, **122**, 799–809.

VOGEL, H. J. & VOGEL, R. H. (1974). Enzymes of arginine biosynthesis and their repressive control. *Adances in Enzymology*, **40**, 65–90.

WANNER, B. L., KODAIRA, R. & NEIDHARDT, F. C. (1978). Regulation of *lac* operon expression: reappraisal of the theory of catabolite repression. *Journal of Bacteriology*, **136**, 947–54.

WEBER, G. (1975). Energetics of ligand binding to proteins. *Advances in Protein Chemistry*, **29**, 1–83.

WEBER, K. & GEISLER, N. (1980). *lac* repressor fragments produced in vivo and in vitro: an approach to the understanding of the interaction of repressor and DNA. In *The Operon*, 2nd edn, ed. J. H. Miller & W. S. Reznikoff, pp. 155–77. Cold Spring Harbor, N.Y.: Cold Spring Harbor Laboratory.

WEGMAN, J. & CRAWFORD, I. P. (1968). Tryptophan synthetic pathway and its regulation in *Chromobacterium violaceum*. *Journal of Bacteriology*, **95**, 2325–35.

WEISS, R. L. & DAVIS, R. H. (1977). Control of arginine utilization in *Neurospora*. *Journal of Bacteriology*, **129**, 866–73.

WEITZMAN, P. D. J. (1980). Citrate synthese and succinate thiokinase in classification and identification. In *Microbial Classification and Identification. Society for Applied Bacteriology Symposium Series No. 8*, ed. M. Goodfellow & R. G. Board, pp. 107–125. London, New York: Academic Press.

WEITZMAN, P. D. J. & DANSON, M. J. (1976). Citrate synthase. *Current Topics in Cellular Regulation*, **10**, 161–204.

WEITZMAN, P. D. J. & JONES, D. (1975). The mode of regulation of bacterial citrate synthase as a taxonomic tool. *Journal of General Microbiology*, **89**, 187–90.

WELCH, G, R. & GAERTNER, G. H. (1976). Coordinate activation of a multienzyme complex by the first substrate. Evidence for a novel regulatory mechanism in the polyaromatic pathway of *Neurospora crassa*. *Archives of Biochemistry and Biophysics*, **172**, 476–89.

WELCH, G. R. & GAERTNER, F. H. (1980). Enzyme organization in the polyaromatic-biosynthetic pathway: the *arom* conjugate and other multienzyme systems. *Current Topics in Cellular Regulation*, **16**, 113–62.

WHEELIS, M. L. (1975). The genetics of dissimilatory pathways in *Pseudomonas*. *Annual Review of Microbiology*, **29**, 505–24.

WOODWARD, M. J. & CHARLES, H. P. (1980). Genes for ribitol and arabitol utilisation, and genes for galactitol utilisation, behave as chromosomal alternatives. *The Society for General Microbiology Quarterly*, **7**, 82.

WYMAN, J. (1972). On allosteric models. *Current Topics in Cellular Regulation*, **6**, 209–26.

ZAKIN, M. M., GAREL, J.-R, DAUTRY-VARSAT, A. COHEN, G. N. & BOULOT, G. (1978). Detection of the homology among proteins by immunochemical cross-reactivity between denatured antigens. Application to the threonine and methionine regulated aspartokinases – homoserine dehydrogenases from *Escherichia coli* K12. *Biochemistry*, **17**, 4318–23.

ZAMENHOF, S. & EICHHORN, H. H. (1967). Study of microbial evolution through loss of biosynthetic function: establishment of 'defective' mutants. *Nature, London*, **216**, 456–8.

ZIEG, J., SILVERMAN, M., HILMEN, M. & SIMON, M. (1980). The mechanism of phase variation. In *The Operon*, 2nd edn, ed. J. H. Miller & W. S. Reznikoff, pp. 411–23. Cold Spring Harbor, N.Y.: Cold Spring Harbor Laboratory.

ZURAWSKI, G., GUNSALUS, R. P., BROWN, K. D. & YANOFSKY, C. (1980). Structure and regulation of *aroH*, the structural gene for the tryptophan-repressible 3-deoxy-D-arabino-heptulosonic acid-7-phosphate synthetase of *Escherichia coli*. *Journal of Molecular Biology*, **145**, 47–73.

ZURAWSKI, G. & YANOFSKY, C. (1980). *Escherichia coli* tryptophan operon leader mutations, which relieve transcription termination, are *cis*-dominant to *trp* leader mutations, which increase transcription termination. *Journal of Molecular Biology*, **142**, 123–9.

THE EVOLUTION OF MEMBRANE-BOUND BIOENERGETIC SYSTEMS: THE DEVELOPMENT OF VECTORIAL OXIDOREDUCTIONS

PETER B. GARLAND

Biochemistry Department, Medical Sciences Institute, University of Dundee, Dundee DD1 4HN, Scotland, UK

INTRODUCTION

Membrane-bound systems of particular bioenergetic interest include porters or permeases driven by electrochemical gradients, ion-translocating ATPases, and proton-translocating respiratory and photosynthetic chains. These subjects were well reviewed at a recent Symposium of this Society (Haddock & Hamilton, 1977) and need not be presented again.

The evolution of such systems has been assessed by several authors from a chemiosmotic viewpoint; i.e. how the selective pressures on evolving organisms were successfully met by the evolution of proton-translocating mechanisms (Raven & Smith, 1976, 1981; Smith & Raven, 1978; Broda & Peschek, 1979; Michel, Michel, Boonstra & Konings, 1979; Gest, 1980). Accordingly I shall give only an outline of the evolution of bioenergetic systems. However, most treatments of this subject have been physiological rather than biochemical, i.e. the emphasis has been on the chronological order in which various parts of the present-day chemiosmotic apparatus appeared in response to environmental challenges. So there remains a need to think, or to start to think, about how the molecular apparatus of bioenergetic systems evolved from whatever preceded it.

EVOLUTION OF BIOENERGETIC SYSTEMS

The evolutionary time scale for the development of biological energy conversion systems is summarized in Fig. 1, of which the main points are as follows.

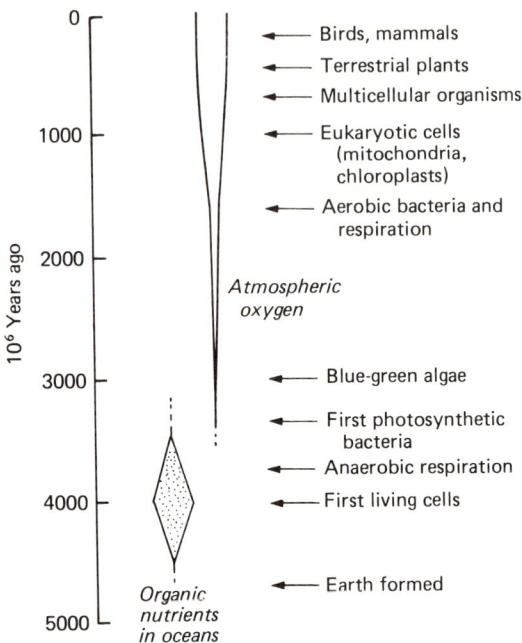

Fig. 1. Evolutionary time scale for the development of photosynthesis and respiration.

The primeval soup

About 4×10^9 years ago the earth had cooled to about its present temperature, the sea and land masses had formed, and the atmosphere was reducing and anoxic. Over the previous $8-9 \times 10^8$ years the exposure of the earth's atmosphere of methane, carbon dioxide, nitrogen, ammonia, hydrogen sulphide and hydrogen to electrical storms, to heat and to ultraviolet radiation had led to the accumulation of many organic molecules; carbohydrates, aminoacids, heterocyclic compounds of carbon and nitrogen, and hydrocarbons. Concentration of these in solution or on surfaces may have favoured their condensation into polymers; primitive polysaccharides, polypeptides, polynucleotides and lipid bilayers (an honorary polymer for the purpose of this discussion). It was in this primeval soup that the earliest life forms are considered to have originated.

Fermentative metabolism and substrate level phosphorylation

The first 'living' cells were presumably self-replicating collections of molecules. For instance, a liposome able to recruit new lipids from the environment and thence to grow and divide would have certain attributes of a living cell. The evolution of enzymes and metabolic pathways is harder to envisage but, given that it happened, the simplest way of generating ATP in an anaerobic but nutrient-rich environment would be by substrate-level phosphorylations, as in glycolysis. The products of fermentation such as lactate or acetate, are acidic, and this has led to two interesting evolutionary hypotheses. The first is that the proton-translocating ATPase of the bacterial cell membrane initially evolved as a device for regulating intracellular pH, and came to be used in ATP synthesis only secondarily and later, when alternative means of proton pumping such as respiration or photosynthesis had been developed (Raven & Smith, 1976). The second is that the efflux of acids (with H^+) from fermenting cells could by itself set up a useful electrochemical H^+ gradient capable of being coupled to the transport of other solutes (Michel et al., 1979).

Anaerobic respiration

A central problem in fermentation is the disposal of reducing equivalents. Terminal oxidants such as oxygen or nitrate were unavailable in the early pre photosynthetic time. But fumarate probably was, in that it could be formed by fermentative pathways. Thus the development of anaerobic oxidoreduction pathways terminating in the reduction of fumarate to succinate was likely. As pointed out by Kröger (1977), additional substrate-level phosphorylations could become available as a consequence of an electron-transport pathway terminating in fumarate reductase. However, an even greater selective advantage would come if the fumarate-reducing electron-transport pathway was not merely some cytoplasmic adjunct to the intermediary metabolism of substrate-level phosphorylation, but additionally was a proton-translocating system of the cytoplasmic membrane. A respiratory proton pump, albeit anaerobic, would assist a cell in two ways; it would spare ATP from consumption by the proton-translocating ATPase, and it could generate ATP by reversal of the ATPase. Thus even before the evolution of photosynthesis or aerobic respiration, the essential features of membrane-bound chemiosmotic systems were probably present; a proton-translocating ATPase and a proton-translocating electron-transport chain. Subsequent evolution probably did little to change the ATPase, which is highly conserved (Postma & Van Dam, 1976). However, the sources and sinks of electrons for electron transport exhibit a great diversity (Haddock & Jones, 1977).

An energy crisis

The early organisms lived on environmental carbon and energy sources accumulated on the surface of the earth over the preceding $8-9 \times 10^8$ years. In effect, they were living on fossil fuels. Then, as now, the depletion of environmental carbon and energy sources at a rate in excess of their replenishment would inevitably have checked the continuation of life unless alternative sources were exploited. The biochemical problem faced by the early heterotrophic bacteria was to find alternative sources of carbon for cell mass and of energy for biosynthesis, transport, etc. Because fermentable organic nutrients were becoming scarce, a selective pressure was created which favoured cells that could convert carbon dioxide, which was plentiful enough, to carbohydrate. This is a reductive process, so the problem therefore was to find suitable hydrogen donors in the environment. As already mentioned, the prebiotic supplies of organic reductants were becoming depleted, both through biochemical conversion to less suitable products and through geological loss of cell material in muds and sediments. In principle three classes of hydrogen donor were available; organic acids such as succinate formed earlier by fermentation of carbohydrate, hydrogen sulphide and water. None of these reductants is of sufficiently low (negative) oxidoreduction potential to reduce carbon dioxide to carbohydrate. The solution to this problem was to develop a photosynthetic apparatus that would enable these relatively high-potential oxidoreductants to reduce a low-potential acceptor which in turn could reduce carbon dioxide. The overall reactions can be represented as

$$\text{succinate} + A \xrightarrow{\text{light energy}} \text{fumarate} + AH_2 \qquad (1)$$

$$H_2S + A \xrightarrow{\text{light energy}} S + AH_2 \qquad (2)$$

$$H_2O + A \xrightarrow{\text{light energy}} \tfrac{1}{2}O_2 + AH_2 \qquad (3)$$

$$6CO_2 + 6AH_2 \longrightarrow C_6H_{12}O_6 + 6A \qquad (4)$$

where A stands for the low-potential hydrogen acceptor, usually NADP.

Early photosynthetic bacteria using organic acids or hydrogen sulphide as a source of reducing equivalents had a limited environmental range that became even more restricted as the available supplies of organic acids and hydrogen sulphide were utilized. The evolution of organisms capable of using water as the hydrogen source occurred some 3×10^9 years ago, with the development of green algae. The biological consequences of this development were threefold. Firstly, the algae had minimal requirements of their environment and, unlike the photosynthetic bacteria, were not restricted to habitats containing hydrogen sulphide or organic acids. Secondly, oxygen entered the atmosphere, allowing the evolution of aerobic pathways of catabolism. Thirdly, the combined presence of oxygen and plant material permitted the evolution of animals.

THE EVOLUTION OF VECTORIAL BIOCHEMISTRY

Vectorial versus scalar reactions

Scalar reactions are those that involve no sense of spatial direction. The reactions of glycolysis as catalyzed by soluble enzymes in a test tube or cytoplasm are scalar. By contrast, an enzyme that resides in a membrane and catalyzes an asymmetric reaction across that membrane, can be said to catalyze a vectorial reaction. An example is the proton-translocating ATPase of the bacterial cytoplasmic membrane, catalyzing the reaction

$$2H^+_{in} + ATP_{in} + H_2O \rightarrow ADP_{in} + Pi_{in} + 2H^+_{out} \qquad (5)$$

where 'in' and 'out' refer to intra- and extra-cytoplasmic phases respectively. The importance of vectorial biochemistry in biological energy conversions underlay the formulation of Mitchell's chemiosmotic hypothesis (Mitchell, 1968, 1977). A diagrammatic representation of the steps in the evolution of bioenergetic systems (Fig. 2) emphasizes again the rôle of the membrane and of vectorial electron and proton movements across the membrane.

Did vectorial enzymes in membrane develop from scalar enzymes of the cytoplasm?

A vectorial enzyme catalyzes the same chemical reaction as its soluble scalar counterpart does. In addition, the vectorial enzyme

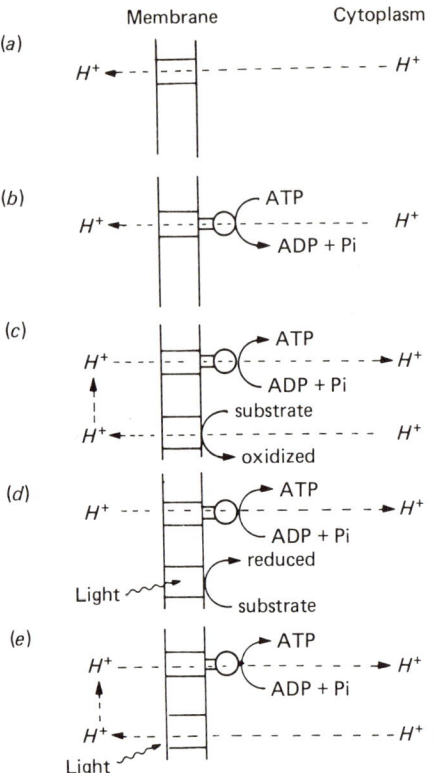

Fig. 2. Development of proton-translocation systems in the cytoplasmic membrane. (a) Early cell, H^+ formed by fermentation exported by passive diffusion or with an acid anion. (b) Proton-translocating ATPase. (c) Proton-translocating anaerobic respiration. Low potential substrates utilized. (d) Non-cyclic photosynthetic electron flow. Low potential substrates generated. (e) Cyclic photosynthetic electron flow. In each instance the movements of H^+ are shown with interrupted lines. No attempt is made here to indicate the stoichiometry of the proton-translocating reactions. Inorganic phosphate is shown as Pi.

has to become inserted into the membrane. Furthermore, this insertion must have the correct polarity. There is now a growing knowledge of the mechanisms of synthesis, assembly and insertion of membrane proteins (Lodish, 1980; Wickner, 1979). One thing is clear; the presence of a membrane protein in a cell requires more genetic information than would its soluble counterpart. It therefore seems a reasonable hypothesis (Gest, 1980) that the first membrane-bound vectorial enzymes responsible for the transmembranous movements of protons and electrons in respiration and photosynthesis developed initially from enzymes that were soluble and already justifying their existence by virtue of their scalar catalysis. This hypothesis leads directly to two interesting questions. Firstly, for a given primitive vectorial membrane enzyme, what was its soluble predecessor? Secondly, what changes in primary structure (i.e. genetic information) were necessary in order for the soluble enzyme to become membrane-bound and vectorial? These questions emphasize the mechanisms whereby energy-conserving mechanisms evolved. They represent a departure from the usual but possibly now-completed practice of asking questions about the chronological order in which various bioenergetic systems developed relative to each other.

In the next section I apply this hypothesis and its attendant questions to the evolution of the proton-translocating ATPase and an anaerobic proton-translocating oxidoreduction enzyme, namely, hydrogenase.

EVOLUTION OF THE PROTON-TRANSLOCATING ATPASE

This enzyme consists of two parts; a transmembranous hydrophobic H^+ channel known as F_0, and a hydrophilic F_1 component attached to the cytoplasmic side of F_0 (Postma & Van Dam, 1976). Raven & Smith (1976) proposed that this enzyme evolved as a proton pump in answer to the problems caused by intracellular acidity resulting from fermentation. However, one might well go back a step or two further and suppose that before there was a pump there was a proton channel, i.e. F_0 preceded F_1. Certainly permeability problems for entry of nutrients and exit of waste products must have arisen as soon as the first cells successfully wrapped themselves in a lipid bilayer; indeed, it is doubtful if the first cells could be called first cells until they had done so. Some necessary degree of leakiness and, with selection and evolution, specific leakiness, could have been achieved by incorporation of proteins into the bilayer. This does not imply any great evolutionary jump. So out of necessity and simplicity it can be anticipated that the first membrane-bound components of bioenergetic systems were transport proteins, for nutrients entering and waste products leaving. Seen in this light, the F_0 proton channel was just that – a proton channel. It is then necessary to postulate some cytoplasmic enzyme that served the cell by hydrolyzing pyrophosphate bonds, not necessarily the terminal one of ATP. A gene fusion or other genetic event might then have given the pyrophosphate-hydrolyzing enzyme the new property of

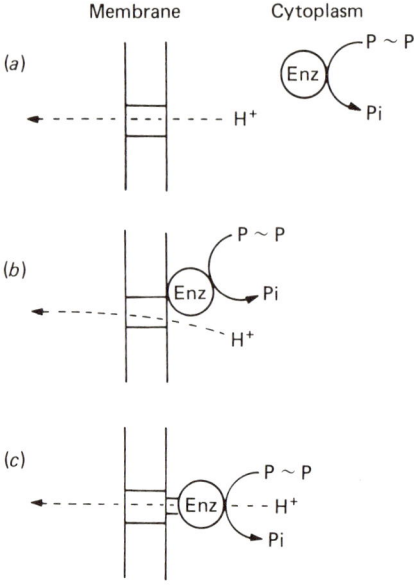

Fig. 3. Hypothetical stages in the evolution of a proton-translocating ATPase. In (*a*), there is a cytoplasmic enzyme (Enz) that catalyses the hydrolysis of ATP or pyrophosphate. There is a proton channel in the membrane, providing a pathway for fermentatively produced H^+ to diffuse out of the cell. In (*b*), the enzyme has become membrane-bound at the site of the proton channel. H^+ diffusion across the membrane is not yet coupled to ATP or pyrophosphate hydrolysis. In (*c*), the enzyme blocks the cytoplasmic end of the proton channel. H^+ movement is now coupled to ATP or pyrophosphate hydrolysis. The proton channel in the membrane corresponds to the F_0 component of the proton-translocating ATP, whereas the enzyme part corresponds to the F_1 component. No attempt is made to indicate the stoichiometry of the reaction. Inorganic phosphate is shown as Pi, and pyrophosphate as $P \sim P$.

attachment to F_0, either as a continuation of the F_0 polypeptide or by addition of a hydrophobic-binding sequence. Thus membrane-bound F_1 could have evolved. This analysis, admittedly speculative, does avoid the difficulty inherent in proposing that the F_0 plus F_1 complex evolved from scratch as a membrane-bound entity with no function other than that currently ascribed to it. Fig. 3 gives a diagrammatic summary of the postulated steps in the evolution of the proton-translocating ATPase.

EVOLUTION OF PROTON-TRANSLOCATING OXIDOREDUCTIONS

The early oxidoreduction enzymes were presumably soluble enzymes of the cell cytoplasm, where they fulfilled a conventional role in intermediary metabolism (Gest, 1980). At least two strong reasons may have encouraged the evolutionary move of some of these enzymes towards membrane attachment. Firstly, there is the ability of membrane-bound multi-enzyme systems to exhibit some of the properties of multi-enzyme complexes. For instance, intermediates have much smaller distances over which to move, and in doing so they do not have to enter the surrounding aqueous bulk phases. Secondly, the membrane offers a hydrophobic environment in which the oxidoreduction properties of quinones can be exploited.

A membrane-attached oxidoreduction system does not necessarily translocate protons, but the changes needed to make it do so need not be extensive. For example, consider the case of hydrogenase in *Escherichia coli*. This iron-sulphur protein is part of the cytoplasmic membrane. The hydrogen/proton couple at pH 7.0 has a standard oxidoreduction potential of -0.42 V, sufficiently negative to drive many, if not all, reductive biosyntheses. Thus a hydrogen-utilizing oxidoreduction enzyme could justify its development in terms of its rôle in intermediary metabolism. And, as discussed above, there could be selective pressures favouring attachment of the enzyme to the membrane. However, if the enzyme acquired a primary structure that inserted it across the membrane, rather than along the membrane, then the outcome could be a vectorial enzyme that now catalyses a trans-membranous electron transfer from hydrogen on one side to a hydrogen acceptor on the other. As shown in Fig. 4, the result would be a proton-translocating vectorial oxidoreduction.

I have taken hydrogenase of *E. coli* as an example for the following reasons. It catalyzes proton translocation (Jones, 1979a), it is functionally and structurally trans-membranous (Jones, 1979b; D. H. Boxer & A. Graham, personal communication), it is relatively simple, containing only one type of polypeptide (Adams & Hall, 1979; Graham *et al.*, 1980), and it is possibly an exceedingly primitive enzyme.

The later evolution of more complex proton-translocating oxidoreduction enzymes in the membrane becomes harder to consider, because new developments could be made not only by recruiting new enzymes from the cytoplasm, but by alteration of enzymes already dedicated to residing in the membrane. Primary-structure studies of membrane proteins have been held up because of the difficulties posed by hydrophobic peptides, but these difficulties can now be by-passed by sequencing, not the whole

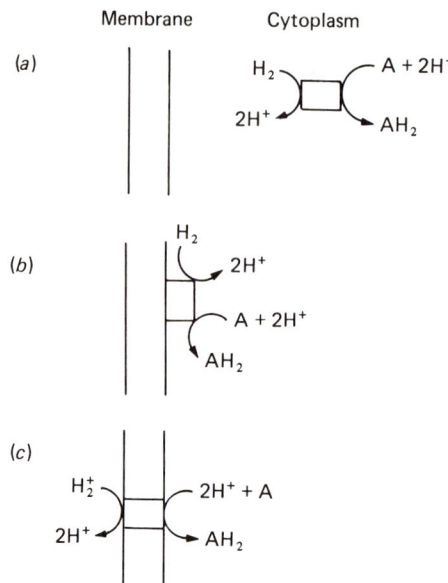

Fig. 4. Hypothetical stages in the evolution of a proton-translocating hydrogenase. In (a) the hydrogenase is shown as a cytoplasmic enzyme, transferring electrons and protons from hydrogen to an acceptor A. In (b) the hydrogenase is membrane-bound but still catalyses a scalar reaction. In (c) the hydrogenase has become trans-membranous, and catalyses a vectorial reaction. Although electrons, not H^+, are the translocated species in this scheme, the end-result is identical to proton-translocation.

polypeptide, but the gene cloned by recombinant-DNA methods. Thus the future prospect is that the evolutionary relationships of many of the oxidoreduction enzymes of proton-translocating respiration or photosynthesis will be studied in terms of their primary structure or gene sequence, with special reference to the acquisition of those sequences that position an enzyme in the membrane in the correct orientation for vectorial rather than scalar catalysis.

I am grateful to John Raven and Wil Konings for preprints, and to David Boxer and Alec Graham for permission to refer to their unpublished work on the structurally transmembranous orientation of hydrogenase in the cytoplasmic membrane of *E. coli*.

REFERENCES

ADAMS, M. W. W. & HALL, D. O. (1979). Purification of the membrane bound dehydrogenase of *Escherichia coli*. *Biochemical Journal*, **183**, 11–22.

BRODA, E. & PESCHEK, G. A. (1979). Did respiration or photosynthesis come first? *Journal of Theoretical Biology,* **81,** 201–12.
GEST, H. (1980). The evolution of biological energy-transducing systems. *FEMS Microbiology Letters,* **7,** 73–7.
GRAHAM, A., BOXER, D. H., HADDOCK, B. A., MANDRAND-BERTHOLET, M.-A. & JONES, R. W. (1980). Immunochemical analysis of the membrane bound hydrogenase of *Escherichia coli*. *FEBS Letters,* **113,** 167–72.
HADDOCK, B. A. & HAMILTON, W. A. (1977). Microbial Energetics. *Symposia of the Society for General Microbiology,* **27,** Cambridge University Press.
HADDOCK, B. A. & JONES, C. W. (1977). Bacterial respiration. *Bacteriological Reviews,* **41,** 47–99.
JONES, R. W. (1979*a*). The role of the membrane-bound hydrogenase in the energy-conserving oxidation of molecular hydrogen by *Escherichia coli*. *Biochemical Journal,* **188,** 345–50.
JONES, R. W. (1979*b*). The topography of the membrane-bound hydrogenase of *Escherichia coli* explored by non-physiological electron acceptors. *Biochemical Society Transactions,* **7,** 724–5.
KRÖGER, A. (1977). Phosphorylative electron transport with fumarate and nitrate as terminal hydrogen acceptors. *Symposia of the Society for General Microbiology,* **27,** 61–93.
LODISH, H. F. (1980). The natural history of transmembrane cell-surface glycoproteins. *Biochemical Society Symposia,* **45,** 105–22.
MICHEL, P. A. M., MICHEL, J. P. J., BOONSTRA, J. & KONINGS, W. N. (1979). Generation of an electrochemical proton gradient in bacteria by the excretion of metabolic end products. *FEMS Microbiology Letters,* **5,** 357–64.
MITCHELL, P. (1968). *Chemiosmotic coupling and energy transduction*. Bodmin, England: Glynn Research Ltd.
MITCHELL, P. (1977). Epilogue: from bioenergetic abstraction to biochemical mechanism. *Symposia of the Society for General Microbiology,* **27,** 383–423.
POSTMA, P. W. & VAN DAM, K. (1976). The ATPase complex from energy transducing membranes. *Trends in Biochemical Sciences,* **1,** 16–17.
RAVEN, J. A. & SMITH, F. A. (1976). The evolution of chemiosmotic energy coupling. *Journal of Theoretical Biology,* **57,** 301–12.
RAVEN, J. A. & SMITH, F. A. (1981). H$^+$ transport in the evolution of photosynthesis. *Biosystems,* (in Press).
SMITH, F. A. & RAVEN, J. A. (1978). The evolution of H$^+$ transport and its role in photosynthetic energy transduction. In *Light Transducing Membranes,* pp. 233–51. New York & London. Academic Press Inc.
WICKNER, W. (1979). The assembly of proteins into biological membranes: the membrane trigger hypothesis. *Annual Review of Biochemistry,* **48,** 23–45.

SHIFT AND DRIFT IN INFLUENZA VIRUSES

W. MIN JOU, M. VERHOEYEN, R.-X. FANG,
R. DEVOS, D. HUYLEBROECK & W. FIERS

*Laboratory of Molecular Biology, State University of Ghent,
Ledeganckstraat 35, B-9000 Ghent, Belgium*

INTRODUCTION

Influenza continues to be an important cause of illness and death in the human population, the latter especially among so-called high-risk patients. Although influenza A virus, the main causative agent of influenza, was isolated from man in 1933 (Smith, Andrewes & Laidlaw, 1933), this was not followed by the rapid development of vaccination programmes to control the disease, as was successfully accomplished with other human viral diseases, such as smallpox, poliomyelitis and measles, once the virus had been isolated and cultured.

The main reason for this apparent paradox lies in the frequent chemical and antigenic changes in the two surface proteins of the virus, the hemagglutinin (HA) and the neuraminidase (NA). The influenza A virus has a divided genome consisting of eight negative-stranded RNA segments (Palese & Schulman, 1976; McGeoch, Fellner & Newton, 1976; Scholtissek *et al.*, 1976), seven of which probably code for one protein each (Palese, 1977), while segment 8 specifies two non-structural proteins (Lamb & Choppin, 1979; Inglis, Barrett, Brown & Almond, 1979). The surface proteins, HA and NA, are coded by genes 4, and 5 or 6 (depending on the strain), respectively. Two types of antigenic variation are observed in influenza A virus, antigenic drift and antigenic shift. Antigenic drift is the gradual change observed in the surface proteins of virus isolates every one or two years. Antigenic shift is a radical change of the antigenic properties of the HA or both the HA and the NA leading to the emergence of new pandemic strains. During this century antigenic shifts have occurred in 1918, 1957 and 1968 (Dowdle, 1976). HA is the antigen against which neutralizing antibodies are primarily directed (Laver & Kilbourne, 1966; Drzeniek, Seto & Rott, 1966), and so it is considered the more important protein involved in these changes. The recently proposed nomenclature for HA antigen subtypes of human influenza A viruses (Schild *et al.*, 1980) is Hl (1918–1957; previously H0, H1 and Hsw), H2 (1957–1968) and H3 (1968–   ). Only two human NA antigen subtypes have been described: N1 (1918–1957) and N2 (1957–   ). In 1977 a virus of the H1N1 subtype (the viral subtype is defined by the combination of the hemagglutinin and neuraminidase symbols) closely resembling the viruses of 1950 strains has reappeared (Kendal, Noble, Skehel & Dowdle, 1978; Nakajima, Desselberger & Palese, 1978; Scholtissek, von Hoyningen & Rott, 1978). Since 1977, influenza A viruses of the H3N2 and H1N1 subtypes have been co-circulating in the human population. While all the human HA and NA subtypes have also been isolated from lower mammals and birds, both in the same combinations as found in man and in other combinations (Hinshaw, Webster & Rodriguez, 1979), nine additional HA subtypes and seven NA subtypes from non-human sources have been described (Schild *et al.*, 1980).

Until now most studies have been directed towards the hemagglutinin molecule, and while in initial studies (at the protein level) practical considerations played a role, this policy was continued even when methods not requiring the use of protein in the study of influenza A antigen variability were introduced (see below), because of the major importance of the hemagglutinin. The present contribution will be largely devoted to the hemagglutinin molecule, as much less information is available on the neuraminidase.

It has been known for many years (Laver & Webster, 1968, 1972) that antigenic changes in the hemagglutinin were associated with major (shift) and minor (drift) changes in the amino-acid sequence by comparison of their peptide maps. To characterize more directly the chemical changes involved, two groups have determined the amino-acid sequence of extensive parts of an H2 hemagglutinin (summarized in Waterfield, Espelie, Elder & Skehel, 1979) and of an H3 hemagglutinin (Dopheide & Ward, 1980; Ward & Dopheide, 1980). Until recently, protein sequencing was the most direct method to study hemagglutinin structure, since an investigation of the genetic material itself was even more laborious for an RNA the size of the HA gene. Procedures for synthesizing nearly full-length DNA copies from RNA and for cloning these copies in bacterial plasmids (Maniatis, Kee, Efstratiadis & Kafatos, 1976; Higuchi, Paddock, Wall & Salser, 1976), in combination with fast DNA sequencing technology (Maxam & Gilbert, 1977; Sanger, Nicklen & Coulson, 1977), seemed a more efficient approach to study these problems. The gene sequence obtained predicts the complete amino-acid sequence of the mature protein and also the regions that are removed during the maturation process. Special problems such as those encountered during the sequencing of hydrophobic regions of the protein (the transmembrane portion of the hemagglutinin, for example), are avoided.

We have successfully cloned and determined the sequences of the hemagglutinin genes of the relatively recent human strain A/Victoria/3/75 (H3N2) and strain A/Aichi/2/68, which circulated at the beginning of the so-called Hong Kong (H3N2) period (Min Jou et al., 1980; Verhoeyen et al., 1980). For practical reasons, the hemagglutinin genes were derived from the high-yielding recombinants X47 (Victoria) and X31 (Aichi), with strain A/PR/8/34. Other groups working along similar lines have reported the hemagglutinin structure of two additional H3 strains, A/Memphis/102/72 (Sleigh et al., 1980) and A/NT/60/68/29C, a laboratory variant of A/NT/60/68 (Both & Sleigh, 1980), of one H2 strain, A/Japan/305/57 (Gething, Bye, Skehel & Waterfield, 1980), and of one H1 strain, A/WSN/33 (Davis, Hiti & Nayak, 1980). The sequence of one avian hemagglutinin gene from fowl plague virus (FPV) is also known (Porter et al., 1979). A partial sequence of the Aichi gene, which is identical to ours, is given in Gething et al. (1980). Ward & Dopheide (1980) and Dopheide & Ward (1980) also reported most of the Memphis sequence and part of the Aichi sequence determined by protein sequencing; these amino-acid sequences differ only in a few residues from the sequences deduced from cloned DNA (Sleigh et al., 1980; Verhoeyen et al., 1980).

CONSTRUCTION OF MOLECULAR CHIMERAS

For the cloning of a double-stranded (ds) DNA copy of the influenza hemagglutinin gene, we essentially followed the methodology described by Devos, van Emmelo, Contreras & Fiers (1979) for the cloning of bacteriophage MS2 genetic information. Like MS2 RNA, the influenza virus RNAs are non-polyadenylated, in contrast to the classical situation where a poly(A)-containing eukaryotic messenger RNA is cloned (Maniatis et al., 1976; Higuchi et al., 1976). Because of the limited amount of starting material used (50–80 μg of the eight viral RNA segments), we did not attempt to isolate the hemagglutinin RNA before polyadenylation, nor did we try to separate the poly(rA)$^+$ material after polyadenylation.

However, the quality of the RNA before and after the polyadenylation reaction was checked analytically on a low per cent polyacrylamide gel in the presence of urea. In both cases the normal pattern of the eight influenza RNA segments was revealed. Moreover, since α-^{32}P-ATP was used during the polyadenylation reaction, the radioautogram proved the physical attachment of a poly(rA) tail to all eight genome segments, including the hemagglutinin gene.

The mixture of eight polyadenylated viral RNAs was converted into dsDNA by a stepwise procedure (Fig. 1) involving avian myeloblastosis virus reverse transcriptase, RNase treatment, *Escherichia coli* DNA polymerase I and S1 nuclease (Devos *et al.*, 1979; Min Jou *et al.*, 1980). The sizes of the molecules of dsDNA in the mixture were measured on a 1.5% agarose gel, and dsDNA corresponding in size to the full-length gene 4 segment was recovered from the gel as a source of DNA for cloning. The poly(dA) · poly(dT) joining procedure of Jackson, Symons & Berg (1972) was used to construct molecular chimeras of the influenza hemagglutinin dsDNA and the bacterial plasmid pBR322 DNA. PstI-cleaved pBR322 DNA was polydeoxyadenylated by means of terminal deoxynucleotidyl transferase, and the purified hemagglutinin dsDNA molecules had poly(dT) tails attached. Poly(dA)-pBR322 DNA (200 ng) and poly(dT)-hemagglutinin gene (~70 ng) were annealed at 65°C, cooled slowly to room temperature over a 4-h period, and used to transform *Escherichia coli* K12 HB101 cells. In one experiment, using the Victoria strain, 47 transformants were obtained, while for the Aichi strain, 39 tetracycline-resistant transformants were found. When hybridized to the corresponding ^{32}P-labelled hemagglutinin RNA probe (Grunstein & Hogness, 1975), four and five colonies, respectively, showed positive hybridization. In each case a plasmid (designated pVHA14 and pXHA14, respectively) containing a presumed full-length hemagglutinin insert (1860 base pairs including the A · T tails) was chosen for sequencing.

NUCLEOTIDE SEQUENCE OF INFLUENZA A HEMAGGLUTININS

The inserts in plasmids pVHA14 and pXHA14 were first characterized by restriction mapping. Fragments were then prepared by digestion with the appropriate enzymes, 5'-terminally labelled and sequenced according to the method of Maxam & Gilbert (1977).

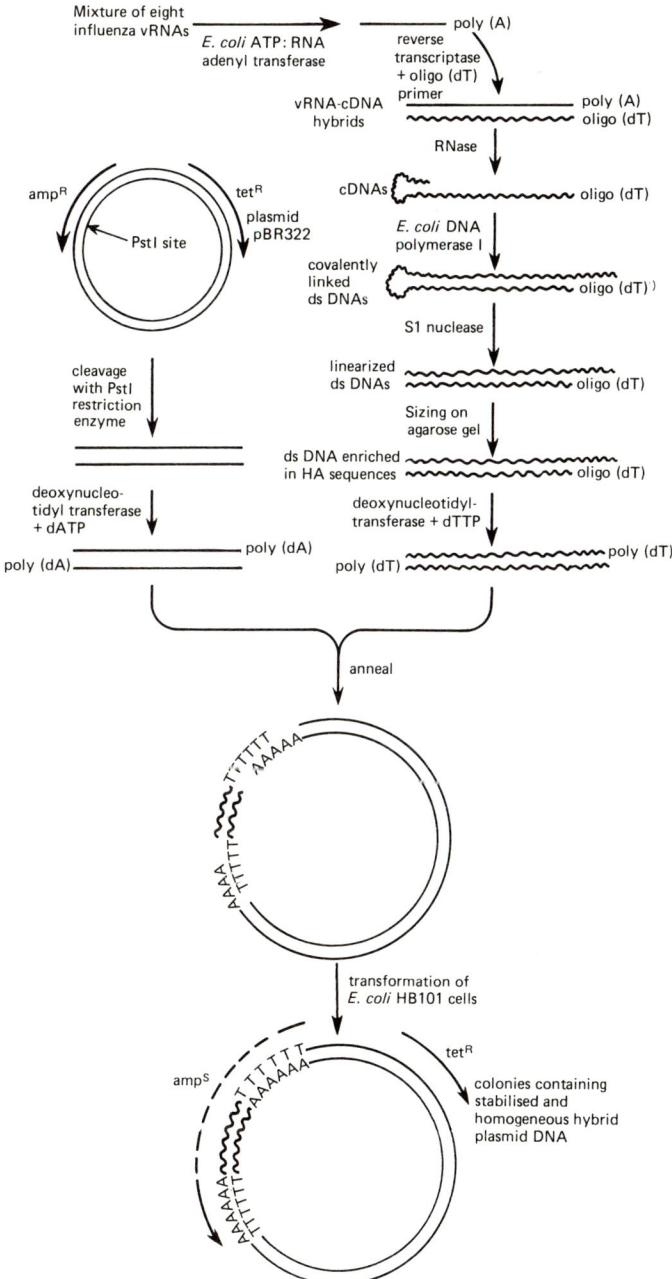

Fig. 1. Strategy for cloning influenza HA sequences. For details see text.

The presence of the sequence 5'–A–G–C–A–A–A–A–G–C–A–G–G– at the 5' end of the insert preceded by the oligo(dT) primer proves that the complete sequence starting from the 3' end of the viral RNA had been cloned, as the complementary sequence is found conserved at the 3' end of all influenza A virus (negative-stranded) RNAs studied so far (Skehel & Hay, 1978; Air, 1979; Robertson, 1979; Both & Air, 1979; Desselberger, Riacaniello, Zazra & Palese, 1980). This sequence is also found at the 5' end of all cloned influenza A genes: the hemagglutinin genes listed in the Introduction, the A/PR/8/34 gene 7 specifying the matrix protein (Winter & Fields, 1980), and gene 8, which codes for the two non-structural proteins, derived from three different virus strains (Porter, Smith & Emtage, 1980; Lamb & Lai, 1980; Baez et al., 1980). The only variable position in this segment of 12 nucleotides is nucleotide 4: it can be A or G. Beyond position 12, the sequence diverges in the different genes, and beyond residue 13 it diverges even in the hemagglutinin gene of different subtypes.

As expected from the cloning strategy used (Efstratiadis, Kafatos & Maniatis, 1977; Devos et al., 1979), initiation of synthesis of the second DNA strand by making use of the self-priming capacity of the first DNA strand arising from a presumed terminal hairpin loop, followed by S1 nuclease treatment, leads to omission of a small portion of the hemagglutinin genetic information corresponding to the 3' terminus of the coding strand. Overlapping information obtained by sequencing the 5' end of the viral RNA by reverse transcription using a labelled restriction fragment as primer, revealed that 23 nucleotides were missing from the pVHA14 insert (and by analogy 27 in pXHA14). The 13 nucleotides at the 5' end of all influenza A viral RNAs are identical (Skehel & Hay, 1978; Robertson, 1979; Desselberger et al., 1980), and are followed by three variable residues and a U-tract of variable length at least till nucleotide 21. The conserved 5' and 3' ends display partial inverted complementarity, and a role for this structure in the initiation of viral RNA replication has been suggested (Desselberger et al., 1980).

While the length and, to a large extent, the sequence of both the 5'- and 3'-untranslated regions are conserved within the H3 subtype (Fig. 2), this is not the case for hemagglutinin genes of different subtype. The length of that part of the cRNA preceding the initiating AUG codon, varies from 21 (H7) to 43 (H2) residues. The 3'-untranslated region (not including the stop codon) varies from 29

291

```
A/Aichi/2/68  (X31)           AGCAAAAGC  AGGGGAUAAU  UCUAUUAAUC  [AUG] AAG,ACC,AUC,AUU,GCU,UUG,AGC,UAC,AUU,UUC,UGU,CUG,GCU,CUC,GGC,CAA,GAC,CUU,CCA,GGA,AAU.
A/Victoria/3/75 (X47)                                              AAC                                                    U  U  U  C
                                                                                                                           U  A
     96
 5   GAC,AAC,     AGC,ACA,GCA,ACA,ACG,CUG,UGC,CUG,GGA,CAU,CAU,GCG,CCA,AAA,ACA,AUC,ACA,GAU,GAU,CAG,AUU,GAA,GUG,ACU,AAU,GCU,ACU.
              AAC                                                    G                 G    A
   198
 19  GAG,CUA,GUU,CAG,AGC,UCC,UCA,ACG,GGG,AAA,AUA,UGC,AAC,AAU,CCU,CAU,CGA,AUC,CUU,GAU,GGA,AUA,GAC,UGC,ACA,GAU,GCU,CUA,UUG,GGG,GAC,CCU,CAU.
                  G  U                                                                  A
   303
 30  UGU,GAU,GUU,UUU,CAA,AAU,GAG,ACA,UGG,GAC,CUU,UUC,GUU,GAA,CGC,AGC,AAA,CCU,UAC,AGC,AUC,UGG,ACU,CGG,GAG,UUC,ACU,UGG,ACU,UGG,GCC,UCC,UUU,AGG,UCA.
                   GA                                                                   A    A   A  C
   408
 40  CUA,GUU,GCC,UCA,GCC,ACU,CUG,GAG,UUU,AUC,UGG,CUU,GUG,AAG,CCA,AAU,GGG,GUA,CGU,CAG,AGC,AAU,GCU,UGC,AAA,AGG,GGA,CCU,GGU,AGC.
                                                                                        G
   513
 51  UUG,UUC,AGU,AGA,CUG,AAC,UGG,UUG,ACC,AAA,UCA,GGA,AGC,ACA,UAU,CCA,GUG,CUG,AAC,ACU,AUG,UCA,GAU,GCA,CCU,AUU,GAU,ACC,UGU,AUU,UCU,GAA,UGC,AUC,ACU,CCA,AAU.
                                                UA                       AA         C                                        GG   C   GC
   618
 61  GGG,AUU,CAC,CAC,CCG,AGC,ACC,AAC,CAA,GAA,CAA,ACC,ACC,CUG,UCA,GCA,GGA,GUC,ACA,GGC,AGA,AGC,CAG,CAA,ACU,AUA,AUC,CCG.
        G                              A              A                      A
   723
 72  AAU,AUC,GGG,UCC,AGA,CCC,UAU,AGG,GGU,UCU,AGU,AGA,AUA,AGC,AUC,UAU,UGG,ACA,AUA,GUU,AAG,CCG,GGA,GAC,GUA,CUG,GUA,AUU,AAU,AGU,AGU,GGG,AAC.
                  G                                                                            CC        A
   828
 82  CUA,AUC,GCU,CCU,CGG,GGU,UAU,UUC,AAA,AUG,CGC,ACU,GGG,AAA,AGC,UCA,AUA,GCU,CCA,UCG,AUU,GAU,ACC,UGU,AUU,UCU,GAA,UGC,AUC,ACU,CCA,AAU.
                  U                                                                            GG    C   GC
   933
 93  GGA,AGC,AUU,CCC,AAU,GAC,AAG,CCC,UUU,CAA,AAC,GUA,AAC,AAG,AUC,ACA,AAC,GUA,AAC,AAG,GCA,UGC,CCC,AAG,UAU,GUU,AAG,CAA,AAC,ACC,CUG,AAG,UUA,GAC,GGG,AUG.
                                                                                    G                       U
  1038
103  CGG,AAU,GUA,CCA,GAG,AAA,CAA,ACU,AGA,GGC,CUA,UUC,GGC,GCA,AUA GCA,AUA GGA,UUC,ACA,AUG,GGU,UGG,UAC,GGU,UUC,AGG,CAU.
                                                         A
                                  HA1 ←—→ ←—→ HA2
  1143
114  CAA,AAU,UCU,GAG,GGC,ACA,GGA,CAA,GCA,GAU,CUU,AAA,AGC,ACU CAA,GCA,GCC,AUC,GAC,CUC,ACC,AUC,AUG,GAA,AAG,GAA,CAC,AAC,AUG,AAG,ACG,AAC,GAG.
                   C
  1248
124  AAA,UUC,CAU,CAA,AUC,GAG,AAG,GAA,AUC,UCA,GAA,GAA,AUU,CAG,AAC,CUC,AAA,GAA,AAG,ACU,CAG,GCU,GCA,AUA,GAU,GGG,GUG,ACC,AAU,AAA,GUG,AAC,UCU,AUC,AUU,GAC.
  1353
135  GAG,CUU,CUU,GUC GCU,CUG,GAG,AAU,CAA,CAU,ACA,AUU,GAC,CUG,ACU,GAC,UCG,GAA,AUG,AAC,AAG,CUG,UUU,GAA,AAA,ACA,AGG,CAA,CAG,CUA,AGG,GAA,AAU,GCU,GAA,GAC.
                                                 U                                A
  1458
145  GAG,AUG,GGC,AAU,GGU,UGC,UUC,AAA,AUA,UGC,CAC,AAA,UGU,GAC,AAU,GCC,UGC,AUA,GAG,UCA,UUC,AGA,AAU,GGU,ACU,UAU,GAC,CAU,GAU,GUA,UAC,AGA,GAC,GAA,GCA.
           C                                            G
  1563
156  UUA,AAC,AAC,CGG,UUC,CAG,AUC,AAA,GAU,GUU,GAA,CUG,AAG,UCU,GGA,UAC,AAA,GAC,UGG,AUC,CUG,UGG,AUU,UCC,UUU,GCC,AUA,UCA,UGC,UUU,UUG,CUU,UGU,GUU,GUU.
                                                          A                              A
  1668
166  UUG,CUG,GGG,UUC,AUC,AUG,UGG,GCC,UUC,AAU,AGG,AAU,AAG,CAA,AAU,AGG,AUA,GGC,AAU,GAC,AAU,AGG,AAU,AGG,AUA,UGA [UGA] GUGAUUUAGU AAUUAAAAAC ACCCUUGUU CUACU
                                                 A  A                                           A
```

Fig. 2. The hemagglutinin genes of influenza strains A/Aichi/2/68 and A/Victoria/3/75. The nucleotide sequence of the Aichi gene is shown as the complementary RNA (coding strand). Only substitutions in the Victoria gene are shown; substitutions leading to an amino acid change are underlined. Start and stop codons are boxed.

(H7) to 41 (H2) nucleotides. Whereas all human influenza A strains sequenced so far have a UGA stop codon followed by one or two UAG and/or UAA stop signals in the same reading frame, the FPV hemagglutinin stop signal is a single UAA terminator. The length of the coding portion of the gene varies only slightly between subtypes: 1686 (H2), 1698 (H3) and 1689 (H7) nucleotides (the H1 sequence has not yet been completed).

THE HEMAGGLUTININ MOLECULE

The hemagglutinin and the neuraminidase appear as glycoprotein spikes attached to the viral membrane. The HA spike is composed of three HA molecules, and each HA molecule of infectious virions consists of two disulphide-bonded polypeptide chains, HA1 and HA2 (Ward & Dopheide, 1979; Waterfield et al., 1979). It is synthesized as a single polypeptide chain which is post-translationally processed, involving removal of an NH_2-terminal signal peptide after the processes of transfer through the membrane and insertion have been completed (McCauley et al., 1979), and proteolytic cleavage into HA1 and HA2 chains to yield mature hemagglutinin and infectious virus (Lazarowitz & Choppin, 1975; Klenk, Rott, Orlich & Blödorn, 1975).

In the H1, the H2 and the four H3 strains sequenced (Figs. 3 and 4) the region excised between HA1 and HA2 (the connecting peptide) consists of a single arginine residue, while in FPV, six residues (Lys–Lys–Arg–Glu–Lys–Arg) are probably removed. In both cases, the amino-acid sequence is compatible with the observation that uninfectious virus containing uncleaved HA (grown in cells where the host protease is absent or inactive) can be activated by trypsin treatment (Lazarowitz & Choppin, 1975; Klenk et al., 1975). The difference in sequence may explain why FPV is able to grow in a wider range of host cells.

Within the H3 subtype the length (16 residues) and hydrophobic character of the signal peptide seem to be maintained, but the exact sequence is not (Fig. 3). In fact, it is the most variable region of the molecule (Table 1). Between subtypes, the length varies slightly (H1, 17 residues; H2, 15; H3, 16; H7, 18) and there is little sequence homology (Fig. 4). Both types of variation presumably reflect the main characteristics of signal peptide sequences, in the sense that it is the hydrophobic nature and length of the peptide that determines

Table 1. *Genetic variation in influenza virus hemagglutinin (H3)*

Comparison	Amino-acid changes in				
	Signal	HA1	Con. Pept.	HA2	Total
Aichi–Memphis	1	15	0	4	19
Memphis–Victoria	2	17	0	2	21
Aichi–Victoria	3	22	0	4	29
Total variable sites	3	27	0	5	35
Number of Residues	16	329	1	221	567
% Variation	18.8	8.2		2.3	6.2

how well it functions rather than the exact amino-acid sequence. In these respects the hemagglutinin signal peptides are similar to those described in other membrane and secreted proteins (Blobel *et al.*, 1979).

A second hydrophobic sequence, at the NH_2-terminus generated in HA2, which may have a rôle in the interaction with, and penetration through, the cell membrane during infection, is highly conserved in all the influenza A strains examined so far (Figs. 3 and 4) (Waterfield *et al.*, 1979), even in the single B-type strain known, and also in the NH_2 terminal sequence of the F1 subunit of Sendai virus fusion protein (Waterfield *et al.*, 1979). A third hydrophobic region, near the carboxy-terminus of the HA2 polypeptide, is probably involved in anchoring the protein in the lipid bilayer (Skehel & Waterfield, 1975). The H3 strains have a completely conserved hydrophobic segment of 26 amino acids in positions 185–210 of the HA2 polypeptide (Fig. 3). A similar region can be discerned in exactly the same position in the H7 strain and also (but shifted one residue towards the carboxy-terminus) in both H1 and H2 subtypes (Fig. 4). The amino-acid homology in this region is very limited between viruses of different subtypes; in fact it is lower than in other regions of the HA2 polypeptide. It is most likely that the 10–11 residues at the carboxyl terminus are inside the virus particle. The precise functioning of the signal interrupting the translocation of membrane proteins (termed the 'stop-transfer' sequence by Blobel *et al.*, 1979) remains to be established, but it may be that, as for signal peptides, hydrophobicity is the main determinant in this interaction. Examination of the known and presumed trans-membrane sequence of other viral (Rose *et al.*,

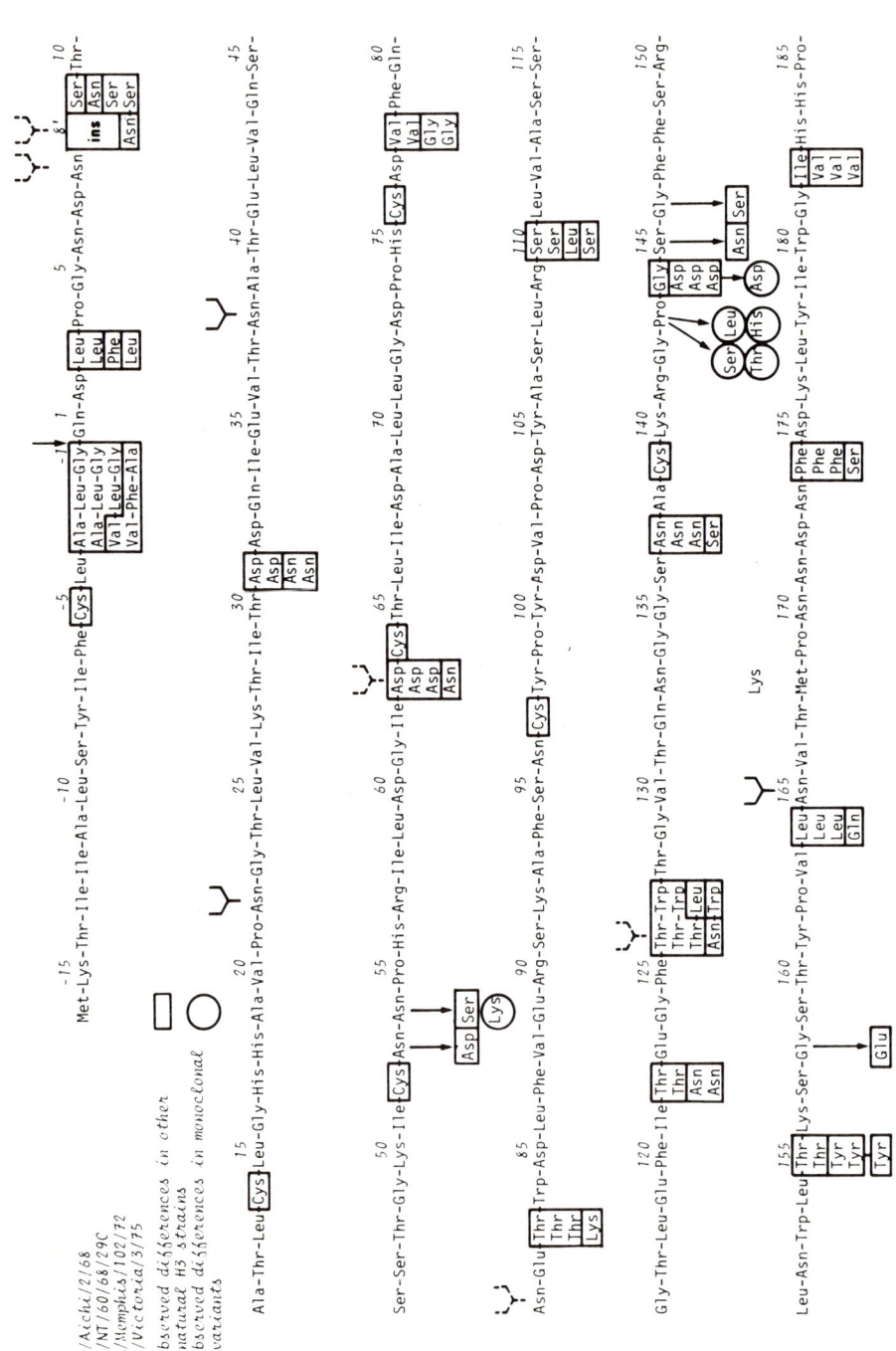

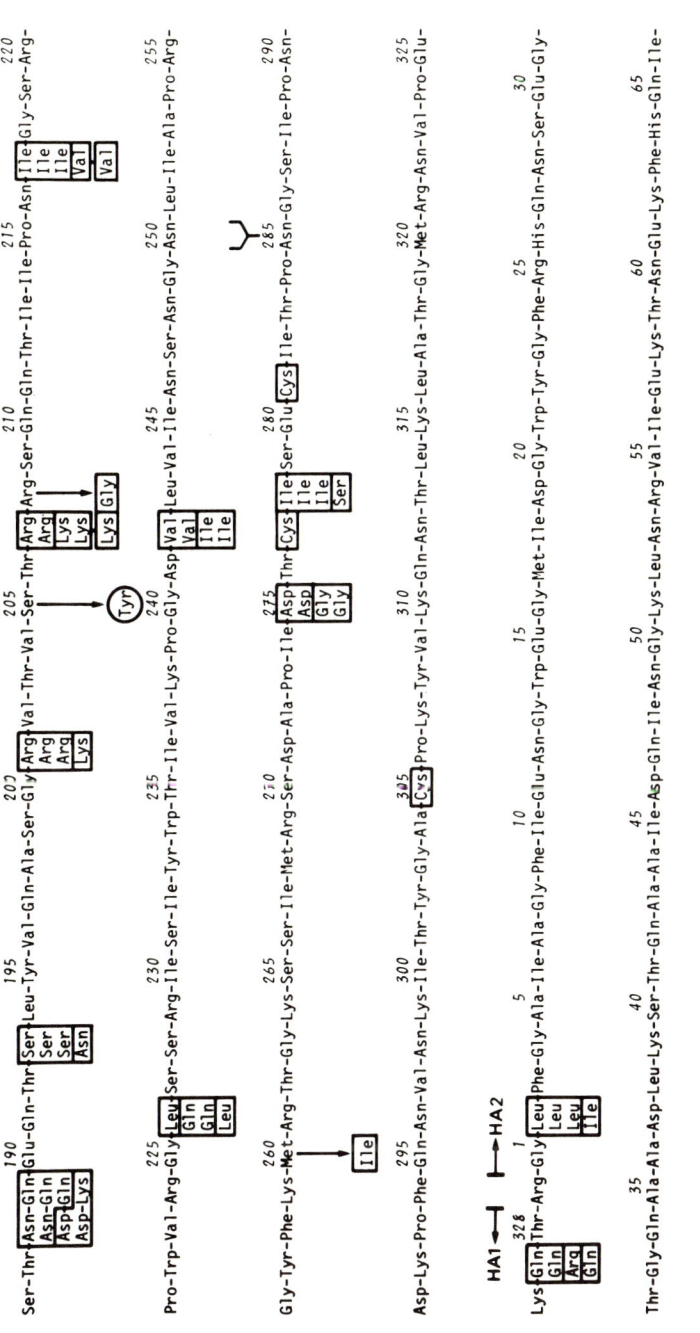

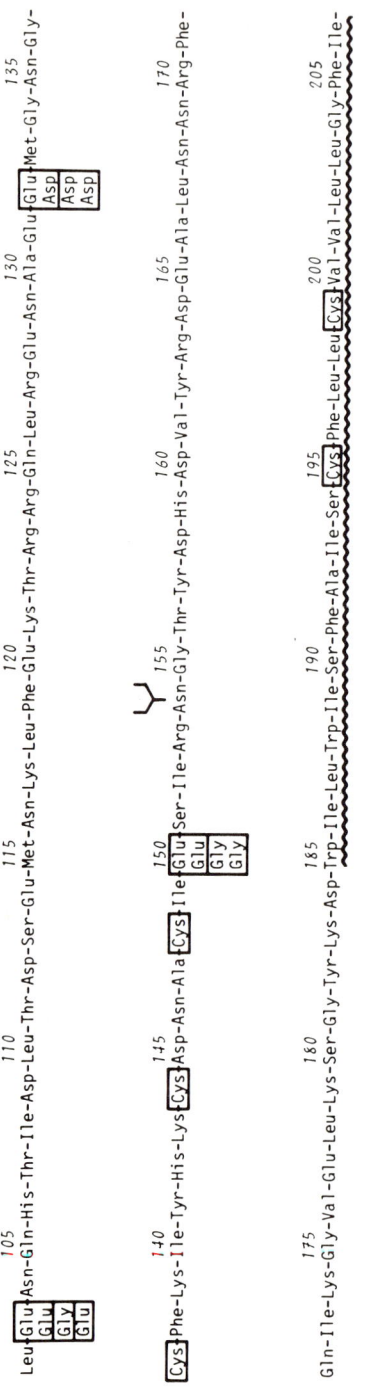

Fig. 3. Comparison of the influenza A hemagglutinin protein sequence within the H3 subtype. The amino-acid sequence of the hemagglutinin derived from A/Aichi/2/68 is shown on top and differences in A/NT/60/68/29C, A/Memphis/102/72 and A/Victoria/3/75 are boxed. These data are all derived from cloned DNA. Differences observed in H3 hemagglutinins by peptide mapping from other field strains as well as from monoclonal variants are also indicated. Constant ⋎ and variable ⋎ potential glycosylation sites are marked above the asparagine residues by full and interrupted lines, respectively. Positions of cysteine residues are constant throughout the molecule and are shown in boxes. The wavy underline indicates the hydrophobic region of HA2 that is embedded in the lipid bilayer of the virion.

1980; Garoff *et al.*, 1980) and cell surface glycoproteins (Springer & Strominger, 1976; Robb, Terhorst & Strominger, 1978; Tomita, Furthmayr & Marchesi, 1978; Rogers *et al.*, 1980) shows that they all contain a hydrophobic segment of 20 to 31 amino acids and a highly variable hydrophilic internal carboxy-terminus of between 2 and 31 residues. The trans-membrane domain in all cases contains one or more serine residues which could be the fatty-acid attachment site(s) observed on the membrane glycoproteins of several enveloped RNA viruses, including influenza (Schmidt & Schlesinger, 1979; Schmidt, Bracha & Schlesinger, 1979).

COMPARISON OF H3 STRAINS: ANTIGENIC DRIFT

The nucleotide sequences of the hemagglutinin genes of the Aichi/68 and Victoria/75 strains are shown in Fig. 2. In total there are 67 nucleotide differences (3.8%), which include a single insertion of three nucleotides close to the NH_2-terminus of the Victoria HA1 (coding for an extra asparagine residue) and 64 nucleotide substitutions: 1 change in the 5'-untranslated region and 63 changes in the coding region. Of the latter changes, 34 are silent nucleotide substitutions and the other 29 cause 28 amino-acid differences (the amino-acid change at position 155 of the HA1 is caused by mutation of two nucleotides). Including the insertion there are 29 amino-acid changes accumulated over a seven-year period (Fig. 3), which corresponds to a 5.1% amino-acid divergence. The complete amino-acid sequence of four H3 hemagglutinins is compared in Fig. 3. The two 1968 isolates (Aichi and the laboratory variant NT/60/29C) differ in five amino acids. In general, once an amino acid has changed it does not change again (except for a few apparent reversions to the original amino acid). The same phenomenon is observed occasionally at the level of silent mutations (not shown). Similar observations were made by comparing maps of soluble tryptic peptides and amino-acid composition data from strains isolated during the period 1968–1977 with the amino-acid sequence of A/Memphis/102/72 (Ward & Dopheide, 1980; Laver, Air, Dopheide & Ward, 1980*a*). These changes are also shown in Fig. 3.

As the antigenic character is thought to reside largely, if not entirely, in HA1 (Jackson, Russell, Ward & Dopheide, 1978; Jackson *et al.*, 1979; Potter & Oxford, 1979), it is not unexpected that most amino-acid changes are in this part of the molecule (see

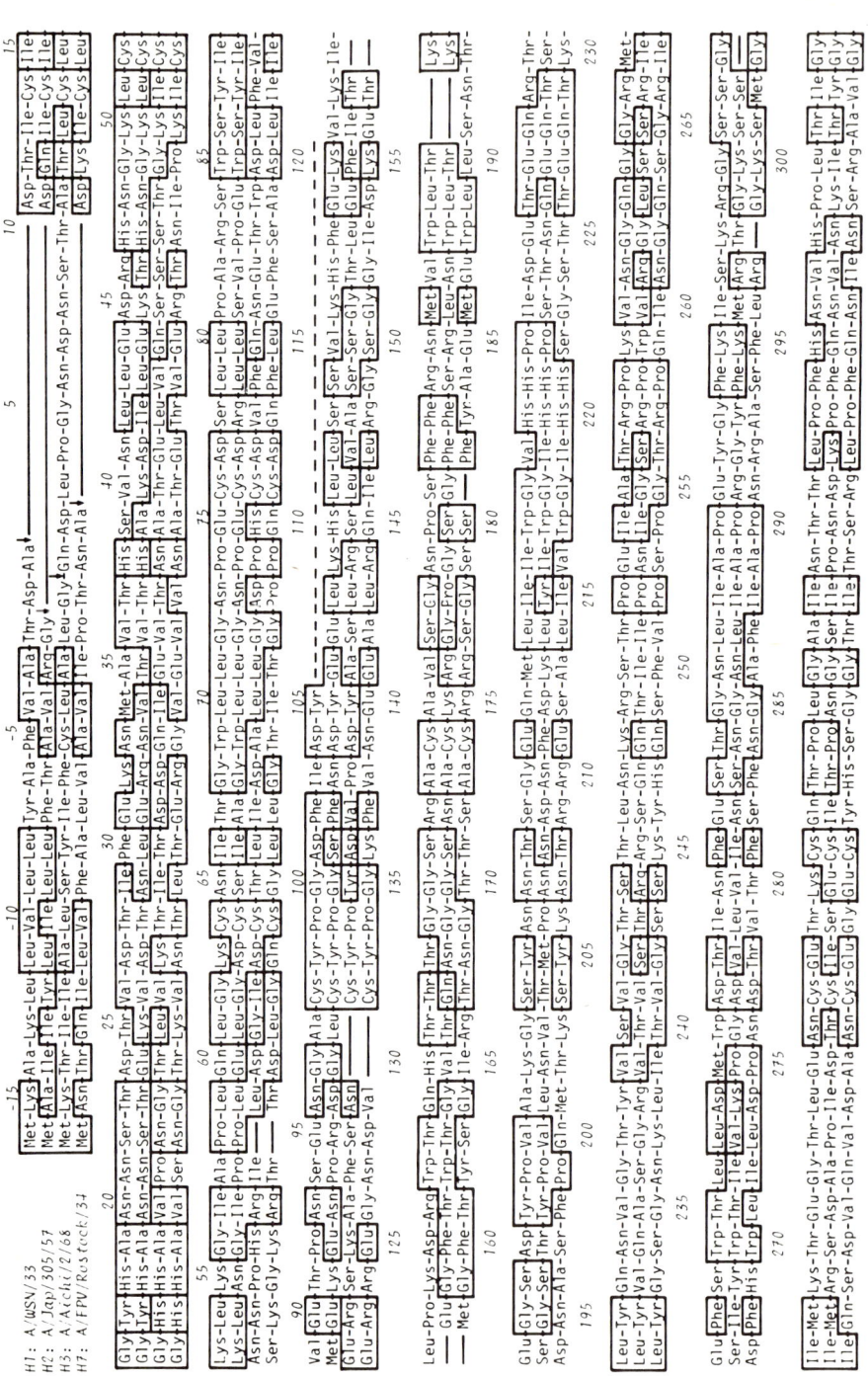

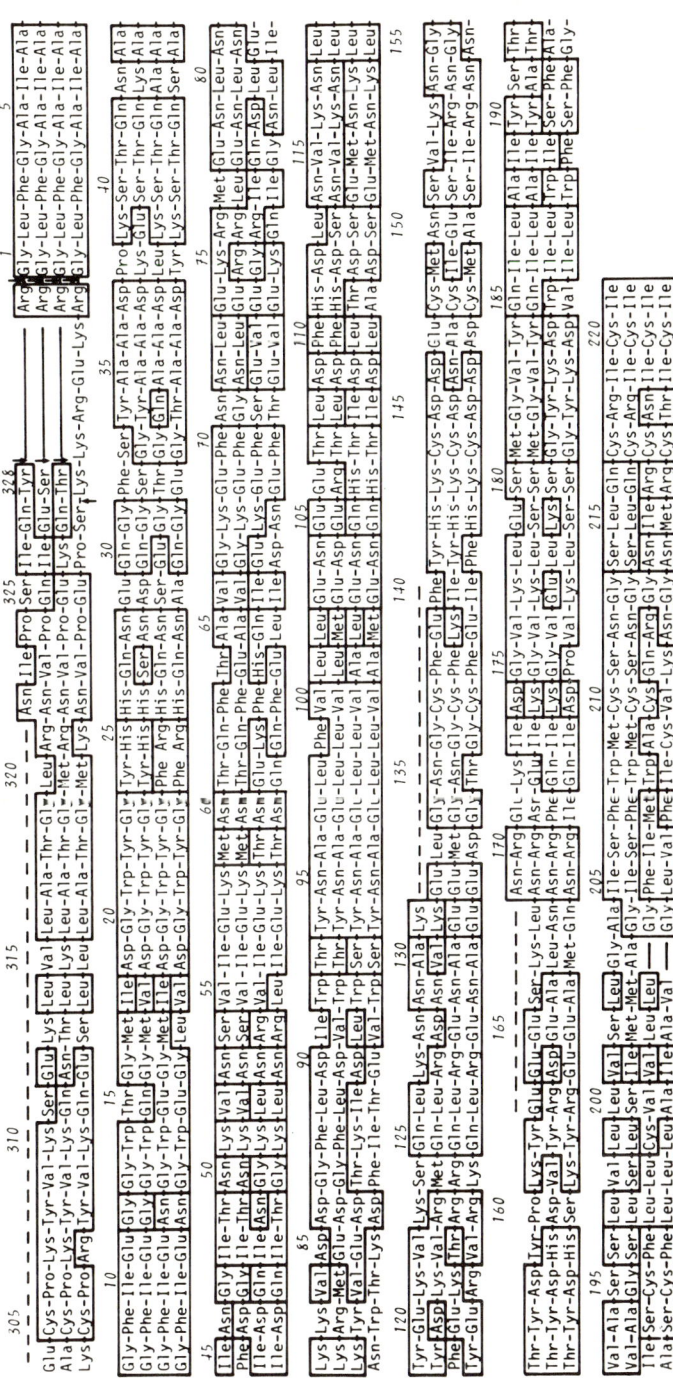

Fig. 4. Comparison of influenza A hemagglutinin protein sequence between different subtypes. The amino-acid sequence of the three hemagglutinin subtypes detected in the human population so far (H1, Davis et al., 1980; H2, Gething et al., 1980; H3, Verhoeyen et al., 1980) and of a hemagglutinin subtype from an avian species (H7, Porter et al., 1979) are compared. Delineation of the different sections of the molecule (signal peptide–HA1–connecting region–HA2) is indicated by arrows. Full lines are used for aligning the different subtypes for maximal sequence homology. Homologous regions are shown in boxes. The numbers refer to the H3 sequence; they correspond only approximately with the other subtypes.

Table 2. *Percentage amino-acid homology between hemagglutinins of different subtypes*

	H1[a]	H2	H3	
H2	64	–	–	HA1
	77			HA2
	69			Total
H3	29	34	93 [b]	HA1
	50	50	98	HA2
	37	40	95	Total
H7 (Havl)	28	36	37	HA1
	47	52	66	HA2
	36	42	48	Total

[a] 32% of HA1 has been sequenced and 83% of HA2.
[b] Drift in the period 1968–1975.

also Table 2). Over the 1968–1975 period of antigenic drift, 22 (6.7%) of the amino acids have been replaced in HA1, whereas only four (1.8%) have changed in HA2. Virus variants have been isolated which are resistant to monoclonal antiserum, and the hemagglutinins of these have probably only one single amino-acid residue changed (Laver *et al.*, 1979; Laver *et al.*, 1980*b*). As these variants are still neutralized by total antiserum against the parent virus, it seems very likely that a more profound change of at least two amino acids would be required to generate a new field strain capable of overcoming the immunity of the population.

For the moment, we cannot tell which amino acids have played a role in the antigenic variability, but it is reasonable to assume that it is a not insignificant fraction of the 22 changes in the HA1 polypeptide observed between Aichi/68 and Victoria/75. Taking into account the number of natural variants isolated and the number of epidemics reported during that period (Pereira, 1976), the number of changes leading to antigenic variations is probably in the range 7 to 15, or even more. If one looks at the numbers of sites at which changes are found in HA1 and HA2, relative to their sizes, there appear to be 16 sites in HA1 likely to be concerned with antigenic variation. A delineation of antigenic sites (defined here as a limited number of amino acids involved in the reaction with the antibody; Laver *et al.*, 1979) by a comparison of amino-acid differences in a limited number of field strains may be complicated further by the fact that changes can occur not only in the 'antigenic sites' but also in other positions in such a way that they influence the antigenic site by a conformational effect.

Laver et al. (1979, 1980b) have analysed Hong Kong variants selected with monoclonal antibodies to study this problem. They found amino-acid changes at positions 54, 143, 205 and in the regions 187–201 and 217–224 of the HA1 (Fig. 3). Both an amino-acid change at position 54 (but to a different amino acid) and at the adjacent residue 53 have been described in field strains (Laver et al., 1980a). There is no counterpart in natural strains for the extreme variability of residue 143 in monoclonal variants, but the adjacent residues, 144, 145 and 146 have undergone changes. Natural mutations around the fourth variant (205) are at positions 207 and 208. Thus, in all instances where a direct comparison is possible, there is only one case (residue 54) where an amino acid has been replaced at the same position in a monoclonal variant and in a field strain, but to a completely different amino acid (to lysine and serine, respectively). The reason for these discrepancies is unknown, but it is quite clear that the immunological pressure in an infected, immune individual is very different from the immune response in a mouse used for isolating the monoclonal hybridoma.

Three amino acids show up as invariable in the natural H3 strains examined so far: cysteine, proline and histidine. However, the hypervariable position 143 in monoclonal variants is occupied by a proline residue and one of the changes observed is to histidine (Fig. 3). Whether the changes involving this proline have to be considered 'natural' or whether they could be due to the atypical selection pressure by a monoclonal antibody remains an open question. Although between subtypes a large fraction of the prolines is constant (see below), the one at position 143 is variable.

Considering the close evolutionary relationship of the H3 hemagglutinins, most of the amino-acid changes have been acquired by single base substitutions and are the type of amino-acid change frequently observed to occur in proteins (Dayhoff, Schwartz & Orcutt, 1978). Apart from their relationship directly determined by the genetic code, the substitutions are often of a chemically conservative nature: Gly↔Ala; Ile↔Val; Leu↔Phe; Leu↔Ile; Glu↔Asp; Arg↔Lys. One amino-acid change (Thr→Tyr, at position 155 of HA1) is both infrequently observed and of a chemically rather drastic nature; it is the only change caused by two nucleotide substitutions. Whether this change is antigenically important is not known.

In conclusion, hemagglutinins of virus isolates of the same (H3) subtype are chemically, surprisingly similar, certainly in contrast to their antigenic divergence (Pereira, 1976). In principle, antigenic drift can be described as caused by the stepwise accumulation of single base substitutions in the nucleotide sequence. Some of these cause antigenically important amino-acid changes and become rapidly fixed under the pressure of the immune system because of their growth advantage. However, we have no realistic picture of the extent of changes required to produce a variant that will escape neutralization and be able to cause an epidemic; also, the delineation of the number and position of the 'antigenic sites' on the HA molecule remains obscure. But the three-dimensional structure of the Aichi/68 hemagglutinin will probably soon be solved (Wilson, Skehel & Wiley, 1980) and this may, in combination with the available and forthcoming sequence information, provide a solid framework for understanding the variability of the virus within a subtype, the mechanism of antigenic drift and the number and distribution of antigenic sites on the hemagglutinin molecule, and for ascribing functions to the constant and variable regions between subtypes (see below).

COMPARISON OF DIFFERENT SUBTYPES: ANTIGENIC SHIFT

It has become possible to compare most of the HA amino-acid sequence of the early human H1 strain A/WSN/33 (previously considered H0) with the complete structure of the HA of strains that represent the onset of the two succeeding subtypes in the human population (causing an antigenic shift), the H2 Japanese (1957) and the H3 Hong Kong (1968) subtypes. Also, the complete sequence of the HA from an avian strain of subtype H7 (previously Hav1), which has never been observed in the human population, is available for comparison. In Fig. 4 the amino-acid sequences have been aligned to show optimal homology. The alignment involves only a few insertions and deletions, with the exception of the NH_2-terminus of the HA1, where the H3 sequence has an extension of 10 (or 11, in strain Victoria) residues and in the H7 sequence in the region connecting HA1 and HA2 (see above). There is considerable sequence homology between the hemagglutinins belonging to the four different subtypes (Fig. 4 and Table 2) illustrating their common evolutionary origin. The homology in the HA1 portion is more limited than in HA2, but usually it is in the same range: between 28 and 37% (it should be remembered that the low values in comparisons involving the HA1 part of H1 are derived from only 32% of the sequence). The homology in HA2 is consistently higher (between 47 and 52% in general), but the conservation amounts to 66% between H3 and H7. This illustrates that the evolutionary relationship of influenza A hemagglutinin subtypes is not a simple one. The two most closely related hemagglutinin subtypes are undoubtedly H1 and H2.

Clearly such large differences in the amino-acid sequences of the hemagglutinins from different human subtypes indicate that antigenic shift cannot have arisen by stepwise mutation of the gene of the previous subtype. It has been proposed that reassortment of hemagglutinin, neuraminidase and probably other genes between a human influenza virus and an influenza virus from lower mammals or birds may lead to an antigenic shift in the human population (Laver & Webster, 1979); variation in the animal reservoir, including the human HA and NA subtypes, both in the combination in which they occur in humans, as well as in many other HA–NA combinations, has been well documented (Hinshaw, Webster & Rodriguez, 1979). The relative differences observed between three human and one avian hemagglutinin subtype would be compatible with such a mechanism. The most convincing evidence for the emergence of the H3N2 subtype from a reassortment event between the H2N2 virus and a virus of unknown origin, which provided probably only the H3 hemagglutinin, comes from RNA · RNA hybridization experiments (Scholtissek, Rohde, von Hoyningen & Rott, 1978). A similar event may have led to the appearance of the H2N2 subtype (Scholtissek et al., 1978).

In 1977 a re-emergence of viruses of the H1N1 subtype was noted. Quite unexpectedly these viruses not only belonged to a subtype previously found in the human population but their surface antigens (Kendal et al., 1978) and even their complete gene constellation (Nakajima et al., 1978; Scholtissek et al., 1978) closely resembled that of 1950 strains. Although there is no explanation for the reappearance of these H1N1 viruses after they had vanished for 27 years, mechanisms other than reassortment must be considered, such as virus latency, persistent infection, etc. Since 1977 we are in a previously undocumented situation of co-circulation of two influenza A subtypes, H1N1 and H3N2. Recently a reassortment event between both subtypes has been described, leading to a 'new' H1N1 variant carrying four genes coding for internal proteins of the virus from a co-circulating H3N2 parent and thus providing the virus with a survival advantage (Young & Palese, 1979).

The almost constant size and the partial amino-acid homology between different hemagglutinins suggest that all hemagglutinins have a similar three-dimensional structure. Cysteine residues are extremely well conserved throughout the HA1 chain and in that part of the HA2 polypeptide outside the viral membrane: there are nine in HA1 and three in HA2. All 12 cysteine residues are believed to be involved in inter- and intra-molecular disulphide bridges, but so far there is no agreement on the bonding pattern (Dopheide & Ward, 1980; Gething *et al.*, 1980). The only part of the molecule where some variation is observed in the location of cysteine residues is the presumed trans-membrane section close to the carboxyl-terminus of HA2, where there are extra cysteines at position 195 (H3 and H7) and 199 (H3). The rôle of the three conserved cysteine residues in the region that is probably on the inner side of the membrane remains to be elucidated.

GLYCOSYLATION SITES

The hemagglutinin molecule is a glycoprotein and from protein studies on the Japanese HA (Gething *et al.*, 1980) and the Memphis HA (Dopheide & Ward, 1980; Ward & Dopheide, 1980) it is known that all glycosylation occurs via *N*-glycosidic linkage to asparagine in the sequence Asn–X–Ser or Asn–X–Thr, the presence of this sequence being a necessary, but not a sufficient condition for the reaction (Neuberger, Gottschalk, Marshall & Spiro, 1972). In the A/Memphis/102/72 (H3) strain, all seven potential sites are glycosylated and, since the same sites are present in the 1968 strains A/Aichi/2/68 and A/NT/60/68/29C, it is probable that the glycosylation pattern has remained constant during the period 1968–1972 (Fig. 3). However, one site (position 81 in HA1) disappears and two potential new sites (at positions 63 and 126) are created in Victoria/75 relative to the 1968 prototype (and Memphis/72). The insertion of an extra asparagine residue at position 8' in strain Victoria modifies the pre-existing site at position 8 into a potential double site 8–8'; but the glycosylation pattern in these cases is unknown. Thus, the pattern of glycosylation in the HA1 is not invariable within the H3 subtype. The effect (if any) of this variation on the antigenic properties remains to be established.

In strain A/Japan/305/57, the HA chain is glycosylated at five positions: two potential double sites are only modified once (as is likely to be the case in strain Victoria), and a sequence Asn–Pro–Ser was found to be unglycosylated. There are seven potential glycosylation sites in the FPV hemagglutinin sequence.

Whereas there is no conservation of glycosylation sites between subtypes in general, their distribution along the chain is more or less similar, probably representing regions on the outside of the molecule. Residue 154 of the HA2 has been found glycosylated in all subtypes examined so far.

CODON USAGE

Codon usage in the HA1 and in the HA2 coding regions of Victoria/75 is shown in Table 3. The most striking difference in amino-acid composition is the complete absence of proline residues in HA2, compared to 20 in HA1, whereas in the other three subtypes only a single proline residue is present in HA2 (Fig. 4). The codon usage is also largely similar in all other aspects.

Vertebrate DNA is known to be very low in the dinucleotide CpG. CpG-containing codons (for serine, proline, threonine, alanine and arginine) are generally less frequent than expected in Victoria HA, but the effect is not as drastic as that observed in other cases, such as simian virus 40 DNA (Fiers et al., 1978). The lower CpG content is reflected not only in codon usage but also in CpG dinucleotides bridging two codons. These observations have been discussed before (Min Jou et al., 1980).

CONCLUSION

In the last few years we have seen a considerable extension in our knowledge about the primary structure of the most important of the influenza A surface antigens, the hemagglutinin, largely due to the application of DNA cloning and fast nucleic-acid sequencing techniques. The phenomena of antigenic drift and antigenic shift have been documented by the elucidation of hemagglutinin structures belonging to the same (H3) subtype as well as to other human (H1

Table 3. *Codons used in the HA1 and HA2 chains of the A/Victoria/ 3/75 hemagglutinin gene. The first number refers to the frequency with which the triplet is used in the HA1 chain (including the connecting arginine residue), the second number indicates the frequency for the HA2 chain and the third number gives the total frequency*

	U	C	A	G	
U	Phe { 4+4=8; 5+7=12 } Leu { 0+1=1; 3+2=5 }	Ser { 4+1=5; 3+2=5; 7+4=11; 1+1=2 }	Tyr { 7+1=8; 4+6=10 } Ochre Amber	Cys { 3+2=5; 6+6=12 } Opal Trp 6+6=12	U C A G
C	Leu { 4+4=8; 0+2=2; 6+0=6; 9+8=17 }	Pro { 6+0=6; 4+0=4; 7+0=7; 3+0=3 }	His { 4+4=8; 2+1=3 } Gln { 9+8=17; 4+2=6 }	Arg { 0+0=0; 2+0=2; 1+0=1; 2+1=3 }	U C A G
A	Ile { 6+6=12; 7+8=15; 8+8=16 } Met 4+4=8	Thr { 10+4=14; 4+0=4; 8+3=11; 5+1=6 }	Asn { 16+9=25; 14+7=21 } Lys { 13+12=25; 6+3=9 }	Ser { 5+0=5; 12+1=13 } Arg { 6+3=9; 3+6=9 }	U C A G
G	Val { 7+4=11; 4+1=5; 4+3=7; 6+0=6 }	Ala { 5+3=8; 2+3=5; 5+6=11; 1+1=2 }	Asp { 8+4=12; 9+10=19 } Glu { 4+11=15; 4+8=12 }	Gly { 4+6=10; 3+5=8; 11+3=14; 10+5=15 }	U C A G

and H2) and nonhuman (H7) subtypes. Since these comparisons are so far solely at the level of primary nucleotide and amino-acid sequence, they are largely quantitative. When the three-dimensional structure of the H3 hemagglutinin is available, it may be possible, in combination with available and forthcoming sequence information, to start making intelligent guesses about (and to devise further experiments to delineate) the different functions of the hemagglutinin and to define the extent and the limitations of its variability.

Hemagglutinins belonging to the same (H3) subtype are chemically surprisingly similar, and antigenic drift can be described as the accumulation of a series of single base changes. Some of these mutations cause antigenically-important amino-acid changes and become rapidly fixed under the pressure of the immune system because of their growth advantage.

The comparison of hemagglutinins from four different subtypes illustrates both their common evolutionary origin and the complexity of that evolutionary relationship. It also suggests that all hemagglutinins have a similar three-dimensional structure. As for the mechanism of shift, the structural data could be compatible with a mechanism of reassortment of hemagglutinin and other genes between a human influenza virus and an influenza virus from wild or domestic animals, as has been suggested (Laver & Webster, 1979). In fact, there is convincing evidence for reassortment phenomena in the history of human influenza (Scholtissek *et al.*, 1978), but as the nature or origin of the donor strain remains obscure the evidence for the animal reservoir remains circumstantial. The relatively recent re-emergence in the human population of viruses of the H1N1 subtype closely resembling 1950 strains (Kendal *et al.*, 1978; Nakajima *et al.*, 1978; Scholtissek *et al.*, 1978) illustrates that reassortment may not be the only way by which antigenic shift can occur and that other mechanisms, such as virus latency or persistent infection, can contribute in this process.

We thank W. Kuziel for help with the manuscript. The research was supported by grants from the Fonds voor Kollektief Fundamenteel Onderzoek and the Gekoncerteerde Akties of the Belgian Ministry of Science. M.V. thanks the N.F.W.O. for a fellowship.

REFERENCES

AIR, G. M. (1979). Nucleotide sequence coding for the 'signal peptide' and N-terminus of the hemagglutinin from an Asian (H2N2) strain of influenza virus. *Virology,* **97,** 468–72.

BAEZ, M., TAUSSIG, R., ZAZRA, J. J., YOUNG, J. F., PALESE, P., REISFELD, A. & SKALKA, A. M. (1980). Complete nucleotide sequence of the influenza A/PR/8/34 virus NS gene and comparison with the NS genes of the A/Udorn/72 and A/FPV/34 strains. *Nucleic Acids Research,* in press.

BLOBEL, G., WALTER, P., CHANG, C. N., GOLDMAN, B. M., ERICKSON, A. H. & LINGAPPA, R. (1979). Translocation of proteins across membranes: the signal hypothesis and beyond. In *Society for Experimental Biology Symposium XXXIII: Secretory Mechanisms,* eds. C. R. Hopkins & C. J. Duncan, pp. 9–36. Cambridge: Cambridge University Press.

Both, G. W. & Air, G. M. (1979). Nucleotide sequence coding for the N-terminal region of the matrix protein of influenza virus. *European Journal of Biochemistry*, **96**, 363–72.

Both, G. W. & Sleigh, M. J. (1980). Complete nucleotide sequence of the haemagglutinin gene from a human influenza virus of the Hong Kong subtype. *Nucleic Acids Research*, **8**, 2561–75.

Davis, A. R. Hiti, A. L. & Nayak, D. P. (1980). Construction and characterization of a bacterial clone containing the hemagglutinin gene of the WSN strain (H0N1) of influenza virus. *Gene*, **10**, 205–18.

Dayhoff, M. O., Schwartz, R. M. & Orcutt, B. C. (1978). A model of evolutionary change in proteins. In *Atlas of Protein Sequence and Structure*, vol. 5, suppl. 3, ed. M. O. Dayhoff, pp. 345–52. Silver Springs, Maryland: The National Biomedical Research Foundation.

Desselberger, U., Racaniello, V. R., Zazra, J. R. & Palese, P. (1980). The 3' and 5' terminal sequences of influenza A, B and C virus genes are highly conserved and show partial inverted complementarity. *Gene*, **8**, 315–28.

Devos, R., van Emmelo, J., Contreras, R. & Fiers, W. (1979). Construction and characterization of a plasmid containing a nearly full-size DNA copy of bacteriophage MS2 RNA. *Journal of Molecular Biology*, **128**, 595–619.

Dopheide, T. A. & Ward, C. W. (1980). Structural studies on a Hong Kong influenza hemagglutinin. The structure of the light chain and the arrangements of the disulphide bonds. In *Structure and Variation in Influenza Virus*, eds. G. Laver & G. Air, pp. 21–26. New York: Elsevier North-Holland.

Dowdle, W. R. (1976). Influenza: Epidemic patterns and antigenic variation. In *Influenza Virus; Vaccines, and Strategy*, ed. P. Selby, pp. 17–21. London: Academic Press.

Drzeniek, R., Seto, J. T. & Rott, R. (1966). Characterization of neuraminidases from myxoviruses. *Biochimica et Biophysica Acta*, **128**, 547–58.

Efstratiadis, A., Kafatos, F. C. & Maniatis, T. (1977). The primary structure of rabbit β-globin mRNA as determined from cloned DNA. *Cell*, **10**, 571–81.

Fiers, W., Contreras, R., Haegeman, G., Rogiers, R., Van de Voorde, A., Van Heuverswyn, H., Van Herreweghe, J., Volckaert, G. & Ysebaert, M. (1978). Complete nucleotide sequence of SV40 DNA. *Nature, London*, **273**, 113–20.

Garoff, H., Frischauf, A.-M., Simons, K., Lehrach, H. & Delius, H. (1980). Nucleotide sequence of the cDNA coding for the semliki forest virus membrane glycoproteins. *Nature, London*, in press.

Gething, M.-J., Bye, J., Skehel, J. J. & Waterfield, M. (1980). Cloning and DNA sequence of double-stranded copies of hemagglutinin genes from H2 and H3 strains elucidates antigenic shift and drift in human influenza virus. *Nature, London*, **287**, 301–6.

Grunstein, M. & Hogness, D. S. (1975). Colony hybridization: a method for the isolation of cloned DNAs that contain a specific gene. *Proceedings of the National Academy of Sciences, U.S.A.*, **72**, 3961–5.

Higuchi, R., Paddock, G. V., Wall, R. & Salser, W. (1976). A general method for cloning eukaryotic structural gene sequences. *Proceedings of the National Academy of Sciences, U.S.A.*, **73**, 3146–50.

Hinshaw, V. S., Webster, R. G. & Rodriguez, R. J. (1979). Influenza A viruses: combinations of hemagglutinin and neuraminidase subtypes isolated from animals and other sources. *Archives of Virology*, **62**, 281–90.

Inglis, S. C., Barrett, T., Brown, C. M. & Almond, J. W. (1979). The smallest genome RNA segment of influenza virus contains two genes that may overlap. *Proceedings of the National Academy of Sciences, U.S.A.*, **76**, 3790–4.

Jackson, D. C., Dopheide, T. A., Russell, R. J., White, D. O. & Ward, C. W. (1979). Antigenic determinants of influenza virus hemagglutinin. II. Antigenic reactivity of the isolated N-terminal cyanogen bromide peptide of A/Memphis/72 hemagglutinin heavy chain. *Virology*, **93**, 458–65.

Jackson, D. C., Russell, R. J., Ward, C. W. & Dopheide, T. A. (1978). Antigenic determinants of influenza virus hemagglutinin. I. Cyanogen bromide peptides derived from A/Memphis/72 hemagglutinin possess antigenic activity. *Virology*, **89**, 199–205.

Jackson, D., Symons, R. & Berg, P. (1972). Biochemical method for inserting new genetic information into DNA of simian virus 40: circular SV40 DNA molecules containing lambda phage genes and the galactose operon of Escherichia coli. *Proceedings of the National Academy of Sciences, U.S.A.*, **69**, 2904–9.

Kendal, A. P., Noble, G. R., Skehel, J. J. & Dopheide, W. R. (1978). Antigenic similarity of influenza A (H1N1) viruses from epidemics in 1977–1978 to 'Scandinavian' strains isolated in epidemics of 1950–1951. *Virology*, **89**, 632–6.

Klenk, H.-D., Rott, R., Orlich, M. & Blödorn, J. (1975). Activation of influenza viruses by trypsin treatment. *Virology*, **68**, 426–39.

Lamb, R. A. & Choppin, P. W. (1979). Segment 8 of the influenza virus genome is unique in coding for two polypeptides. *Proceedings of the National Academy of Sciences, U.S.A.*, **76**, 4908–12.

Lamb, R. A. & Lai, C.-J. (1980). Sequence of interrupted and uninterrupted mRNAs and cloned DNA coding for the two overlapping nonstructural proteins of influenza virus. *Cell*, **21**, 475–85.

Laver, W. G., Air, G. M., Dopheide, T. A. & Ward, C. W. (1980*a*). Amino acid sequence changes in the haemagglutinin of A/Hong Kong (H3N2) influenza virus during the period 1968–1977. *Nature, London*, **283**, 454–7.

Laver, W. G., Air, G. M., Webster, R. G., Gerhard, W., Ward, C. W. & Dopheide, T. A. (1970). Antigenic drift in type A influenza virus: sequence differences in the hemagglutinin of Hong Kong (H3N2) variants selected with monoclonal hybridoma antibodies. *Virology*, **98**, 226–37.

Laver, W. G., Air, G. M., Webster, R. G., Gerhard, W., Ward, C. W. & Dopheide, T. A. (1980*b*). The antigenic sites on influenza virus hemagglutinin. Studies on their structure and variation. In *Structure and Variation in Influenza Virus*, ed. G. Laver & G. Air, pp. 295–307. New York: Elsevier North-Holland.

Laver, W. G. & Kilbourne, E. D. (1966). Identification in a recombinant influenza virus of structural proteins derived from both parents. *Virology*, **30**, 493–501.

Laver, W. G. & Webster, R. G. (1968), Selection of antigenic mutants of influenza viruses. Isolation and peptide mapping of their hemagglutinating proteins. *Virology*, **34**, 193–202.

Laver, W. G. & Webster, R. G. (1972), Studies on the origin of pandemic influenza. II. Peptide maps of the light and heavy polypeptide chains from the hemagglutinin subunits of A2 influenza viruses isolated before and after the appearance of Hong Kong influenza. *Virology*, **48**, 445–55.

Laver, W. G. & Webster, R. G. (1979). Ecology of influenza viruses in lower mammals and birds. *British Medical Bulletin*, **35**, 29–33.

Lazarowitz, S. G. & Choppin, P. W. (1975). Enhancement of the infectivity of influenza A and B viruses by proteolytic cleavage of the hemagglutinin polypeptide. *Virology*, **68**, 440–54.

McCauley, J., Bye, J., Elder, K., Gething, M.-J., Skehel, J. J., Smith, A. & Waterfield, M. D. (1979). Influenza virus hemaglutinin signal sequence. *FEBS Letters*, **108**, 422–6.

McGeoch, D. J., Fellner, P. & Newton, C. (1976). Influenza virus genome consists of eight distinct RNA species. *Proceedings of the National Academy of Sciences U.S.A.*, **73**, 3045–9.

MANIATIS, T., KEE, S. G., EFSTRATIADIS, A. & KAFATOS, F. C. (1976). Amplification and characterization of a β-globin gene synthesized in vitro. *Cell*, **8**, 163–82.

MAXAM, A. M. & GILBERT, W. (1977). A new method for sequencing DNA. *Proceedings of the National Academy of Sciences, U.S.A.*, **74**, 560–4.

MIN JOU, W., VERHOEYEN, M., DEVOS, R., SAMAN, E., FANG, R.-X., HUYLEBROECK, D., FIERS, W., THRELFALL, G., BARBER, C., CAREY, N. & EMTAGE, S. (1980). Complete structure of the hemagglutinin gene from the human influenza A/Victoria/3/75 (H3N2) strain as determined from cloned DNA. *Cell*, **19**, 683–96.

NAKAJIMA, K., DESSELBERGER, U. & PALESE, P. (1978). Recent human influenza A (H1N1) viruses are closely related genetically to strains isolated in 1950. *Nature, London*, **274**, 334–9.

NEUBERGER, A., GOTTSCHALK, A., MARSHALL, R. D. & SPIRO, R. G. (1972). Carbohydrate-peptide linkages in glycoproteins and methods for their elucidation. In *The Glycoproteins*, ed. A. Gottschalk, pp. 450–90. Amsterdam: Elsevier North-Holland.

PALESE, P. (1977). The genes of influenza virus. *Cell*, **10**, 1–10.

PALESE, P. & SCHULMAN, J. L. (1976). Mapping of the influenza virus genome: identification of the hemagglutinin and the neuraminidase genes. *Proceedings of the National Academy of Sciences U.S.A.*, **73**, 2142–6.

PEREIRA, M. S. (1976). Strain surveillance in man. In *Influenza Virus, Vaccines, and Strategy*, ed. P. Selby, pp. 25–31. London: Academic Press.

PORTER, A. G., BARBER, C., CAREY, N. H., HALLEWELL, R. A., THRELFALL, G. & EMTAGE, J. S. (1979). Complete nucleotide sequence of an influenza virus haemagglutinin gene from cloned DNA. *Nature, London*, **282**, 471–7.

PORTER, A. G., SMITH, J. C. & EMTAGE, J. S. (1980). Nucleotide sequence of influenza virus RNA segment 8 indicates that coding regions for NS1 and NS2 proteins overlap. *Proceedings of the National Academy of Sciences U.S.A.*, **77**, 5074–8.

POTTER, C. W. & OXFORD, J. S. (1979). Determinants of immunity to influenza infection in man. *British Medical Bulletin*, **35**, 69–75.

ROBB, R. J., TERHORST, C. & STROMINGER, J. L. (1978). Sequence of the COOH-terminal hydrophilic region of histocompatibility antigens HLA-A2 and HLA-B7. *The Journal of Biological Chemistry*, **253**, 5319–24.

ROBERTSON, J. S. (1979). 5′ and 3′ terminal nucleotide sequences of the RNA genome segments of influenza virus. *Nucleic Acids Research*, **6**, 3745–57.

ROGERS, J., EARLY, P., CARTER, C., CALAME, K., BOND, M., HOOD, L. & WALL, R. (1980). Two mRNAs with different 3′ ends encode membrane-bound and secreted forms of immunoglobulin μ chain. *Cell*, **20**, 303–12.

ROSE, J. K., WELCH, W. J., SEFTON, B. M., ESCH, F. S. & LING, N. C. (1980). Vesicular stomatitis virus glycoprotein is anchored in the viral membrane by a hydrophobic domain near the COOH terminus. *Proceedings of the National Academy of Sciences U.S.A.*, **77**, 3884–8.

SANGER, F., NICKLEN, S. & COULSON, A. R. (1977). DNA sequencing with chain-termination inhibitors. *Proceedings of the National Academy of Sciences U.S.A.*, **74**, 5463–7.

SCHILD, G. C., NEWMAN, R. W., WEBSTER, R. G., MAJOR, D. & HINSHAW, V. S. (1980). Antigenic analysis of influenza A virus surface antigens: considerations for the nomenclature of influenza virus. *Archives of Virology*, **63**, 171–84.

SCHMIDT, M. F. G., BRACHA, M. & SCHLESINGER, M. J. (1979). Evidence for covalent attachment of fatty acids to Sindbis virus glycoproteins. *Proceedings of the National Academy of Sciences U.S.A.*, **76**, 1687–91.

SCHMIDT, M. F. G. & SCHLESINGER, M. J. (1979). Fatty acid binding to vesicular stomatitis virus glycoprotein: a new type of post-translational modification of the viral glycoprotein. *Cell,* **17,** 813–19.

SCHOLTISSEK, C., HARMS, E., ROHDE, W., ORLICH, M. & ROTT, R. (1976). Correlation between RNA fragments of Fowl Plague virus and their corresponding gene functions. *Virology,* **74,** 332–44.

SCHOLTISSEK, C., ROHDE, W., VON HOYNINGEN, V. & ROTT, R. (1978). On the origin of the human influenza virus subtypes H2N2 and H3N2. *Virology,* **87,** 13–20.

SCHOLTISSEK, C., VON HOYNINGEN, V. & ROTT, R. (1978). Genetic relatedness between the new 1977 epidemic strains (H1N1) of influenza and human influenza strains isolated between 1947 and 1957 (H1N1). *Virology,* **89,** 613–17.

SKEHEL, J. J. & HAY, A. J. (1978). Nucleotide sequences at the 5'-termini of influenza viral RNAs and their transcripts. *Nucleic Acids Research,* **5,** 1207–19.

SKEHEL, J. J. & WATERFIELD, M. D. (1975). Studies on the primary structure of the influenza virus hemagglutinin. *Proceedings of the National Academy of Sciences U.S.A.,* **72,** 93–7.

SLEIGH, M. J., BOTH, G. W., BROWNLEE, G. G., BENDER, V. J. & MOSS, B. A. (1980). The haemagglutinin gene of influenza A virus: Nucleotide sequence analysis of cloned DNA copies. In *Structure and Variation in Influenza Virus,* ed. G. Laver & G. Air, pp. 69–79. New York: Elsevier North-Holland.

SMITH, W., ANDREWES, C. H. & LAIDLAW, P. P. (1933). A virus obtained from influenza patients. *Lancet,* **2,** 66–8.

SPRINGER, T. A. & STROMINGER, J. L. (1976). Detergent-soluble HLA antigens contain a hydrophilic region at the COOH-terminus and a penultimate hydrophobic region. *Proceedings of the National Academy of Sciences U.S.A.,* **73,** 2481–5.

TOMITA, M., FURTHMAYR, H. & MARCHESI, V. T. (1978). Primary structure of human erythrocyte glycophorin A. Isolation and characterization of peptides and complete amino acid sequence. *Biochemistry,* **17,** 4756–70.

VERHOEYEN, M., FANG, R.-X., MIN JOU, W., DEVOS, R., HUYLEBROECK, D., SAMAN, E. & FIERS, W. (1980). Antigenic drift between the hemagglutinin of the Hong Kong influenza strains A/Aichi/2/68 and A/Victoria/3/75. *Nature, London,* **286,** 771–6.

WARD, C. W. & DOPHEIDE, T. A. (1979). Primary structure of the Hong Kong (H3) haemagglutinin. *British Medical Bulletin,* **35,** 51–6.

WARD, C. W. & DOPHEIDE, T. A. (1980). The Hong Kong (H3) hemagglutinin. Complete amino acid sequence and oligosaccharide distribution for the heavy chain of A/Memphis/102/72. In *Structure and Variation in Influenza Virus,* ed. G. Laver & G. Air, pp. 27–38. New York: Elsevier North-Holland.

WATERFIELD, M. D., ESPELIE, K., ELDER, K. & SKEHEL, J. J. (1979). Structure of the haemagglutinin of influenza virus. *British Medical Bulletin,* **35,** 57–63.

WILSON, I. A., SKEHEL, J. J. & WILEY, D. C. (1980). Structural studies on the haemagglutinin glycoprotein of influenza virus. In *Structure and Variation in Influenza Virus,* ed. G. Laver & G. Air, pp. 339–48. New York: Elsevier North-Holland.

WINTER, G. & FIELDS, S. (1980). Cloning of influenza cDNA into M13: the sequence of the RNA segment encoding the A/PR/8/34 matrix protein. *Nucleic Acids Research,* **8,** 1965–74.

YOUNG, J. F. & PALESE, P. (1979). Evolution of human influenza A viruses in nature. Recombination contributes to genetic variation of H1N1 strains. *Proceedings of the National Academy of Sciences U.S.A.* **76,** 6547–51.

EVOLUTION OF ANTIGENIC VARIATION IN THE SALIVARIAN TRYPANOSOMES

M. J. TURNER AND J. S. CORDINGLEY

MRC Biochemical Parasitology Unit, The Molteno Institute, University of Cambridge, Downing Street, Cambridge CB2 3EE, UK

INTRODUCTION

The pathogenic African trypanosomes are parasitic protozoa which inhabit the blood and tissue fluids of their mammalian hosts, and evade elimination by the host's immune system through a process of antigenic variation. Infection is characterized by a relapsing parasitemia, in which each remission is a consequence of destruction of trypanosomes by host antibodies, and each recrudescence is due to the reappearance of antigenically distinct variants. Antigenic variation is a feature of other protozoan infections (reviewed in Turner, 1980), most notably in members of two other genera, *Plasmodium* and *Babesia*, but it is best characterized in the salivarian trypanosomes which constitute four of the seven subgenera of the genus Trypanosoma (Fig. 1). The species comprising this group are all transmitted cyclically by tsetse flies or are descendants of such species but are now non-cyclically transmitted either venereally or mechanically outside the tsetse fly belt of Africa. Most of our information on antigenic variation comes from *Trypanosoma brucei*, two infraspecific categories of which, *Trypanosoma brucei rhodesiense* and *Trypanosoma brucei gambiense* are responsible for human sleeping sickness. These are often referred to as subspecies but are probably better regarded as nosodemes. *T. brucei* also infects lower mammals including laboratory rodents, and is convenient for research.

The life cycle of *T. brucei* is complicated (Fig. 2). When an infected tsetse fly (*Glossina* sp.) takes a blood meal, trypanosomes are transferred from the salivary glands into the lymph and blood of the mammalian host. Trypanosome numbers rapidly increase, and during this ascending phase of the parasitemia, the trypanosomes occur predominantly as long, slender, multiplicative forms. After five to seven days, antibodies capable of both lysing and agglutinating the trypanosomes are detectable in the serum, the disease enters a remission phase and the parasitaemia drops rapidly to a low, even

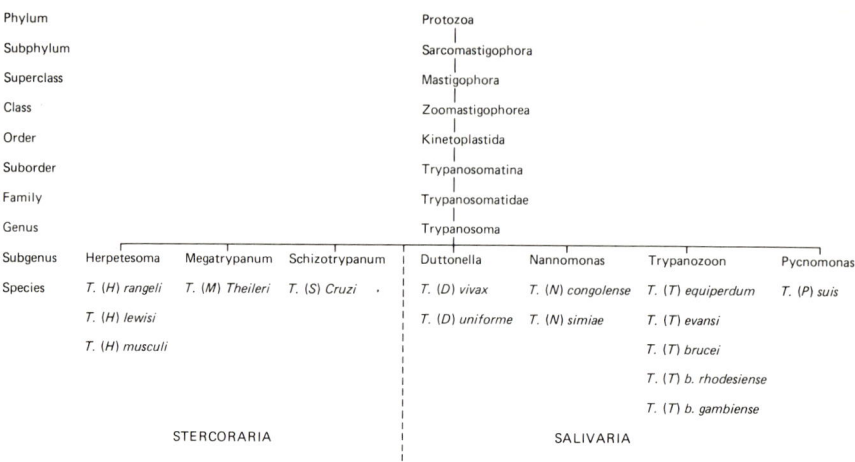

Fig. 1. Classification of trypanosomes. After Hoare (1972).

subpatent level. During this phase an increasing proportion of short, stumpy, non-replicative forms appear. These are apparently non-infective to new mammalian hosts by mechanical infection, and therefore seem to be in a condition incapable of maintaining the bloodstream infection. Survivors of this crisis are slender trypomastigotes of a new antigenic type, unaffected by the circulating antibody, and multiply until they in turn are eliminated by the host's immune system. This cycle apparently can continue indefinitely, and usually leads to the death of the host. The existence of distinct morphological forms (pleomorphism) is an essential feature of these bloodstream trypomastigotes, because monomorphic strains comprising only slender forms generated in the laboratory are incapable of cyclical transmission. The change from long, slender to short, stumpy forms is accompanied by an alteration in respiratory metabolism (Vickerman, 1965). In the long, slender forms ATP is generated by glycolysis and terminal respiration is mediated by reoxidation of NADH in the α-glycerophosphate cycle (Grant, Sargent & Ryley, 1961), and pyruvate is excreted. The change to short, stumpy forms is accompanied by the production of some of the enzymes of the TCA cycle and the appearance of a functional mitochondrial electron transport chain. These changes are thought to facilitate establishment of infection in the fly midgut after ingestion by the tsetse fly, since all stages occurring in the insect have normal mitochondrial terminal respiration (Bowman, 1974). After multiplication in the midgut, trypanosomes migrate to the

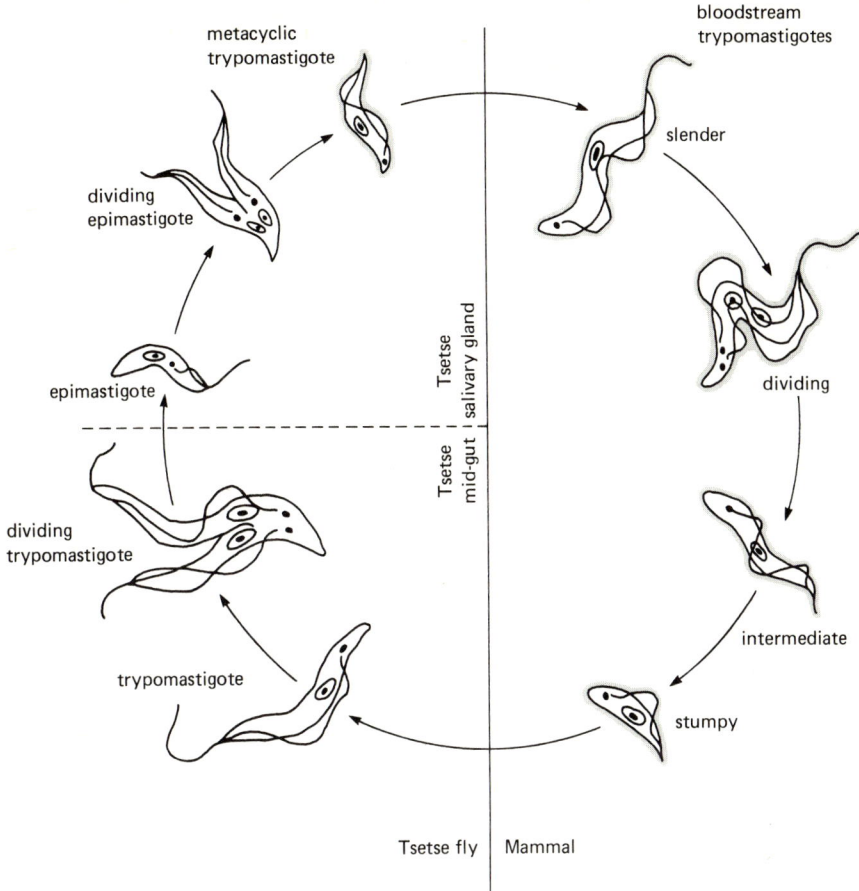

Fig. 2. Diagrammatic representation of the life cycle of *T. brucei*. After Vickerman (1971). Forms which possess a surface coat are shaded.

salivary glands, undergoing further multiplication while attached to the salivary epithelium and ultimately developing into infective metacyclic trypomastigotes. The cycle of development within the fly takes seventeen to fifty days.

All bloodstream forms (Fig. 3) are distinguished by the presence of an electron-dense surface coat, 12–15 nm thick, external to the limiting membrane bilayer (Vickerman, 1969). The surface coat is not present in fly midgut forms (Vickerman, 1969) and only reappears with the development of infective metacyclic trypomastigotes (Vickerman, 1969; Steiger, 1971). The evidence that antigenic variation is associated with changes in the composition of the surface coat may be summarized as follows. First, in suitable media

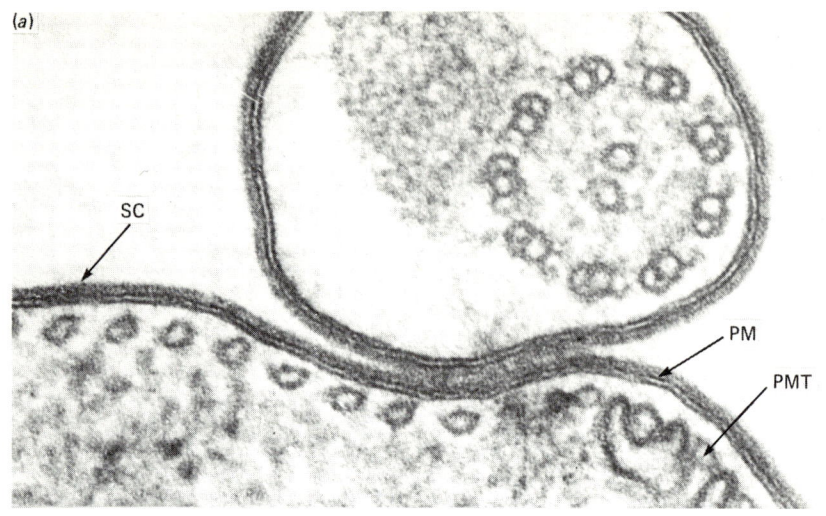

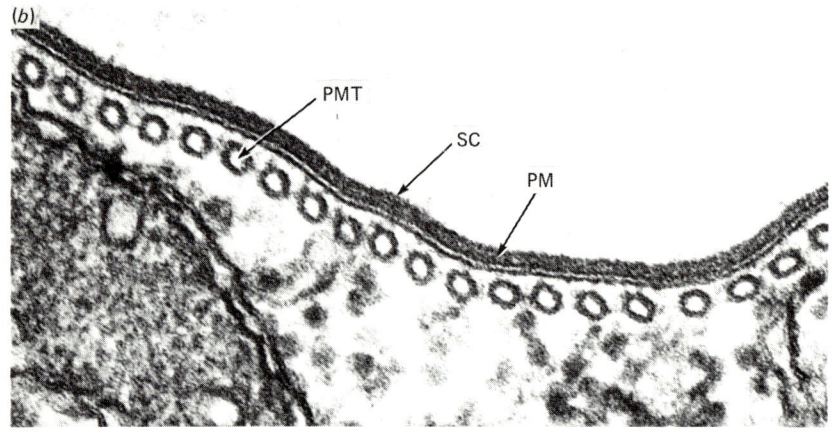

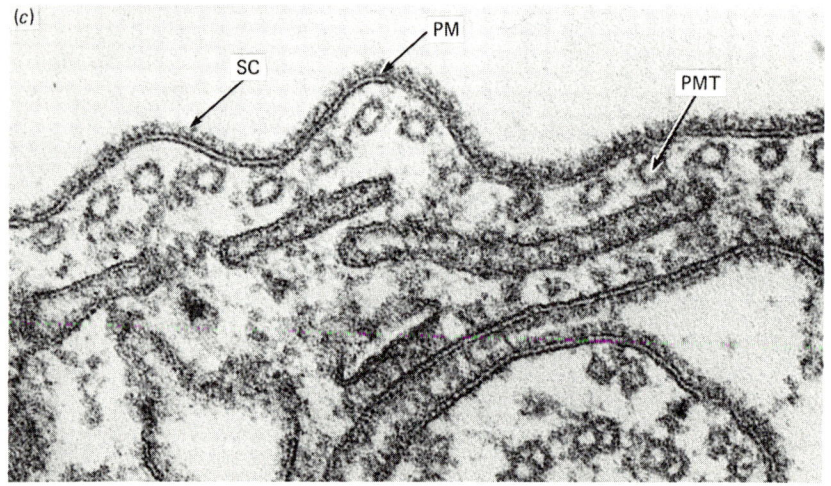

bloodstream trypomastigotes differentiate into a form synonymous with that found in the insect midgut, the procyclic trypomastigote. Such procyclic forms are incapable of undergoing antigenic variation and irrespective of the antigenic type from which they are derived all share a common surface antigenicity (Seed, 1964). This transformation is accompanied by the loss of the surface coat (Brown, Evans & Vickerman, 1973). Secondly, variant specific antibodies, prepared by infection with cloned trypanosomes, have been shown to bind to the surface coat of homologous but not heterologous trypanosomes (Vickerman & Luckins, 1969). Finally, surface labelling of bloodstream trypanosomes with formyl-methionine sulphone methyl phosphate results in the labelling of a single glycoprotein which may be readily purified to homogeneity. The composition of the glycoprotein is variant-specific and if used as an immunogen produces variant-specific immunity (Cross, 1975).

Before any attempt can be made to describe how such a system may have evolved, it is essential to discuss the nature of the phenomenon more completely and to characterize the antigens. Such a description is best divided into discussion of the serology, biochemistry and molecular biology of antigenic variation.

SEROLOGY OF ANTIGENIC VARIATION

Ritz (1916) demonstrated that infections initiated with a single trypanosome produced the characteristic relapsing parasitemia showing that each trypanosome possesses the information necessary to generate new variants. The number of variants which may be generated from a clone is not known, although 101 variants have been derived from a single clone of *Trypanosoma equiperdum* (Capbern, Giroud, Baltz & Mattern, 1977). Groups of clones which generate the same families of variants, such that all variants are directly or indirectly interconvertible, are termed serodemes. The number of different serodemes within each subgenus or species is unknown, but certainly different field isolates commonly express widely differing repertoires although some overlap (~5%) is often

Fig. 3. Transverse section of part of bloodstream trypanastigotes of (*a*) *Trypanosoma (N.) congolense*, (*b*) *Trypanosoma (T.) brucei* and (*c*) *Trypanosoma (D) vivax*, showing surface coat (SC), plasma membrane (PM) and pellicular microtubules (PMT). Note the more granular appearance of the surface coat of *Trypanosome vivax* (*c*). Previously unpublished micrograph kindly supplied by L. Tetley and K. Vickerman. (×118 000.)

detectable. Variants common to different serodemes are known as isotypic variant antigen types, or iso-VATs, and are frequently found to cross species barriers. The nature of such isotypes is clearly of interest from the evolutionary standpoint and will be discussed later. From the above, it is clear that the size of the gene pool cannot be estimated with any certainty, but it seems likely to run into thousands. The difficulties of analysing such a complex system have been magnified by the apparent lack of any sexual process in the life cycle and the failure to propagate bloodstream trypomastigotes *in vitro*. Thus it has been impossible to utilize classical genetics in the way used in analysis of antigenic variation in *Paramecium*. However, a suitable culture system has now been developed (Hirumi, Doyle & Hirumi, 1977) and recent work on the distribution of isozymes within the population has suggested the existence of a sexual stage (Tait, 1980) so more rapid progress may be expected in the future.

One of the most important features of antigenic variation is that it is not a random process. This was first demonstrated in a classic series of experiments (Gray, 1965a,b). Gray reported that after cyclical transmission, certain variants of a serodeme tended to appear very early in the course of infection, irrespective of the antigenic composition of the population ingested by the vector. Gray called these variants the 'basic antigens' of the strain, and Cunningham (1966) claimed that these were expressed on the metacyclic trypanosomes in the salivary glands of the fly. Gray also noted that on syringe passage of variants arising later in infection, reversion to antigens seen earlier in the series often occurred. Such variants were termed the predominant antigens, and again seemed to develop early in the course of infection after cyclical transmission. This gave rise to the concept of an ordered sequence of expression, starting with a basic antigen expressed both on metacyclics and within the first bloodstream population, proceeding through the predominant types until cyclical transmission produced reversion to the basic antigenic type (Fig. 4). Later work has shown this to be an over-simplification. Two experimental approaches have shown the picture to be more complex. Because of the difficulties in controlling the antigenic composition of trypanosomes ingested and transmitted by the fly, experiments have been described in which infection has been initiated with a small number (10–1 000) of cloned trypanosomes, and the antigenic composition of the first relapse population has been analysed (van Meirvenne,

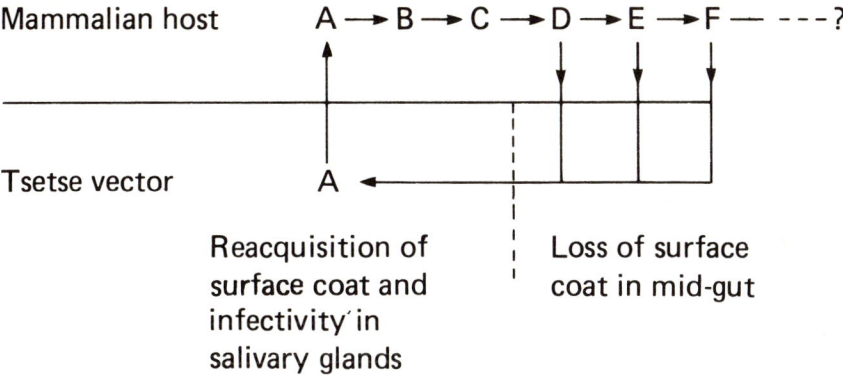

Fig. 4. Sequential expression model, in which A, B, C, D, E and F represent different antigenic types. After Cross (1978).

Janssens & Magnus, 1975; Miller & Turner, 1980). This population is never homogeneous. Indeed, in our own experiments in which antisera to seven variants were available, these antisera identified on average only about 65% of each relapse population. The remainder could not be identified, and it is a problem inherent in this approach that the unidentified fraction could represent a further one or one hundred antigenic types. There is no demonstrable requirement for any antigenic type to relapse to any other in such experiments (Fig. 5). Rather, there appears to be a statistically definable order of priority which is different for each parent variant. The 'predominant types' described by Gray represent a group of variants which have relatively high probabilities of expression by all other variants. Analysis of growth rates show that this is not a factor in determining the frequency of expression of different variants in the relapse population, although it does have an effect in determining the major type in each wave of parasitemia (Miller & Turner, 1980). Recently, a computer analysis of published data concluded that a mechanism based on random generation of new variants followed by selection by growth rate alone could not produce the degree of variant orderliness reported in the literature (Kosinski, 1980).

What is the rôle of the immune response in antigenic variation? All the available evidence points to a purely selective effect, based on the following observations. First, both immunofluorescent staining (van Meirvenne *et al.*, 1975) and neutralization experiments (Doyle, 1977) have demonstrated that minor heterotypes are spontaneously generated at a continuous low level of approximately

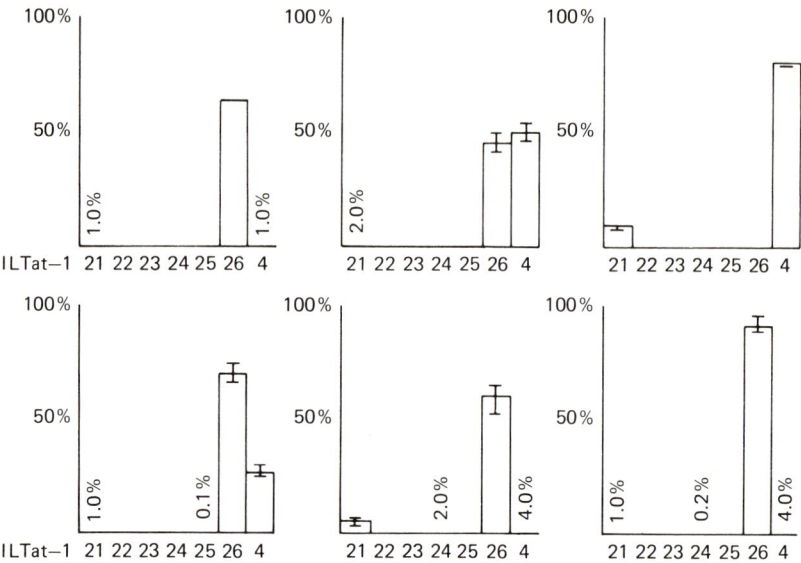

Fig. 5. Non-random expression of variants in a population of trypanosomes following a relapse. Infections of rats were initiated with approximately ten trypanosomes from cloned populations, in this case of antigenic type ILTat 1-25. Heterotypes are found in this cloned population at a level of less than 1 in 1 000, hence the probability of having included a mixture of antigenic types in the inoculum is small. After the first remission in the parasitemia, the succeeding relapse population was sampled, and the antigenic composition monitored by immunofluorescence. All members of this relapse population are descendants of an antigenically homogeneous population, and therefore reflect any preferential expression of new variant types. Each graph represents the results obtained in one rat. The data are expressed as a histogram, in which the percentage in the relapse population of each of seven different antigenic types is shown. In this case, ILTat 1-25 preferentially expresses ILTat 1-26 on relapse, but the preference is neither absolute nor obligatory. After Miller & Turner (1981).

1 in 10^4 trypanosomes in a cloned population. Such a level is consistent with the timing of appearance of new variants *in vivo*. Second, antigenic types expressed in relapse populations are the same as those present as minor heterotypes in the clones from which the relapses were derived (Miller, 1980). Miller also found that heterotypes expressed during the course of a relapsing infection show no changes in proportion or antigenic type attributable to the influence of the immune system. Finally, Doyle *et al.* (1980) have convincingly demonstrated that new variants can be generated *in vitro* in the complete absence of stimulation by the immune response. In conclusion, then, if induction by antibody does take place, then it occurs at a rate less than or equal to the spontaneous generation of new variants.

There are two further complications in this analysis. First it has been reported (Seed, 1978; Miller & Turner, 1981), that growth rates of variants differ when measured alone and in mixtures. Secondly, in *Trypanosoma vivax* the repertoire of variants expressed has been found to differ in different host species (de Gee, Shah & Doyle, 1979). These two findings strongly suggest that there are additional selective environmental effects operating on all variants, irrespective of the mechanisms by which they are generated.

Another approach used in analysis of antigenic variation has been to analyse the antigenic composition of the metacyclic trypanosomes, and to compare this with the first bloodstream population to develop. This has proved a rather controversial area of research. Jenni (1977a, b) described the preparation of an antimetacyclic serum which reacted with 100% of metacyclic trypanosomes of one strain of *T. brucei*, but not other strains. Within 36 h of infection, only 10–20% of the bloodstream trypanosomes reacted with this antiserum, suggesting that the majority of the trypanosomes had changed antigenic type already. Jenni interpreted his data as showing that the metacyclic population was antigenically homogeneous, a view supported by similar data obtained using *Trypanosoma congolense* (Nantulya, Doyle & Jenni, 1980). However, other workers have reported (Le Ray, Barry & Vickerman, 1978; Barry, Hajduk, Vickerman & Le Ray, 1979) that metacyclic populations contain more than one antigenic type, and that these types include variants found in bloodstream populations. Barry & Hajduk (1979) showed that antisera to one variant produced in the blood stream could recognize 11–20% of metacyclics of one strain. However, pooled antisera against a large collection of bloodstream variants did not increase the proportion recognised, suggesting that while metacyclic trypanosomes are antigenically heterogeneous, they do not contain a large selection of the antigenic types expressed later in the infection. These conflicting results may be reconcilable on the assumption that Jenni's antimetacyclic serum was polyspecific.

The finding that antigenic change takes place early following cyclical transmission is undisputed. Additional evidence for this antigenic instability of trypanosomes in the early stages of a cyclically transmitted infection comes from the observation that it has been found impossible to generate antigenically homogeneous trypanosome populations from infections initiated with a single metacyclic trypanosome (Doyle, personal communication; Hudson, Taylor & Elce, 1980).

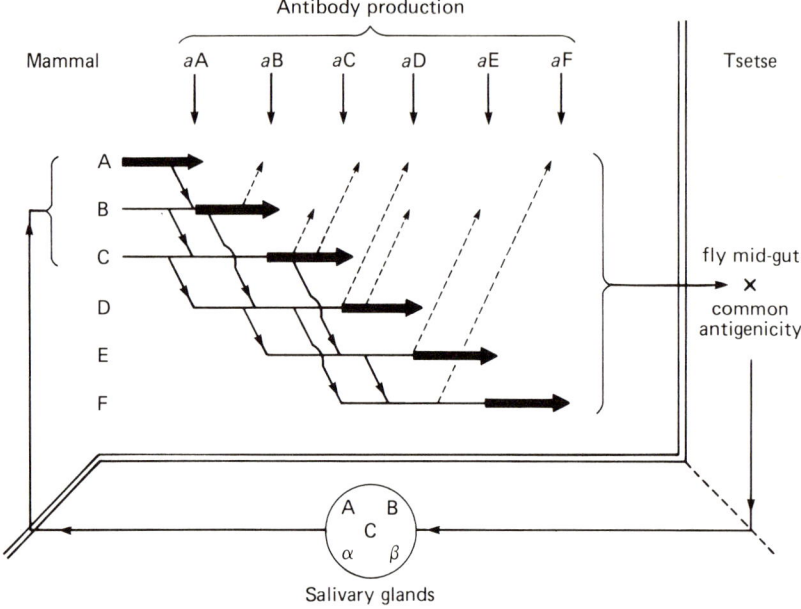

Fig. 6. A model of the nature of antigenic variation, based on serological observations. Infection is initiated with a mixture of antigenic types (A, B, C). It remains controversial whether all or only some of the bloodstream variants which occur in the course of infection are present at this early stage. This model assumes that not all are present. In each wave of parasitemia, one antigenic type may predominate (thick arrow), and may suppress the levels of minor heterotypes (thin arrow) or induce their sequestration in tissue spaces. Elimination by antibodies of the major variant in each wave of parasitemia allows its replacement by development of a minor heterotype. As infection proceeds, new variants are produced spontaneously and at a low level in a non-random way (oblique arrows). Re-expression of variants appearing earlier in the sequence is allowed (dashed arrows) but because of the presence of circulating antibody such events are only observed on mechanical passage to a non-immune host. 'Predominant' antigens are those whose re-expression occurs at a comparatively high frequency in all variants, hence they are often detected after such mechanical passage. On entry into *Glossina sp.* all variants lose their surface coat and adopt a common surface antigenicity. In the salivary glands, infectivity is regained with the reacquisition of surface coat by the metacyclic trypomastigotes. The metacyclic population is antigenically heterogeneous, and in addition to expressing antigens detectable in bloodstream populations (A, B, C) may express antigens which are either unique to the salivary glands or only very rarely found in the bloodstream (α, β). On re-entry into the mammal, selection occurs of those variants best able to establish a bloodstream infection (A, B, C).

Thus the biology of antigenic variation demonstrates the evolution of a remarkably complex switching mechanism. Cyclical transmission results in a dedifferentiation to a form in which the antigen genes are inactivated. Reactivation occurs with the development of infective metacyclic forms. Once activated, new variants are expressed at a fixed rate in some apparently non-random way (Fig. 6). Development of different phenotypes may be environmentally

Table 1. *Amino-acid compositions and isoelectric points of surface glycoproteins (VSGs). Data from Miller (1980)*

VSG	1–21	1–22	1–23	1–24	1–25	1–26	1–4
Amino-acids[a]							
Asx	66.3	68.1	69.1	57.5	68.2	79.7	56.9
Thr	56.4	60.4	54.3	51.3	44.7	58.3	53.0
Ser	33.6	45.7	30.4	39.3	37.8	31.3	35.8
Glx	81.4	73.3	78.2	81.1	73.4	65.8	78.9
Pro	27.7	21.6	21.0	22.3	21.2	15.9	15.2
Gly	57.3	40.4	53.2	47.5	48.2	41.3	59.9
Ala	91.2	86.9	83.5	74.4	119.9	82.0	93.8
Cys	7	n.d.[b]	11.0	14.0	n.d.	11.4	n.d.
Val	18.1	20.6	17.8	20.7	13.5	26.3	21.8
Met	1.4	8.0	7.2	6.5	2.7	7.1	7.6
Ile	23.7	31.4	18.3	18.5	24.3	20.8	24.9
Leu	36.7	46.0	51.2	56.2	46.8	46.0	48.9
Tyr	10.0	13.4	10.3	11.6	6.0	10.7	5.9
Phe	9.2	9.6	7.4	14.6	7.4	16.1	9.5
His	5.4	11.1	13.5	5.8	4.7	7.5	9.3
Lys	40.4	50.1	59.9	63.0	63.8	64.9	68.4
Arg	34.4	12.8	14.8	15.3	17.4	15.2	10.6
Isoelectric point							
(pI)	6.8	5.93	6.7	5.49	6.2	8.6	5.8

[a] Number of residues per 600 amino-acids, excluding tryptophan, which was not determined.
[b] n.d., not determined.

controlled (host or vector; different host species). Can the biochemistry of the antigens provide any clue to their evolutionary origin?

BIOCHEMISTRY OF ANTIGENIC VARIATION

Much progress has been made over the past five years in characterizing the variant surface glycoproteins (VSGs). Cross (1975) described the purification of antigen from *T. brucei* in a simple three-step procedure. Cell rupture releases about 75% of the VSGs in an apparently water soluble form. Following DEAE-cellulose chromatography, the VSGs can be purified to homogeneity by isoelectric focussing. The isoelectric points vary markedly in the range 5.8–9.1. The VSGs have molecular weights of around 60 000, and contain 7–17% carbohydrate, comprising four sugars only – *N*-acetylglucosamine, mannose, galactose and glucose (Johnson & Cross, 1977). The glucose content varies from preparation to preparation, and its absence from purified glycopeptide suggests

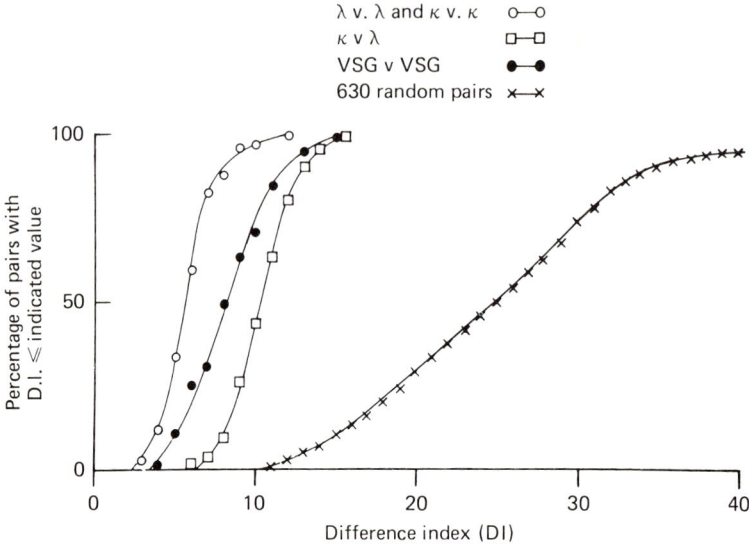

Fig. 7. Analysis of amino-acid composition data by Difference Index (D.I.) method. Seventy-eight pairwise comparisons of the amino-acid compositions of immunoglobulin ϰ light chains of known sequence gave a mean D.I. of 5.5. Forty-five such comparisons of immunoglobulin λ light chains gave a mean D.I. of 5.7. One hundred and thirty comparisons of the ϰ chains with the λ chains gave a mean D.I. of 10.5. Fifty-five pairwise comparisons of VSG amino-acid compositions gave a mean D.I. of 8.0. Different ϰ chains contain 80% sequence homology, as do λ chains, whereas ϰ and λ share 40% homology. This suggests therefore that VSGs share about 50% homology. The distribution curves for the D.I.s within each population are plotted, and the distribution curve for D.I.s obtained from 630 pairs of randomly chosen proteins. (After Metzger et al. (1968).)

that it is an artefact (Holder & Cross, 1980). The different isoelectric points are reflected in differences in the amino-acid compositions which are otherwise unremarkable (Table 1). Bridgen, Cross & Bridgen (1976) presented N-terminal sequence data on four VSGs obtained from *T. brucei* clones isolated at roughly weekly intervals during a chronic infection of a rabbit. There were no sequence homologies within the first 30–40 residues.

Given that the antigens are different, it is more fruitful from the evolutionary standpoint to consider those features of the VSG molecules that are shared. The antigens are obviously functionally related. They all must be inserted through the trypanosome membrane, remain bound to the cell surface, and form a coherent surface coat, and these properties must be reflected in structural similarities. Evidence that the VSGs of *T. brucei* are conformationally related comes from proteolysis experiments, in which it was shown that brief exposure to trypsin cleaves VSGs into an

N-terminal fragment (about two thirds of the molecule) and a C-terminal fragment (one third) (Johnson & Cross, 1979). This cleavage can be effected both on purified antigen, and by treatment of viable trypanosomes with trypsin. In the latter case, the C-terminal fragment remains bound to the trypanosome, and can be released by rupturing the cell (Cross & Johnson, 1976). This portion of the molecule, which therefore appears to contain the membrane-binding site, also carries carbohydrate. At least two carbohydrates are attached to the C-terminal fragment. We have shown (Auffret, Allan & Turner, unpublished) that treatment of VSGs with endo-glycosidase-H releases only about 50% of the VSG carbohydrate. The released N-glycosidically linked oligosaccharide contains N-acetyl glucosamine and mannose, plus a small amount of galactose. The remaining carbohydrate contains N-acetylglucosamine, a small amount of mannose and high levels of galactose. Both these carbohydrates seem to be in the C-terminal domain. Recent work (Holder & Cross, 1980) has shown the high mannose oligosaccharide to be internal, whereas in five cases examined, the high galactose carbohydrate is within the C-terminal tryptic glycopeptide. In every case, the C-terminal amino acid, whether asparagine or serine, was glycosylated. Both the internal and the C-terminal tryptic peptides have been sequenced, and show a marked degree of sequence conservation. Additionally, the high-galactose C-terminal tryptic peptides carry the common determinant previously detected in tests for immunological cross-reaction (Barbet & McGuire, 1978; Cross, 1979).

Other than the glycoprotein sequences reported above, there has been little direct proof of sequence homology. We have compared the amino-acid composition of VSGs using the difference index method (Metzger, Shapiro, Mosiman & Vinton, 1968) and by comparison with similar analyses performed on immunoglobulin light chains of known sequence, we have predicted the existence of about 50% sequence homology (McConnell, Cordingley & Turner, 1979) (Fig. 7). However, this has proved impossible to detect using one and two-dimensional peptide mapping techniques (Fig. 8). This is consistent with the existence of small sequence homologies scattered throughout the length of the molecule, rather than the existence of large and distinct variable and constant region domains. The proteolysis experiments referred to above suggest some sort of domain structure and peptide mapping studies on C-terminal glycopeptides comprising about one third of the molecule support the

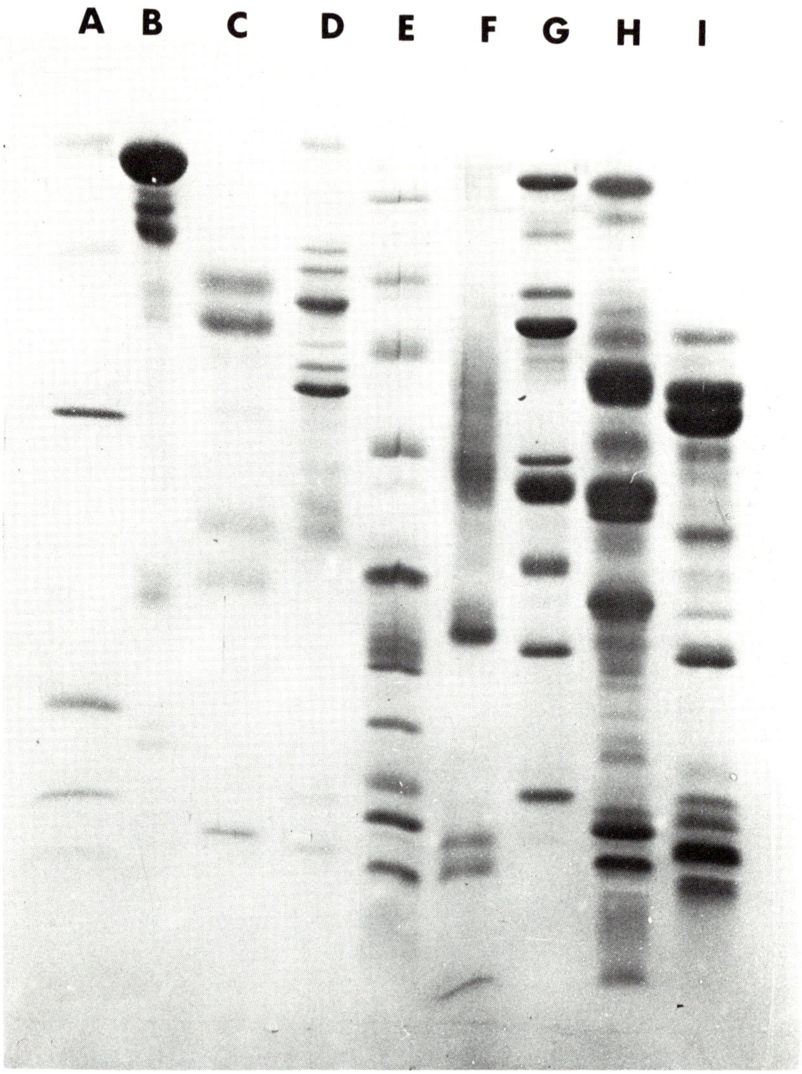

Fig. 8. Cyanogen bromide (CNBr) maps of eight VSGs, isolated from two different stocks. After cleavage with CNBr, and electrophoresis in a sodium dodecylsulphate–polyacrylamide gel, samples were stained with Coomassie brilliant blue. CNBr cleaves proteins at methionine residues, so it can be readily seen that there is no conservation of methionine in readily identifiable domains. Similar results are obtained in both one- and two-dimensional maps with proteolytic enzymes. Track A, molecular weight markers; B, ILTat 1–25; C, MITat 1–3; D, ILTat 1–23; E, MITat 1–2; F, ILTat 1–21 G, MITat 1–6; H, MITat 1–5; I, MITat 1–4. Molecular weight ($\times 10^{-3}$) of molecular weight markers is indicated to the left of track A.

Table 2. *Properties of* Trypanosoma brucei *antigens (VSGs) in solution. Data from Auffert & Turner (1981)*

Antigen	Cross-linking	$10^{-3} \times$ mol.wt of monomer	$10^{-3} \times$ mol.wt of cross-linked product
MITat 1.1	80	56	115
MITat 1.2	90	58	118
MITat 1.4	94	61	120
MITat 1.5	78	59	120
MITat 1.6	95	63	150
ILTat 1.21	49	62	130; 200
ILTat 1.23	20	65	140; 200
ILTat 1.25	10	63	135

The eight different antigens are from two different serodemes, MITAR and ILTAR 1. Cross-linking was accomplished with the bifunctional reagent dimethyl suberimidate at pH8.5. The degree of cross-linking was estimated by scanning a coomassie brilliant blue-stained SDS polyacrylamide gel electrophoretogram of the reaction products and integrating peak areas. The antigens exist as dimers and trimers in solution. ILTat 1.21 and ILTat 1.23 gave both dimers and trimers.

idea of homologies within this region (Auffret, Cross & Turner, 1978). If the prediction of 50% sequence homology is accurate, however, there must be homologies within the N-terminal sequences. Such sequences are presumably not exposed to the host's immune system but are concealed within the interior of the molecule and may be essential for the maintenance of the appropriate three dimensional configuration. With the accumulation of sequence data, the situation will become clearer in the near future.

What other features of the antigens are conserved? First, the VSGs appear capable of fixing complement directly, albeit at a low level, both by the classical and the alternate pathways (Musoke & Barbet, 1977). It is not known whether this represents a conservation of protein or carbohydrate but by analogy with other complement-fixing antigens it seems likely to be the latter. Secondly, the purified VSGs of *T. brucei* exist as dimers and trimers (Table 2) in solution (Auffret & Turner, 1981). Thirdly, all soluble VSGs examined in this laboratory show (Fig. 9) a cholesterol-dependent binding to lipid monolayers (Klein, Turner & Infante, 1980). The significance of this binding is not clear, although it may be related to the mode of attachment to the plasma membrane. Antigen can be released from trypanosomes very easily. Distruption by homogenization (Cross, 1975), extraction at pH 5.5 (Baltz, Baltz & Pautrizel, 1976) and shaking with glass beads (Reinwald,

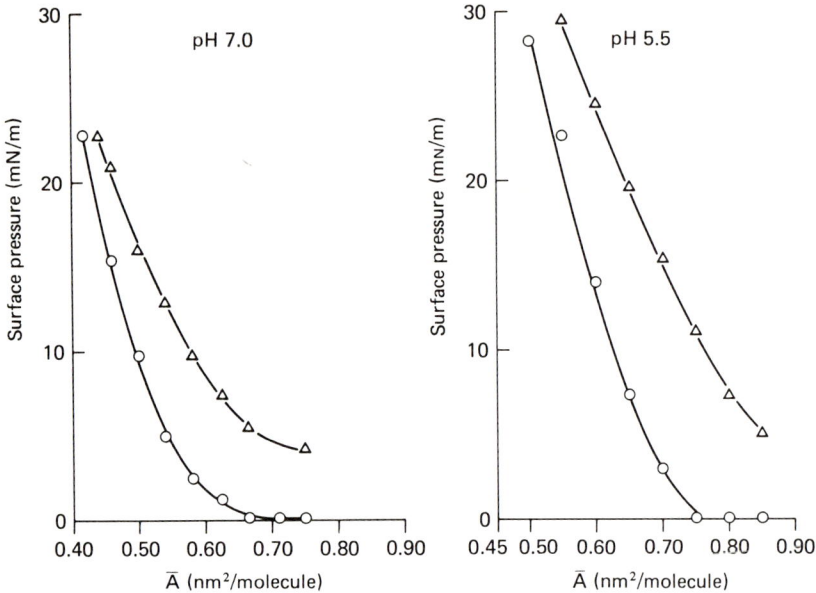

Fig. 9. Interaction of purified VSG with cholesterol-containing lipid monolayers. The results are presented as pressure–area isotherms for lipid monolayers in a Langmuir trough. A monolayer is established on the surface of an aqueous buffer, and the surface pressure measured using a dipping platinum wire force balance. A moveable barrier allows measurements of surface pressure to be made as a function of surface area. On the addition of substrate to the buffer phase, penetration of the monolayer by substrate is measured as an increase in the surface pressure at a given surface area. VSG MITat at 1.6 has been added to a phosphatidyl choline–cholesterol (1:1) monolayer at pH 7.0 and pH 5.5 at a molar ratio of 6:1 (lipid:antigen). Penetration occurs readily at pH 7.0, but is enhanced at pH 5.5. Circles, lipid alone; triangles; lipid plus antigen. Data from R. A. Klein, M. J. Turner & R. B. Infante (unpublished).

Rautenberg & Risse, 1979) have all been successfully employed to solubilize VSGs from different species. Sequencing of cDNA clones encoding the C-terminal region of the molecule has shown that about 20 hydrophobic residues are encoded at the C-terminus which are not present on the solubilized antigen (Boothroyd, Cross, Hoeijmakers & Borst, 1980) but to date all attempts to isolate antigen carrying this hydrophobic tail have failed. If cholesterol-binding is not related to membrane attachment, then it may have some other function as yet not understood, or be an evolutionary relic of the origin of VSGs. The fact that all VSGs examined to date have retained the trait suggests that there is still selection for cholesterol-dependent binding.

As mentioned earlier, there are some antigenic types which are common to different serodemes and cross species barriers. Some biochemical characterization of the VSGs obtained from such 'iso-VATs' have been reported. (Vervoot et al., 1981; Barbet et al., 1981). Two groups of iso-VATs were examined. Group one contained four members, one from each of the three clones of T. brucei (brucei, rhodesiense and gambiense) and one from T. evansi. Group two consisted of one clone of T. b. gambiense and one clone of T. evansi. All are therefore members of the subgenus Trypanozoon but isolated from different locations and different hosts. The T. evansi clones, in fact, were isolated from South American stocks. In each case, all members of each group were serologically indistinguishable by immunofluorescence and by immunolysis. Are they therefore expressing identical VSGs? Apparently not. Isolated VSGs from each family of iso-VATs were indistinguishable on SDS gels but differed in their isoelectric points (pIs between 6.2 and 7.9 in group one and 5.3 and 5.9 in group two). Cross-reaction between isolated VSGs was tested by immunodiffusion and radioimmunoassay. VSGs of group one iso-VATs were indistinguishable in immunodiffusion, whereas group two VSGs gave reciprocal lines of partial identity. Non-isotype VSGs give no detectable cross-reaction by this technique. In radioimmunoassay, heterologous combinations of antigen and antibody within each group gave 100% precipitation which could be fully inhibited by the addition of unlabelled homologous or heterologous antigen. Immunoprecipitation of homologous combinations of antigen and antibody within each group could be inhibited by heterologous antigen from the same group by an amount varying between 19% and 93%. In each case, the cross-reactions could be distinguished from those detectable between non-isotypic VSGs which have been shown to be due to the recognition of common carbohydrate side chains (Holder & Cross, 1981). Peptide maps showed extensive overlaps, suggesting a degree of structural homology commensurate with the results obtained in radioimmunoassay.

Such iso-VATs must have arisen before speciation in the subgenus Trypanozoon occurred and may be the closest surviving relatives of the ancestral gene or genes which gave rise to the family of variant genes.

MOLECULAR BIOLOGY OF ANTIGENIC VARIATION

The study of the molecular biology of antigenic variation is still in its infancy but already a large amount of information has been

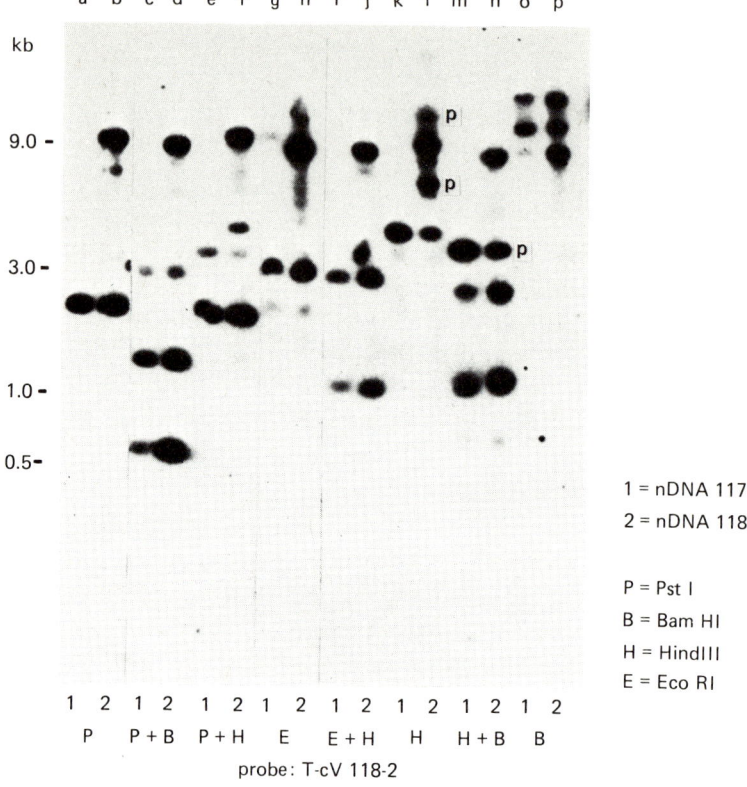

Fig. 10. Hybridisation of DNA complementary to VSG from MITat 1.5 (118) to nuclear DNA digests of variants MIT at 1.5 and 1.4. The restriction digests were electrophoresed through a 0.7% Agarose gel, blotted onto nitrocellulose filter strips and hybridised with nick-translated DNA from recombinant plasmid T-cV 118-2. The p marks persistent partial *Hind* III digestion products, absent in other experiments. Lane 1, nuclear DNA from MITat 1.4; Lane 2, nuclear DNA from MITat 1.5; P, restriction endonuclease *Pst* 1; B, *Bam* H1; H, *Hind* III, E, *Eco*R1. In each digest, the MITat 1.5 probe sees one DNA fragment more in the homologous nuclear DNA than in the heterologous MITat 1.4 nuclear DNA. The fragments common to the MITat 1.4 and 1.5 digests are attributed to a basic copy of the 1.5 VSG gene present in both variants. Results of double digestion experiments with two restriction enzymes are consistent with the presence of only one basic gene copy, lacking intervening sequences. The extra band in the homologous MITat 1.5 nuclear DNA is attributed to the presence of an additional expression-linked copy of the MITat 1.5 gene only present in the MITat 1.5 nuclear DNA. From Hoeijmakers, Frasch, Bernards, Borst & Cross (1980). Reprinted with permission from *Nature*, **284**, 80. Copyright © 1980 Macmillan Journals Ltd.

obtained of relevance both to the evolution of the VSGs and the mechanism of gene switching. Three groups of workers have now reported on results of Southern blotting experiments, using cloned VSG-specific DNA sequences. Hoeijmakers *et al.* (1980a,b) prepared variant specific cDNA clones from four different variant antigen types. Southern blots of nuclear DNA treated with a number of different restriction enzymes showed that each variant antigen gene was present in every VAT belonging to that serodeme and in all cases was surrounded by the same DNA sequences. However in each case an additional gene copy was found in the VATs expressing that gene (Fig. 10). The simplest interpretation of these data is that an expression-linked gene copy is transcribed and that a recombinational event is necessary to activate the gene. The data are consistent with either activation of the gene on only one of a diploid pair or insertion of an extra copy of the gene into an expression-linked locus, by analogy with the gene-cassetting model of yeast-mating type interconversion (Hicks, Strathern & Klar, 1979). Since recent evidence is consistent with diploidy in trypanosomes (Tait, 1980), we may also be looking at a protozoan example of allelic exclusion. These observations have been largely confirmed (Pays et al., 1980) in a different stock of trypanosomes. An expression linked copy of the gene was again observed, but the arrangement of unexpressed sequences did show some slight variations from VAT to VAT. An additional important observation was that when bloodstream trypomastigote forms were transformed to uncoated insect midgut procyclic forms by culture at 26°, the expression-linked copy of the variable antigen gene for the VAT put into culture was lost.

Results from a third laboratory contradict these findings (Williams, Young & Majiwa, 1979; Williams et al., 1980). These workers used cDNA clones of one VAT from a third trypanosome stock to look at the arrangement of that gene in several VATs. Not only was no expression-linked copy of the gene found, but the arrangement of the gene differed in all the variants examined. Indeed the sequences surrounding the gene were different in different isolates of the same VAT. It was argued that the data were consistent with the presence of two copies of the gene, perhaps reflecting alleles in diploid stock. Rearrangements seem to affect both copies independently but not in an expression-linked fashion. At present, these conflicting views cannot be reconciled.

Recombinant DNA technology has also been used to look at the evolution of VSG genes. Frasch et al. (1980) showed that in addition to hybridizing to the basic and expression-linked copies of the gene, each cDNA probe recognized a number of other weakly hybridizing bands which could be removed under more stringent washing conditions. This suggested the existence of families of related genes evolving by gene duplication, followed by deletion and mutations within individual members of the pool. The same cDNA clones, prepared from *T. brucei* VATs recognized DNA sequences in different strains of *T. brucei, T. evansi* and *T. equiperdum*, but not in *Trypanosoma cruzi*, suggesting that the sequence recognized evolved before speciation within the subgenus *Trypanozoon* but after divergence of the salivaria from the stercoraria.

EVOLUTION OF THE SALIVARIAN TRYPANOSOMES

This subject has been well reviewed in the past (Baker, 1974) and only the central points will be discussed in detail here. The prevailing view expressed by Baker (1974) is that the dixenous *Trypanosoma* evolved from monoxenous uniflagellate parasites of the gut of invertebrates. Adoption of haematophagy by the invertebrate host provided the opportunity for adaptation to a second host. At what stage did association with *Glossina* species develop? The salivarian trypanosomes share a number of characteristics suggesting a common evolutionary origin. As stated above all are either cyclically transmitted by *Glossina* species or are believed to be descendants of species that are. In contrast to other trypanosomes which parasitize mammals, they are all transmitted via the proboscis of their vector. However, none of the species are very efficient at infecting their vectors, suggesting a comparatively recent association with *Glossina*. To account for these and other shared features, Woo (1970) suggested that the salivarian trypanosomes evolved relatively recently from species which infected amphibious reptiles and were previously transmitted by leeches. Adaptation to *Glossina* species could have occurred when these insects fed upon reptiles basking on the land, as they are known to do on *Crocodilus niloticus*. Such a hypothesis could account for development of trypanosomes in the proboscis and salivary glands, for in all cases adequately studied dixenous trypanosomes of leeches and aquatic vertebrates are transmitted via the mouthparts of the leech. The apparently recent association between the salivarian trypanosomes and their vector is also explained. In support of the hypothesis, *T. brucei* can be adapted to growth both in turtles (at 37°) and in cayman (at 23°) with retention of infectivity to mice (Woo & Soltys, 1969). The hypothesis tells us nothing about the evolution of the surface coat. If we accept the hypothesis, we must ask whether the ancestral reptilian trypanosomes carry a surface coat. Two leech-transmitted fish haemoflagellates have been reported to possess surface coats. Bloodstream forms of *Trypanosoma danilewski* have a rather thin and fuzzy surface coat (Lom, 1979), resembling those found in stercorarian trypanosomes such as *T. cruzi*, rather than *T. brucei*. In contrast to the salivarian trypanosomes, carbohydrate staining in fixed thin sections shows polysaccharides throughout the surface coat and the membrane. Culture forms lack the surface coat. *Trypanoplasma borrelli* has a fuzzy, rather granular surface coat up to 20 nm thick. Again, carbohydrate staining reveals polysaccharide throughout the surface coat and the cell membrane (Lom & Nohýnková, 1977). Such piscine haemoflagellates can establish chronic infections in their hosts, but nothing is known about the interaction, if any, between the parasite and the host's immune system.

EVOLUTION OF THE SURFACE COAT

There are two distinct requirements in considering the evolution of antigenic variation. The first requirement is the acquisition of a set of VSG genes. The second is the development of a 'leak-proof' switching mechanism – the simultaneous expression of more than one VSG on the surface would be wasteful and perhaps lethal. The rate at which the switching mechanism operates must also be compatible with the rate of development of new antibodies. If it is too slow the parasite will be eliminated.

What was the ancestral VSG? Where was it acquired and how? Vickerman (1969) first suggested that the coat is an adaptation to life in the bloodstream of the host. Clearly its function is now to evade the host's immune sustem, but this requires the existence of a set of VSG genes. The original selective advantage of a surface coat to a trypanosome must have been in an increased probability of achieving cyclical transmission through an increase in the survival time in the mammalian host's bloodstream. The first barrier to successful colonization of the blood would not be the host's acquired immunity, but the mechanisms of non-specific or innate immunity, including both humoral factors, such as properdin, conglutinin, complement and possibly serum lipoproteins, and phagocytosis. It has been suggested that resistance to such effects was the primary function of the surface coat (Cross, 1978). Coated trypanosomes are not readily phagocytosed, nor are they easily lysed by complement.

In this respect, there is a parallel with encapsulated bacteria such as *Streptococcus pneumoniae, Klebsiella pneumoniae, Bacillus anthracis* and *Haemophilus influenzae* in which the polysaccharide or protein capsule provides resistance both to phagocytosis and complement lysis (Robbins *et al.*, 1980). In support of this idea, it has been pointed out that both procyclic trypomastigote forms and trypsinized bloodstream forms of *T. brucei* are readily lysed by fresh serum. Coated trypomastigotes are not susceptible to this effect, which involves activation of complement by the alternate pathway. The coat therefore seems to be exercising a barrier function as suggested earlier for *T. brucei* on the basis of the inaccessibility of lectin receptors in the presence of a surface coat (Cross & Johnson, 1976). Is this a valid model? The complement system is extremely ancient and has changed relatively little in phylogeny. We have observed that procyclic trypomastigotes of *T. brucei* which are susceptible to the action of fresh guinea-pig serum are also lysed by crocodile serum through activation of the alternate pathway (Miller & Turner, unpublished). If the validity of the model is accepted, then either the complement-fixing structure developed after transfer to a new host or some protection against lysis must have been needed when ancestral salivarian trypanosomes first entered the bloodstream of reptiles. The former interpretation is favoured by some recent results (Vickerman & Tetley, 1981) on *Trypanosoma (Duttonella) vivax*. Uncoated procyclic forms of this species do not activate complement, nor does this species acquire a surface coat in the metacyclic stage, immediately before entering the bloodstream of the host. It also possesses the most diffuse and ill-defined surface coat in the bloodstream trypomastigote form of any of the salivaria (Vickerman, 1974). Selection for a complement-impermeable barrier would not be needed, hence its surface coat may more closely resemble that of the ancestral precursor. The generally higher infection rate of the vector observed with this species (Baker, 1974) may also reflect the longest association with *Glossina*. *Nannomonas*, which has an intermediate infection rate, and *Trypanozoon*, which has the lowest, can be seen as successive adaptations to development in the tsetse (Hoare, 1948; reviewed in Baker, 1974). *Duttonella* develops only in the proboscis, *Nannomonas* in the midgut and proboscis, and *Trypanozoon* in the midgut, proboscis and salivary glands. The decreasing infection rates reflect the comparatively recent nature of these adaptations. Perhaps the development of the complement-fixing structure was associated with these changes, and this produced selective pressure to 'tighten up' the surface coat, and to produce coated metacyclic forms. Unfortunately, less information is available on the surface coat of *T. (Duttonella) vivax* than on that of *T. (Nannomonas) congolense* or the *Trypanozoon*. We are not aware of any data published on the nature of the VSG in *T. vivax*.

If complement was not the barrier to colonization, then as indicated above, other humoral or cellular factors may have been, and we propose that the surface coat was selected for on the basis of resistance to such non-specific factors. There are then three basic possibilities for the derivation of the ancestral coat protein gene. First, adaptation of a pre-existing membrane protein. Second, acquisition from the vertebrate or invertebrate host. Third, acquisition from a co-parasite of the mammalian or intermediate host. The first of these possibilities is conceptually the simplest, but requires the presence of a membrane protein of unknown function whose synthesis can be increased sufficiently to provide a confluent surface coat without altering the function of other membrane proteins. An increase in gene number in a manner similar to the amplification of dihydrofolate reductase genes in methotrexate resistant cell lines (Alt, Kellems, Bertino & Schimke, 1978) is the simplest explanation for such increased synthesis. Accumulation of mutations in this gene pool followed by the imposition of an accurate control mechanism could account for the generation of the system we see today. What sort of membrane protein could this ancestral gene have encoded? Presumably, the original function has been lost – no enzymatic activity has ever been described for the VSGs. One possible function is that of a glucose carrier protein. Trypanosomes take up glucose by a process of facilitated diffusion down a concentration gradient, implying the existence of a carrier molecule, and the suggestion has been made that VSGs may function as part of the glucose uptake system (Cross, 1977). The evidence is circumstantial, based on the observation that glucose is associated with VSGs after purification, and that much, perhaps all of this glucose can be removed by gel filtration. The possibility of VSG involvement in glucose transport is amenable to experimental analysis.

VSG-specific recombinant DNA probes do not hybridize to nuclear DNA from *T. cruzi* (Frasch *et al.*, 1980), so either the postulated functional protein is absent in this stercorarian species or the sequence has diverged so rapidly that no detectable homology now exists. In the phylogenetic tree proposed by Baker (1974), the salivaria and *T. cruzi* are quite distantly related. It might be informative if the DNA probe analysis could be extended to some of the trypanosomes of reptiles or fishes.

A second possible derivation is by acquisition from the vertebrate or invertebrate host. It is difficult to imagine how this could occur in an organism which lives entirely extracellularly. The third possibility is acquisition from a virus or other intracellular parasite of the vertebrate or invertebrate host. Several families of enveloped viruses are interesting from this point of view, since they are characterized by a lipid envelope, coated with glycoprotein spikes or knobs. On entering the host cell, virus-directed proteins are rapidly synthesized and envelope glycoproteins are exported to the cell surface. Nucleocapsids are assembled in the cytoplasm and diffuse to the cell surface where they interact with the virus glycoproteins. The nucleocapsid then buds out through the cell membrane to produce mature virions. Could the VSGs have arisen by incorporation into the genome of genes specifying viral envelope glycoproteins?

It is worth considering the mechanisms by which a virus transfers its genome into the cytosol of the cell it is infecting. Penetration occurs at the plasma membrane or intracellularly after the virus has been taken up by endocytosis. Penetration occurs through fusion at the cell surface for most paramyxoviruses (Klenk, 1980), whereas endocytosis is the most common pathway for other envelope viruses. This pathway has been well characterized for the entry of Semliki Forest Virus (SFV) into baby hamster kidney cells (Helenius, Kartenbeck, Simons & Fries, 1980). Virus is endocytosed in coated vesicles, which fuse ultimately with primary lysosomes to form secondary lysosomes. Fusion of the viral and lysosomal membranes then effects transfer of the genome into the cytoplasm. This fusion is both pH-dependent, with an optimum of pH 5.5–6.0, and dependent on the presence of cholesterol in the lysosomal membrane (Helenius *et al.*, 1980; White & Helenius, 1980). As yet it is not known if this pathway is common to all endocytosed envelope virus, but certainly increasing lysosomal pH from 5.5 to about 7.0 inhibits infection by several myxo, oncorna, rubella and paramyxoviruses (Helenius *et al.*, 1980). In SFV, fusion is believed to be associated with one of the spike glycoproteins, E1. The hydrophilic external domain contains a 16 amino-acid hydrophobic sequence which may represent the fusion promoting sequence (Garoff *et al.*, 1980). Does this represent the origin of the cholesterol-dependent interaction of water-soluble VSG with lipid monolayers (Klein, Turner & Infante, 1980)? In preliminary experiments we have shown a 2–3 fold enhancement in this interaction on lowering the pH from 7.5 to 5.5. Sequencing of VSGs will determine the existence or otherwise of similar hydrophobic sequences.

Are there analogies between the biochemistry of viral envelope glycoproteins and VSGs? Broadly speaking, yes. The properties of viral glycoproteins from a number of alphaviruses, flaviviruses and oncornaviruses are summarized in Table 3 and compared with the properties of the VSGs.

Oncornaviruses provide an interesting model. Type C viruses have been isolated from birds, mammals, and reptiles, and all contain a surface coat of 'spikes' or 'knobs' comprised of a glycoprotein or glycoproteins. The major glycoproteins of various viruses vary in molecular weight from 51 800 to 85 000 (Altstein & Zhdanov, 1979). The mode of attachment of this glycoprotein varies from virus to virus, but some portion is usually disulphide linked to a hydrophobic protein or glycoprotein which serves as anchor. However, the remainder is bound only loosely to the envelope through protein–protein or protein–lipid interactions. For example, in Friend murine leukaemia virus about 90% of the 71 000 Dalton major glycoproteins is loosely attached to the virus, and may be removed by washing under the appropriate conditions (Montelaro, Sullivan & Bolognesi, 1978), behaviour analogous with the VSGs.

Envelope viruses are distributed throughout the plant and animal kingdom, hence there can be no restrictions placed on the site of interaction between virus and trypanosome. From considerations of the relationship between *Duttonella*, *Nannomonas* and *Trypanozoon* discussed earlier, we suggest that if such an association occurred, then it was coincident with the transfer into a new host range. Is there any evidence for the association of viruses with trypanosome infections, or with Tsetse flies? Virus-like particles (VLPs) have been found in cytoplasmic vesicles in the salivary glands of *Glossina morsitans centralis* (Jenni, 1973) and in the nucleolar region of midgut muscle of *Glossina fuscipes fuscipes* (Jenni & Steiger, 1974). Whether such particles have any relevance to the parasite–vector relationship is not known. Similar virus particles have been identified in another member of the Trypanosomatidae, *Leishmania hertigi*, a parasite of porcupines (Molyneux,

Table 3. *Properties of viral glycoproteins and trypanosoma VSGs. Data from Alstein & Zhdanov (1979) and Nayak (1977)*

Microbial group	Organism	Major glycoproteins (mol.wt × 10⁻³)	Form on surface	Form in solution	Held to surface by	Solubilised by
Oncornaviruses	Avian Type C	85, 37	85 bound to 37		Hydrophobic 37	Detergent
	Murine Type C –Friend	70, 15	10% 70 bound to 15 90% 70		Hydrophobic 15 ?	Detergent Osmotic stock
	–Rauscher	70; 15	Some 70 bound to 15 Some 70		Hydrophobic 15 ?	Detergent Osmotic stock
	–AKR	70; 15	Some 70 bound to 15 Some 70		Hydrophobic 15 ?	Detergent Osmotic stock
	Other mammalian type C	70	70			
Alphaviruses	Semliki Forest Sindbis	50; 50; 10 50; 50	Complex of 50+50+10 Complex of 50+50		Hydrophobic C-termini Hydrophobic C-termini	Detergent Detergent
Flavivirus	Japanese equine encephelomyelitis virus	63				
Myxovirus	Influenza A	45–70 (Neuraminidase) 80 (55+25) (Hemagglutinin)	2–4 molecules/unit	Dimers/tetramers	Hydrophobic C-terminus	Detergent
			3 molecules/spike	Trimers	Hydrophobic C-terminus	Detergent
Paramyxovirus	Sendai	50 (Fusion protein) 75 (Hemagglutinin/ Neuraminidase)	2–3 molecules/spike	Dimers/trimers	Hydrophobic C-terminus Hydrophobic C-terminus	Detergent Detergent
Trypanosome	VSG	~60	2–3 molecules/spike?	Dimers/trimers	Hydrophobic C-terminus(20%) Hydrophobic C-terminus(80%)	Detergent Osmotic stock

1974). No such observations have been made in any species of Trypanosoma.

MOLECULAR MECHANISMS OF ANTIGENIC VARIATION: AN EVOLUTIONARY HYPOTHESIS

In this section we advance an hypothesis for the evolution of antigenic variation at the molecular level. The scheme is speculative and we have relied heavily upon data from other systems. It is possible that antigenic variation in trypanosomes operates by a mechanism found only in these organisms. However, it is our opinion that the process of molecular changes will use mechanisms found widely in eukaryotes, but adapted to the trypanosome's particular needs, and our hypothesis starts from this concept.

Earlier in this article, we proposed that ancestral trypanosome 'developed' or acquired a membrane protein that gave partial protection against non-specific immunity. Under selective pressure for an increase in the amount of this membrane protein, the ancestral VSG became amplified leading to a higher gene dosage and an increased synthesis of the gene product, in a manner analogous to the amplification of the dihydrofolate reductase genes in methotrexate-resistant mouse cell lines (Alt *et al.*, 1978). The evolution of a non-leaky switching mechanism is harder to envisage. However, the mechanism as it exists today suggests that either the genes became mobile, or that they became associated with 'hot spots' for the insertion of a mobile promoter. It is unlikely that this mechanism is peculiar to trypanosome surface antigen genes, and it is possible that similar switching events occur during development in other eukaryotes. In the following sections, we outline two alternative evolutionary pathways; one which involves a moveable promoter, and one which assumes that the VSG genes are mobile in a manner analogous to 'gene cassetting' in yeast mating type changes (Hicks *et al.*, 1979).

The mobile promoter hypothesis

Both hypotheses start with the assumption that an ancestral VSG gene became amplified as described earlier. Iso-VATs may represent the closest surviving relatives of such amplified ancestral genes. We also propose that the amplified unit contain a non-coding

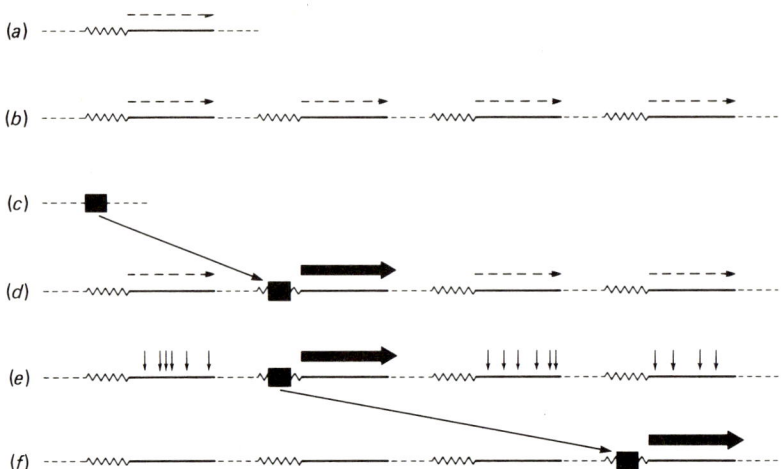

Fig. 11. The mobile promoter hypothesis. (a). An ancestral membrane protein gene is transcribed at a low level (broken arrow). The solid line indicates the structural gene for the membrane protein, the wavy line 5' non-coding sequence and the dotted line DNA of non-specified function but not duplicated. (b). Amplification of the gene occurs in response to selection for denser surface coat, but each gene is still only transcribed at a low level. (c) and (d). A mobile DNA element arises that can insert into the 5' non-coding region, and fortuitously produces high level transcription. (e). The other genes are now free to switch off and to freely mutate (vertical arrows). (f). Movement of the mobile element to an homologous site at the 5' end of a mutated gene causes antigenic variation.

segment at the 5' end of the gene. At some point subsequent to the increase in gene dosage a mobile DNA element became capable of inserting with high probability into the 5' non-coding region of the genes, and is further assumed to have acted as a high-level promoter. This newly acquired high-activity gene was transcribed at a sufficiently high level to render unnecessary the low-levels synthesis of the other copies. These subsequently lost their low-level promoter sites and began to drift in their amino-acid sequence to become the family of related VSG genes we observe today. The mobile element can subsequently activate any of the ancestral genes by insertion in their 5' non-coding region. The existence of predominant antigens can be explained by assuming that the insertion of the promoter is specific for the sequence at the 5' end of all the ancestral genes, but that some genes have diverged in this region more than others, or that the predominant antigens have developed multiple copies of the insertion site. Further refinements of this basic mechanism must be envisaged to explain the apparent allelic exclusion observed in antigen expression. The basic scheme is presented diagrammatically in Fig. 11.

The mobile gene hypothesis

The duplication of the ancestral gene may have taken place in such a way as to generate a block of tandemly repeated genes, comparable to ribosomal cistrons or other clustered repeats such as satellite DNA. The tandemly repeated ancestral VSG genes subsequently became interspersed with another family of tandem repeats. Computer simulations suggest that such interspersion of repetitive DNA can be a consequence of unequal crossing over (Smith, Brown & Dover, unpublished observations). The VSG genes thus became flanked by direct repeats, and in this configuration they would closely resemble other mobile DNA elements, such as Ty1 in yeast (Cameron, Loh & Davis, 1979; Roeder & Fink, 1980); Copia, 412, and 297 elements in Drosophila (Potter, Brorein, Dunsmuir & Rubin, 1979); and retroviruses (Hughes et al., 1978) and consequently developed the ability to transpose gene copies to other locations in the genome. One of the sites at which they were capable of inserting was in, or close to, a high-level promoter. After one of the genes came under the control of such a promoter, the low-level promoters present in each copy of the ancestral gene were no longer necessary and were lost. The gene family thus created would diverge as mutations occurred in and around the genes. Only genes which became inserted at the promoter site became active, thus in part accounting for the exclusion phenomenon of only a single antigen being expressed at a time. Allelic exclusion would occur if the insertion of the VSG gene inactivated an essential gene close to the insertion site. Occupation of the insertion site in both chromosomes would inactivate both of the diploid pair of essential genes and could thus be a lethal event. Once again, the occurrence of predominant antigens can be accounted for by postulating the existence of genes whose terminal repeats have sequences which give that gene a higher probability of transposing to the promoter site. In addition, the continual unequal exchange that may be occurring in the repetitive terminal elements could possibly explain the variations in gene structure observed in some laboratories (Williams et al., 1980; Pays et al., 1980). Such alterations would not be correlated with expression, but would reflect the constant shuffling of repetitive elements in the genome. This hypothesis assumes that trypanosomes undergo sexual reassortment of their genes, and that the cell is diploid, both of which are strongly supported by recent data (Tait, 1980). The mechanism is illustrated in Fig. 12. Neither mechanism

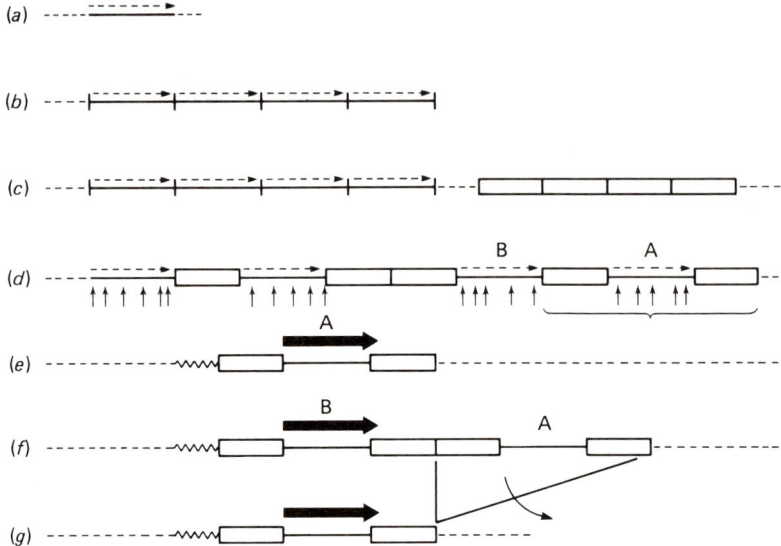

Fig. 12. The mobile gene hypothesis. The symbols are as in Fig. 11. (a). An ancestral membrane protein gene is transcribed at a low level. (b). Tandem reduplication occurs to give a repetitive array of genes. (c). Adjacent repetitive DNA evolves or pre-exists. (d). Unequal crossing over leads to interspersion. A and B represent different VSGs. (e). The presence of such direct terminal repetitive sequences fortuitously turns the ancestral genes into transposing DNA elements, leading to insertion in a locus at which high level expression occurs. (f). A new insertion of a mutated gene followed by (g). deletion creates antigenic variation.

fully accounts for the apparent differences observed in metacyclic and bloodstream antigen repertoires. However, if an environmental effect on phenotype selection is allowed, then such behaviour can be rationalized. The nature of the environmental effect, if any, is not known.

CONCLUSION

We have presented a number of hypotheses on both the nature and origin of the VSGs, and on the evolution and mechanism of the gene switching events which lead to antigenic variation in the salivarian trypanosomes. What are lacking at present are the experimental data needed to test these hypotheses. However, the last five years have seen an explosive growth in research on this problem. This interest is not simply a reflection of the concern to find new methods for the control of bovine and human trypanosomiasis, urgent though this is. It is also a recognition of the unique qualities of these organisms which makes antigenic variation in trypanosomes an ideal, experimentally amenable model for other, wider aspects of eukaryotic differentiation. We may therefore expect in a short time the data which will lead to the acceptance or rejection of the ideas we have put forward.

REFERENCES

ALT. F. W., KELLEMS, R. E., BERTINO, J. R. & SCHIMKE, R. T. (1978). Selective multiplication of dihydrofolate reductase genes in methotrexate-resistant variants of cultural murine cells. *Journal of Biological Chemistry*, **253**, 1357–70.

ALTSTEIN, A. D. & ZHDANOV, V. M. (1979). Envelope antigens of oncoviruses. In *Advances in Virus Research*, **25**, ed. M. A., Lauffe, F. B. Bang, K. Maramorosch & K. M. Smith, pp. 451–88.

AUFFRET, C. A., CROSS, G. A. M. & TURNER, M. J. (1978). Common peptides in the C-terminal regions of variant specific surface antigens from *Trypanosoma brucei*. *Parasitology*, **77**, xxxvii.

AUFFRET, C. A. & TURNER, M. J. (1981). Variant specific antigens of *T. brucei* exist in solution as glycoprotein dimers. *Biochemical Journal*. **193**, 647–50.

BAKER, J. R. (1974). The evolutionary origin and speciation of the genus *Trypanosoma*. In *Evolution in the Microbical World. Symposium of the Society for General Microbiology*, **24**, ed. M. J. Carlile & J. J. Skehel, pp. 343–66. Cambridge University Press.

BALTZ, T., BALTZ, D. & PAUTRIZEL, R. (1976). Affinity of concanavalin A for *Trypanosoma equiperduum:* purification of the antigenic type specific glycoprotein fraction. *Annales d'immunologie*, **127**, 761–74.

BARBET, A. F. & MCGUIRE, T. C. (1978) Cross reacting determinants in variant specific surface antigens of African trypanosomes. *Proceedings of the National Academy of Sciences, U.S.A.*, **75**, 1989–93.

BARBET, A. T., MUSOKE, A. J., VERVOORT, T., VAN MEIRVENNE, N., MPIMBAZA, G. & MAGNUS, E. (1981). Cross reacting antigenic determinants in variable surface glycoproteins from isotypic African trypanosoma. *Journal of Experimental Medicine*. In press.

BARRY, J. D. & HAJDUK, S. L. (1979). Antigenic heterogeneity of bloodstream and metacyclic forms of *Trypanosoma brucei*. In *Pathogenicity of Trypanosomes*, ed. Losos, G. & Chouinard, A, pp 51–56. Ottowa: International Development Research Centre.

BARRY, J. D., HAJDUK, S. L., VICKERMAN, K. & LERAY, D. (1979). Detection of multiple variable types in metacyclic populations of *Trypanosoma brucei*. *Transactions of the Royal Society of Tropical Medicine and Hygiene*, **73**, 205–8.

BOOTHROYD, J. C., CROSS, G. A. M., HOEIJMAKERS, J. H. J. & BORST, P. (1980). A variant surface glycoprotein of *Trypanosoma brucei* is synthesised with a hydrophobic carboxy terminal extension absent from purified glycoprotein, *Nature, London,* **288**, 624–5.

BOWMAN, I. B. R. (1974). Intermediary metabolism of pathogenic flagellates. In *Trypanosomiasis and Leishmaniasis. CIBA Symposium 20 (New Series)*, pp. 255–71. Amsterdam: Elsevier.

BRIDGEN, P. J., CROSS, G. A. M. & BRIDGEN, J. (1976). N-terminal amino acid sequences of variant specific surface antigens from *Trypanosoma brucei*. *Nature, London,* **273**, 613–14.

Brown, R. C., Evans, P. A. & Vickerman, K. (1973). Changes in oxidative metabolism and ultrastructure accompanying differentiation of the mitochondrion in *Trypanosoma brucei. International Journal of Parasitology.* **3**, 691–704

Cameron, J. R., Loh, E. Y. & Davis, R. W. (1979). Evidence for transposition of dispersed repetitive DNA families in yeast. *Cell,* **16**, 739–51.

Capbern, A., Giroud, C., Baltz, T. & Mattern, P. (1977). *Trypanosoma equiperdum:* Etude des variations antigenique au cour de la trypanosome experimentale du lapin. *Experimental Parasitology,* **43**, 6–13.

Cross, G. A. M. (1975). Identification, purification and properties of clone-specific glycoprotein antigens constituting the surface coat of *Trypanosoma brucei. Parasitology,* **71**, 393–417.

Cross, G. A. M. & Johnson, J. G. (1976). Structure and organisation of the variant specific antigens of *Trypanosoma brucei.* In *Biochemistry of parasites and host-parasite relationships,* ed. H. Van den Bossche, pp. 413–20. Amsterdam: Elsevier/North Holland Biomedical Press.

Cross, G. A. M. (1977). Isolation, structure and function of variant specific surface antigens. *Annales de La Société belge de médicine tropicale,* **57**, 389–99.

Cross, G. A. M. (1979). Cross reacting determinants in the C-terminal region of trypanosome variant surface antigens. *Nature, London,* **277**, 310–12.

Cunningham, M. P. (1966). The preservation of viable metacyclic forms of *Trypanosoma rhodesiense* and some studies on the antigenicity of the organisms. *Transactions of the Royal Society for Tropical Medicine and Hygiene,* **56**, 48–59.

deGee, A. L. W., Shah, S. D. & Doyle, J. J. (1979). *Trypanosoma vivax:* sequence of antigenic variants in mice and goats. *Experimental Parasitology,* **48**, 352–8.

Doyle, J. J. (1977). Antigenic variation in the salivarian trypanosomes. In *Immunity to blood parasites of animals and man,* ed. L. H. Miller, J. A. Pine, J. J. McKelvey, pp. 31–63. London and New York: Plenum Press.

Doyle, J. J., Hirumi, H., Hirumi, K., Lupton, E. N. & Cross, G. A. M. (1980). Antigenic variation in clones of animal-infective *Trypanosoma brucei* derived and maintained *in vitro. Parasitology,* **80**, 359–69.

Frasch, A. C. C., Hoeijmakers, J. H. L., Bernards, A., Borst, P. & Cross, G. A. M. (1980). The genes for the variant surface glycoproteins (VSGs) of *Trypanosoma brucei.* In *The host/parasite interface.* ed. H. Van den Bossche. Amsterdam: Elsevier/North Holland Biomedical Press.

Garoff, H., Frischauf, A. M., Simons, K., Lehrach, H. & Delius, H. (1980). Nucleotide sequence of the cDNA coding for the Semliki forest virus membrane glucoproteins. *Nature, London.* **288**, 236–41.

Grant, P. T., Sargent, J. R. & Ryley, J. F. (1961). Respiratory systems in the Trypanosomatidae. *Biochemical Journal,* **81**, 200–6.

Gray, A. R. (1965a). Antigenic variation in clones of *Trypanosoma brucei* I. Immunological relationships of the clones. *Annals of Tropical Medicine and Parasitology,* **59**, 27–36.

Gray, A. R. (1965b). Antigenic variation in a strain of *Trypanosoma brucei* subgroup transmitted by *Glossina morsitans* and *G. palpalis. Journal of General Microbiology,* **41**, 195–214.

Hoare, C. A. (1972). *The trypanosomes of mammals.* Oxford: Blackwell Scientific Publications.

Helenius, A., Kartenbeck, J., Simons, K. & Fries, E. (1980). On the entry of Semliki Forest Virus into BHK-21 cells. *Journal of Cell Biology,* **84**, 404–20.

Hicks, J., Strathern, J. N. & Klar, A. J. S. (1979). Transposable mating type genes in *Saccharomyces cerevisiae. Nature,* **282**, 478–84.

HIRUMI, H., DOYLE, J. J. & HIRUMI, K. (1977). African trypanosomes: culture of animal-infective *Trypanosoma brucei in vitro. Science*, **196**, 992–4.
HOARE, C. A. (1948). The relationship of the hemoflagellates. *Proceedings of the 4th International Congress on Tropical Medicine and Malaria*, **2**, 1110–16.
HOEIJMAKERS, J. H. L., BORST, P., VAN DEN BURG, J., WEISSMANN, C. & CROSS, G. A. M. (1980). The isolation of plasmids containing DNA complementary to mRNA for VSGs of *T. Brucei. Gene.*, **8**, 391–417.
HOEIJMAKERS, J. H. J., FRASCH, A. C. C., BERNARDS, A., BORST, P. & CROSS, G. A. M. (1980). Novel expression linked copies of the genes for variant surface antigens in trypanosomes. *Nature, London*, **284**, 78–80.
HOLDER, A. A. & CROSS, G. A. M. (1981), Glycopeptides from variant surface glycoproteins of *Trypanosoma brucei*. C-terminal location of antigenically cross-reacting carbohydrate moieties. *Molecular and Biochemical Parasitology*, **2**, 135–50.
HUDSON, K. M., TAYLOR, A. E. R., & ELCE, B. J. (1980). Antigenic changes in *Trypanosoma brucei* on transmission by tsetse fly. *Parasite Immunology*, **2**, 57–69.
HUGHES, S. H., SHANK, P. R., SPECTOR, D. H., KUNG, H. E., VOGT, P. K., & GRETMAN, M. L. (1978). Proviruses of avian sarcoma viruses are terminally redundant, coextensive with unintegrated linear DNA and integrated at many sites. *Cell*, **15**, 1397–410.
JENNI, L. (1973). Virus-like particles in a strain of *G. morsitans centralis*, Machado, 1970. *Transactions of the Royal Society of Tropical Medicine and Hygiene*, **67**, 295.
JENNI, L. (1977*a*). Comparison of antigenic types of *Trypanosoma (T. Brucei)* strains transmitted by *Glossina m. morsitans, Acta Tropica*, **34**, 35–41.
Jenni, L. (1977*b*). Antigenic variants in cyclically transmitted strains of the *T. brucei* complex. *Annales de la Société Belge de médicine tropicale*, **57**, 383–6.
JENNI, L. & STEIGER, P. (1974) Virus-like particles of *Glossina fuscipes fuscipes* Newst, 1910. *Acta Tropica*, **31**, 177–180.
JOHNSON, J. G. & CROSS, G. A. M. (1977). Carbohydrate composition of variant specific surface antigen glycoproteins from *Trypanosoma brucei. Journal of Protozoology*, **24**, 587–91.
JOHNSON, J. G. & CROSS, G. A. M. (1979). Selective cleavage of variant surface glycoproteins from *Trypanosoma brucei. Biochemical Journal*, **178**, 689–97.
KLEIN, R. A., TURNER, M. J. & INFANTE, R. B. (1980). The interaction of variant specific antigen from *Trypanosoma brucei* with lipid monolayers. *Parasitology*, **81**, xxvii–xxviii.
KLENK, H.-D. (1980). Viral glycoproteins: initiators of infection and determinants of pathogenicity. In *The molecular basis of microbial pathogenicity*, ed. H. Smith, J. Skehel & M. J. Turner, pp. 135–8. Dahlem Konferenzen. Weinheim: Verlag Chemie Gmbh.
KOSINSKI, R. J. (1980). Antigenic variation in trypanosomes: a computer analysis of variant order. *Parasitology*, **80**, 343–57.
LE RAY, D., BARRY, J. D., EASTON, C. & VICKERMAN, K. (1977). First tsetse fly transmission of the Antat serodeme of *Trypanosoma brucei. Annales de la Société Belge de medicine tropicale*, **57**, 369–81.
LOM, J. & NOHÝNKOVÁ, E. (1977). Surface coat of the bloodstream phase of *Trypanoplasma borelli. Journal of Protozoology*, **24**, 52A.
LOM, J. (1979). Biology of the trypanosomes and trypanoplasms of fish. In *Biology of the Kinetoplastida*, Volume 2, eds. W. H. R. Lumsden & D. A. Evans, pp. 269–337. London: Academic Press.
MCCONNELL, J., CORDINGLEY, J. S. & TURNER, M. J. (1979). The extent of

variability of variant specific antigens of *Trypanosoma brucei brucei*. *Parasitology*, **79**, vi.

METZGER, H., SHAPIRO, M. B., MOSIMAN, J. E. & VINTON, J. E. (1968). Assessment of compositional relatedness between proteins. *Nature*, **219**, 1166–1188.

MILLER, E. N. (1980). Biological aspects of antigenic variation in African trypanosomes. Ph.D. thesis, University of Cambridge.

MILLER, E. N. & TURNER, M. J. (1980). Analysis of antigenic types appearing in first relapse populations of clones of *Trypanosoma brucei*. *Parasitology*. In press.

MOLYNEUX, D. H. (1974). Virus-like particles in Leishmania parasites. *Nature*, **249**, 588–589.

MONTELARO, R. C., SULLIVAN, S. J. & BOLOGNESIK, D. P. (1978). An analysis of Type-C retrovirus polypeptides and their associations in the virion. *Virology*, **84**, 19–31.

MUSOKE, A. J. & BARBET, A. F. (1977). Activation of complement by variant specific surface antigens of *Trypanosoma brucei*. *Nature*, **270**, 438–440.

NANTULYA, V. M., DOYLE, J. J. & JENNI, L. (1980). Studies on *Trypanosoma (nannomonas) congolense*. III. Antigen variation in three cyclically transmitted stocks. *Parasitology*, **80**, 123–131.

NAYAK, D. P. (1977). *The Molecular Biology of Annual Viruses* Vol. 1. New York: Marcel Dekker Inc.

PAYS, E., VAN MEIRVENNE, N., LE RAY, D. & STEINERT, M. (1981). Gene duplication and transposition linked to antigenic variation in *Trypanasoma brucei*. *Proceedings of the National Academy of Sciences*, in press.

POTTER, S. S., BROVEIN, W. J., DUNSMUIR, P. & TUBIN, G. M. (1979). Transposition of elements of the 412 copia, and 297 dispersed repeated gene families in Drosophila. *Cell*, **17**, 415–27.

REINWALD, E., RAUTENBERG, P. & RISSE, H. J. (1979). *Trypanosoma congoleuse*: Mechanical removal of the surface coat *in vitro*. *Experimental Parasitology*, **48**, 384–97.

RITZ, H. (1916). Über Rezidive bei experimenteller Trypanosomiasis II. *Mitteilung Archiv für Schifts und Tropenhygiene*, **20**, 397–420.

ROBBINS, J. B., SCHNEERSON, R., EGAN, W. B., VANN W. & LIU, D. T. (1980). Virulence properties of bacterial capsular polysaccharides – unanswered questions. In *The Molecular basis of microbial pathogenicity*, ed. H. Smith, J. Skehel & M. J. Turner. pp. 115–132. Dahlem Konferenzen 1980. Weinheim: Verlag Chemie Gmbh.

ROEDER, G. S. & FINK, G. R. (1980). DNA rearrangements associated with a transposable element in yeast. *Cell*, **21**, 239–50.

SEED, J. R. (1964). Antigenic similarity among culture forms of the 'brucei' group of trypanosomes. *Parasitology*, **54**, 593–6.

SEED, J. R. (1978). Competition among serologically different clones of *Trypanosoma brucei gambiense in vivo*. *Journal of Protozoology*, **25**, 526–9.

STEIGER, R. (1971). Some aspects of the surface coat formation in *Trypanosoma brucei*. *Acta Tropica*, **28**, 341–6.

TAIT, A. (1980). Evidence for diploidy and mating in trypanosomes. *Nature, London*, **287**, 536–8.

TURNER, M. J. (1980). Antigenic variation. In *The Molecular basis of Microbial Pathogenicity*. ed. H. Smith, J. Skehel & M. J. Turner, pp. 135–8. Dahlem Konferenzen. Weinheim: Verlag Chemie Gmbh.

VAN MEIRVENNE, N., JANSSENS, P. G. & MAGNUS, E. (1975). Antigenic variation in syringe-passaged populations of *Trypanosoma (Trypanozoon) brucei* I Rationalisation of the experimental approach. *Annales de la Société Belge de médicine tropicale*, **55**, 1–23.

VERVOORT, T., BARBET, A. F., MAGNUS, E., MUSOKE, A. J., MPIMBAZA, G. & VAN MEIRVENNE, N. (1981). Identification and isolation of isotypic surface glycoproteins from African trypanosomes. *Journal of Experimental Medicine. In press.*

VICKERMAN, K. (1965). Polymorphism and mitochondrial activity in sleeping sickness trypanosomes. *Nature,* **208,** 762–766.

VICKERMAN, K. (1969). On the surface coat and flagellar adhesion in trypanosomes. *Journal of Cell Science,* **5,** 163–193.

VICKERMAN, K. & LUCKINS, A. G. (1969). Localisation of variable antigens in the surface coat of *Trypanosoma brucei* using ferritin conjugated antibody. *Nature.* **224,** 1125–1126.

VICKERMAN, K. (1971). Morphological and physiological considerations of extracellular protozoa. In *Ecology and Physiology of Parasites,* ed. A. M. Fallis, pp. 58–91. Toronto: University Press.

VICKERMAN, K. (1974). Antigenic variation in African Trypanosomes. In *Parasites in the Immunised host. Mechanisms of survival.* Ciba Foundation Symposium 25 (New Series). Amsterdam: Associated Scientific Publishers.

VICKERMAN, K., TETLEY, L. & MOLOO, S. K. (1981). Absence of a surface coat from metacyclic *Trypanosoma vivax:* possible implications for vaccination against *vivax* trypanosomias. *Transactions of the Royal Society of Tropical Medicine and Hygiene. In press.*

WHITE, J. & HELENIUS, A. (1980). pH-dependent fusion between the Semliki Forest virus membrane and liposomes. *Proceedings of the National Academy of Sciences of the U.S.A.,* **77,** 3273–3277.

WILLIAMS, R. O., YOUNG, J. R. & MAJIWA, P. A. O. (1979). Genomic rearrangements correlated with antigenic variation in *Trypanosoma brucei. Nature,* **282,** 847–849.

WILLIAMS, R. O., YOUNG, J. R., MAJIWA, P. A. O., DOYLE, J. J. & SHAPIRO, S. Z. (1980). Contextural genomic rearrangements of variable antigen genes in *Trypanosoma brucei. Cold Spring Harbor Symposium on Quantitative Biology,* **45,** *In press.*

WOO, P. T. K. & SOLTYS, M. A. (1969). The experimental infection of reptiles with *Trypanosoma brucei. Annals of Tropical Medicine and Parasitology,* **63,** 35–38.

WOO, P. T. K., (1970). Origin of mammalian trypanosomes which develop in the anterior-station of blood sucking anthropods. *Nature,* **228,** 1050–1062.

SUMMARIZING REMARKS

J. R. S. FINCHAM

Department of Genetics, University of Edinburgh, Edinburgh EH9 3JN, UK

This Symposium has illustrated various forms of evolutionary enquiry. For purposes of summary and discussion I will pick out three of these.

The first approach assumes that which we can hardly doubt, namely that evolution has happened and that its main directions have been determined by natural selection. It seeks to explain why the features of morphology and metabolism which we see in living species should have been selected. This involves the analysis of the cellular machinery with the hope of being able to show how it all functions for the good of the organism. This is not an unprejudiced approach, but we know from experience that it works. Teleology is a most valuable tool in helping us to 'make sense' of what might otherwise be a baffling maze of phenomena and to find our way to new discoveries. Professor Krebs' functional analysis of metabolic cycles provides us with a fine example. Yet we should beware of assuming that everything has to be adaptive. Prokaryote genomes are, indeed, so far as we know them, tightly organized, with little room to waste, and we do not find in them the profusion of wreckage and rubbish which we seem to see when we are able to get glimpses of the more spacious genomes of higher eukaryotes, including our own. All the same (to take an example given us by Dr Baumberg), when we find that *Escherichia coli* strain B has apparently abandoned the rational feedback control of arginine synthesis which prevails in other strains it seems much more likely to be a piece of negligence (the selective discipline having lapsed for a while) than a subtle example of adaptation to environment.

The second approach to evolution involves the construction of hypothetical phylogenetic 'trees'. The traditional way of doing this is though comparative studies of structure and/or metabolism, placing the different forms in graded series such that each member of the series may be plausibly regarded as having been derived, at least in respect of the character under study, from something resembling the previous one. By reference to such series one can gain some idea of the way evolution may have gone and guess at the selective forces which have driven the observed changes. In this Symposium, Dr Dawes and Professor Garland, dealing respectively with spores and membranes, exemplify the comparative functional approach. Their reasonable speculations carry us over a huge expanse of time, from the 'primeval soup' onwards. This can be an enlightening exercise and undoubtedly has heuristic value. The difficulty is, of course, that in the absence of any fossil record to speak of, we are restricted to looking at the outermost twigs of the phylogenetic tree and are left to make the best guesses we can about the branching structure from which they sprang. The concept of 'living fossils' is always a risky one and it is best to assume that there are no such things.

We are in a much better position to learn something about the extinct branches of the phylogenetic tree if we have data on nucleic acid sequence, since we can quantify the degree of relatedness of different base sequences in a way that is not possible for any other character. Given the sequence organization, the job of deducing the most probable lines of descent can be delegated to a computer. This kind of exercise has been practised for some years with amino-acid sequences of proteins such as the cytochromes, and we have heard at this symposium of other, very suggestive, results with the ferredoxins and superoxide dismutases. But nucleic-acid sequence clearly gives us a more accurate measure of relatedness. During the last year the results of at least two large surveys have been published, one on 16S rRNA, which we have heard about here, and another, which extends across the prokaryote–eukaryote divide, on the glycine/valine family of tRNA molecules (Cedergren *et al.*, 1980). This approach clearly opens a new chapter in evolutionary studies and comparable data on other nucleic acid families will be awaited with keen interest. Detailed agreement between the evolutionary trees deduced from different kinds of sequences would carry great conviction. One limitation of the method is that, to be useful on the longest time scale, the chosen sequences must be very conservative ones, retaining almost exactly the same function over great phylogenetic distances and fixing mutations at an extremely slow rate. Ribosomal RNAs fill the bill; there is, indeed, no obvious reason why they should ever change and nothing to show that such changes as have occurred are in any way adaptive as opposed to very rare accidents of mutation and 'drift'.

Such sequences may well hold the answers to many evolutionary puzzles. One, not mentioned here, concerns the affinities of that mysterious group, the Dinoflagellates (Beam *et al.*, 1977; Rizzo & Burghardt, 1980). Another, which is in everyone's mind, is the question of the relationships of mitochondrial and plastid genomes both of which, as Dr Cavalier-Smith has explained, may very well have a prokaryotic origin. Unfortunately the tRNA data to which I referred above give no clear information on mitochondrial origins because mitochondrial genomes apparently fix mutations fast enough effectively to cover their own evolutionary tracks. Sequences with relatively rapid rates of change, either because they are effectively 'neutral' or because they are subject to strong selection for change in function, may of course be of value for establishing the rates and kinds of mutational divergence between relatively similar species (see Lee, Bertrand & Yanofsky, 1978; Struhl & Davis, 1980; Miozzari & Yanofsky, 1979; the last reference documents a particularly nice example of two monofunctional genes fusing to give a bifunctional one). But they give little information about evolution on the grand scale.

A third line of enquiry represented in this Symposium relates to the origin of the variation which provides the material for evolution. Point mutations are, of course, well documented and ubiquitous – any DNA sequence which is not stringently selected for constancy, accumulates them. I think, however, that the suspicion is growing that significant evolutionary innovations usually require more drastic restructuring of the genome. In this connection the new world of transposable DNA elements, discovered by P. Starlinger and H. Saedler and described to us here by Dr Saedler, may be of immense significance. In themselves the IS and such-like elements of *E. coli* are no doubt as purely 'selfish', as are phage genomes. They are a constant source of what must be mostly deleterious mutations and genome rearrangements (deletions, inversions, transpositions) and they may well be a major source of the differences which have been found between gene arrangements in different species. A minority of the innovations may turn out to be lucky rather than unlucky accidents. More important, especially in the short run, is the role of pairs of IS elements in effecting the migration, sometimes across species boundaries, of blocks of genes bracketed between them. The significance of these transposons for the spread of drug resistance is well known. A transposon is, in effect, a mobile group of closely co-ordinated 'selfish genes' seeking their collective fortune as a team. The idea of peripatetic gene teams seems, in fact, to be a useful one even apart from the existence of special mobilization elements. Old-fashioned homologous general recombination can do a great deal in this regard. Recent work by Ryan and McConnell (1980) has, for instance, shown that several *E. coli* bacteriophages of the T3–T7 group have very likely originated by recombination of blocks of genes, each block concerned with one set of closely integrated functions – head components or tail fibres and so on. The fact that these groups of mutually adapted genes occur in linked blocks (in the T-even phages also) can be viewed as an adaptive feature from the point of view of the genes, allowing them to stick together as teams through cycles of recombination and enabling them to try their luck in many different recombinant phage types.

The importance of recombination in the origin of new influenza antigenic types has been emphasized here by Min Jou *et al.* By constantly reshuffling genes (or blocks of genes), respectively determining host range and antigenic type, these viruses are able constantly to out-flank their hosts' immunological defences. The capacity for switching its antigens and/or host range is of vital importance to a parasite and we now have several examples where this capacity is a built-in property of the genome. Some involve controlled segmental inversions; these cases include phase-variation of flagellar antigens in *Salmonella* reported by M. Simon and his colleagues (Zieg, Hilmen & Simon, 1978), and the host-range switch determined by the invertible G segment of bacteriophage Mu (Van de Putte, Cramer & Giphart-Gassler, 1980). Controlled transpositions are also becoming familiar, notably in the now well-known mating-type switch of yeast (Saccharomyces), which is based on transpositions involving three partly-homologous segments. Now, in the Trypanosomes (as explained to us by Drs Turner and Cordingley), we have a possibly similar system of controlled transposition governing antigenic phase in a eukaryotic parasite. Such programmed switches represent, like the perfect metabolic adaptations that this review started with, an impressive end result of evolution rather than evolution itself. In a sense, indeed, they are an escape from the necessity to evolve. In evolution, as in life, the taking of one option implies the closing of others. The fortunate trypanosome, like Jekyll and Hyde, is able to take the new without losing access to the old. You probably need to be a single cell in order to do that.

REFERENCES

Beam, C. A., Himes, M., Himelfarb, J., Link, C. & Shaw, K. (1977). Genetic evidence of unusual meiosis in the dinoflagellate *Crytothecium cohnii*. *Genetics*, **87**, 19–32.

Cedergren, R. J., La Rue, B., Sankoff, D., Lapalne, G. & Grosjean, H. (1980). Convergence and minimal mutation criteria for evaluating early events in tRNA evolution. *Proceedings of the National Academy of Sciences, USA* **77**, 2791–5.

Lee, F., Bertrand, G. & Yanofsky, C. (1978). Comparison of the nucleotide sequences of the initial transcribed regions of *Escherichia coli* and *Salmonella typhimurium*. *Journal of Molecular Biology*, **121**, 193–217.

Miozzari, G. F. & Yanofsky, C. (1979). Gene fusion during the evolution of the tryptophan operon in Enterobacteriaceae. *Nature*, **277**, 486–9.

Rizzo, P. J. & Burghardt, R. C. (1980). Chromatin structure in the unicellular algae *Olisthodiscus luteus, Crypthecodinium cohnii,* and *Pendinium balticum*. *Chromosoma (Berlin)* **76**, 91–9.

Ryan, T. & McConnell, D. M. (1980). Evolution of the T7–like phage involved recombination at sites between genetic modules in T7–like and T3–like ancestors. *Heredity* **45**, 149.

Struhl, K. & Davis, R. W. (1980). Conservation and DNA sequence arrangement of the DNA polymerase I gene region from *Klebsiella aerogenes, Klebsiella pneumoniae* and *Escherichia coli*. *Journal of Molecular Biology*, **141**, 343–68.

Van De Putte, P., Cramer, S. & Giphart-Gassler, M. (1980). Invertible DNA determines host specificity of bacteriophage Mu. *Nature,* **286**, 218–22.

Zieg, J., Hilmen, M. & Simon, M. (1978). Regulation of gene expression by site-specific inversion. *Cell,* **15**, 237–44.

INDEX

acetate, oxidation of, 215–17
acetic acid bacteria, 225
acetyl coenzyme A, 216, 217, 218–19, 224
N-acetylglucosamine, peptidoglycan and chitin polymers of, 48
Acholeplasma, 12
acids: production of, by primitive fermenting organisms, 275, 276–7
Acinetobacter calcoaceticus, 248, 251
actin, in all eukaryotes, 37, 47, 65
actinomycete-coryneform group of bacteria, with DNA having high G + C, 7, 8, 11
Actinomycetes, 9, 10
  sporulation in, 88, 94
Actinoplanes group, 9
activation of enzyme synthesis, 256, 260
active transport, in both prokaryotes and eukaryotes, 41
Agrobacterium tumefaciens, 192
L-alanine, in sporulation and germination in *Bacillus* spp., 104, 111
Alcaligenes eutrophus, 194
algae, 277
  ferredoxins of, 180, 190, 193, 206
  sporulation in, 86, 87, 90
alkaline phosphatase, in developing spore, 115–16
amino acids, 218
  controls in synthesis of: aromatic, 231–5; aspartic acid family, 235–6
  see also individual amino acids
ammonia
  assimilation of, by coliform bacteria, 257
  repression of enzymes of nitrogen metabolism by, 252
  subject to decomposition by ultra-violet light, 99
amoeboid motion, in eukaryotes, 37
  not in earliest eukaryotes, 39, 47
cAMP, effector for catabolite activator and repressor proteins, 256, 260
amphibolic metabolic pathways, 231, 237
Anacystis nidulans, 196
anaerobic bacteria, superoxide dismutase in, 196, 200
anaerobic respiration, 275–6
ancestor, universal common (Progenote), 26–7
animals
  date of divergence of, from plants, 176
  small size of mitochondrial DNA of, 73
anthranilate synthase, 231, 232, 234, 238

antibiotics
  production of, an ancestral characteristic of Gram-positive bacteria?, 12–13
  spread of factors carrying resistance to, 145
Aphanocapsa, 24–5
Aphanotheca sacrum, 192
archaebacteria (methanogens, extreme halophiles, extreme thermacidophiles), 6, 19–21
  phylogeny of, 18–19, 23
arginine
  pathway of synthesis of pyrimidine and, 239–40
  regulation of metabolism of, 250, 253, 259, 261
  as reserve in spores, 103
Arthrobacter group, 8, 9, 10
Arthrobacter simplex, 9, 11
Arthromictaceae, 116
Arthromitus, 88, 93
Ascomycetes, 69, 89
aspartate aminotransferase, in controls of citric acid cycle, 218
aspartate carbamoyl transferase, 239, 240, 244
aspartokinases, 235–6, 242, 247
Aspergillus, 73, 225
atmosphere, early anaerobic, 23
  addition of oxygen to, 92, 277
ATP
  energy transduction by, in both prokaryotes and eukaryotes, 41
  required for ligation, after excision of introns from tRNA, 156–8, 162
  yield of, from various reactions, 215–16, 217, 223
ATPase
  membrane-bound, proton-transporting, 276, 277, 278, 279–80
  membrane-bound, transporting secretory vesicles, 65
  mitochondrial, 73
attenuation of enzymes (modulation of efficiency of transcription termination), 256, 257–8
  speculations on origin of, 259
autogenous regulation of enzyme synthesis, 257, 260
auxotrophy: for tryptophan in *Shigella*, mutation causing, 255–6
Azotobacter cysts, 87, 91, 94

355

Azotobacter vinelandii, 16
  ferredoxins of, 182, 186, 187, 190, 191

Babesia, antigenic variation in, 313
Bacillariophyceae, sporulation in, 90
Bacillus, *see*
  Clostrium–Bacillus–Streptococcus group
Bacillus acidocaldarius, 25
Bacillus anthracis, 334
Bacillus cereus, 101
Bacillus coagulans, 93
Bacillus fastidiosus, 104
Bacillus-Lactobacillus-Streptococcus group, 12, 13
Bacillus licheniformis, 236
Bacillus megaterium, 104, 110–11
Bacillus polymyxa, 181, 236
Bacillus spp., 24–5, 102, 104
  endospore formation in, 88, 91, 93; postulates for evolution of, 96–7; stages in, 95–6
  ferredoxins of, 165, 190
Bacillus sphaericus, 93, 100
Bacillus stearothermophilus, 25
  ferredoxins of, 179, 182, 185, 188, 189
  superoxide dismutase of, 196, 199, 200
Bacillus subtilis, 93, 102, 246
  spores of: coat protein, 101; sulpholactic acid in, 103
  sporulation in, 95, 98–9; mutations in germination and outgrowth, 104
  tryptophan pathway in, 138, 251, 254
bacteria
  anaerobic forms of, early; aerobic, perhaps developed several times, 23
  concept of species in, 146
  current classification of, not phylogenetically valid, 21, 23–4; not supported by cytochrome structures, 177
  encapsulated, resistant to phagocytes and complement, 334
  morphology not a reliable phylogenetic indicator for, 1, 11, 21, 22
  proteins of, 176
  reactions catalysed by ferredoxins of, 180–1
  sporulation in, 87, 88, 92–104
  supposed origin of, from anaerobic heterotrophs, questioned, 22–3
  see also archae-, eu-, Gram-negative, *and* Gram-positive bacteria
bacteriophages, 102, 262, 352
baeocytes, of Pleurocapsalean cyanobacteria, 88, 91
Basidiomycetes, 89
Beijerinckia cysts, 88, 94
Bifidobacteria, 9, 10, 11
Blastocladiella, 91

Botanical Myth, of origin of eukaryotes, 49
Brassica campestris, superoxide dismutase of, 196, 201
Brevibacterium linens, 9

Calvatia gigantea, number of spores produced by, 112
carbamoyl phosphate synthetase, 239, 240
carbohydrates, in surface glycoprotein of trypanosomes, 323, 325
carbon catabolite repression of enzymes, 251-2, 253
carbon dioxide fixation
  ferredoxin in, 181, 190
  in photosynthetic bacteria, 185
  selection of organisms capable of, on depletion of carbon sources in 'primeval soup', 276
catabolite activator proteins, 256, 260
catabolite repressor proteins, 256
cell cycle
  in eukaryotes, 35, 39
  length of, 51–3, 54, 58
  meiosis in relation to, 59
  sporulation in relation to, 95, 96, 98
cell division, and sporulation
  differences between, 97–8
  mutually exclusive, 110
cell wall composition, important in phylogeny, 22, 46, 110
  in eubacteria, 6, 10, 11, 19
  in fungi, 77; in different stages of Mucorales, 109–10
  variety of, in archaebacteria, 19
cells
  compartmentation of, in eukaryotes, 36–7, 44, 49, 77
  fusion of, 56, 58
  interaction between components of, in evolution, 33
  size of, 40, 49, 58
Cellulomonas group, 8, 9, 10, 11
centrioles, 9-triplet: widespread in eukaryotes, absent from prokaryotes, 39, and from Eufungi, 49
centromeres: anaphase separation of, in origin of meiosis, 59–60
centrosomes, in development of mitosis, 56, 57
channel-shuttle model (for intermediates with alternative metabolic fates), 234, 241
Chara, 201
chemoautotrophy, not found in eukaryotes, 40
chitin: change from peptidoglycan to, in cell wall of first eukaryotes, 48, 77
chitosomes, 49n

Chlamydomonas, 87, 91
chloramphenicol, *E. coli* transposon carrying resistance to, 141
Chlorella, 87
Chlorobium, 8, 17, 184
Chlorobium limicola, 188, 189
Chlorobium thiosulfatophilum, 188, 191, 200
Chloroflexus, 8, 17
chlorophylls
  a and *b*, in higher plants, 17, 72, and in *Prochloron*, 17
  b, in Euglenophytes, 73n
  c, in plastids of Chromophyte and Cryptophyte algae, and Dinoflagellates, 72, 73n
Chlorophyte algae, 90, 201, 277
  photosynthetic apparatus of, 45, 73n
chloroplasts, 6, 16–17
  ferredoxins of, 190
  five independent symbioses of, with different groups of eukaryotes, 74
  genes of: introns in, 151; split, 38
  see also plastids
chorismate mutases, 232–3
  inhibitors of different activities of, 233, 234, 242
  repression of synthesis of, 251
chromatin: cycle of condensation of, lacking in some Hemiascomycetes and Zygomycetes, 49–50
Chromatium group, 14
Chromatium vinosum, 24–5
  ferredoxins of, 179, 182, 184, 186, 189, 191
  superoxide dismutase of, 190, 200, 206
Chromobacterium violaceum, 250
Chromophyte algae, 72, 73n
chromosomes
  of eukaryotes, several linear, 35; directed transport of, 37; increase in number of, after development of centrosomes, 57; origin of, 60, 61; origin of pairing of, in meiosis, 59–60; processes in evolution of, 50–3
  of prokaryotes, single circular, attached to plasma membrane, 35
Chrysophyte algae, 90
Chytridiomycetes, 75, 110
cilia, absent from prokaryotes, 39, and Eufungi, 49; widespread in eukaryotes, 39, but not originally present, 39, 47
  evolution of, in ancestor of Chytridiomycetes?, 75
  theory of origin of, in ectosymbiotic spirochaetes, 41
circadian rhythms: in eukaryotes, but not originally present, 39
citrate, as catabolite repressor, 252
citrate synthase, 231, 236
  properties of, used in taxonomy, 236, 243

citric acid cycle
  advantages of oxidation of acetate by, over direct oxygenation, 215–18
  in *E. coli* growing anaerobically, 231
  reversed, in carbon dioxide fixation by photosynthetic bacteria, 185
  in trypanosomes, in part of life cycle, 314
  use and regeneration of oxaloacetate in, 217
Clostridium, not a genus, but a major phylogenetic unit, 12, 13, 23
  ferredoxin of, 190, 204
Clostridium acidiurici, 13, 188
Clostridium aminovalericum, 24–5
Clostridium–Bacillus–Streptococcus group, with DNA having low G + C, 7, 8
Clostridium butyricum, 188, 204
Clostridium innocuum, 13
Clostridium pasteurianum, 13, 179, 182, 183–4, 185, 187, 191
Clostridium ramosum, 12
Clostridium spp., 88, 93, 102
coccoid forms
  groups of, not valid phylogenetic units, 21
  as morphologically degenerate forms, 25–6, often having relatives with more complex morphology, 12
coenzyme A, 216, 217–18, 224
coenzymes: unusual, in archaebacteria, 20
collagen: introns in gene for, in chick, 151
complement, trypanosomes and encapsulated bacteria resistant to, 333–4
copper, catalyses superoxide dismutation, 195, 196
co-repressors of enzymes, 249, 255
coryneform bacteria
  animal pathogen group of, 9, 11
  plant pathogen group of, 8, 9, 10
  see also actinomycete-coryneform group
crossing-over, in transition from circular to linear chromosomes, 60
Cryptophyte algae, 72, 73n
Cyanidium caldarium, 192, 193
cyanobacteria, cyanophytes, 8, 16–17
  chromosome of, 50
  ferredoxins of, 190, 193, 206
  fossils of, 17, 92
  lack usual bacterial range of controls of enzyme synthesis, 253
  photosynthetic apparatus of, 45–6, 72, 75
  respiratory system of, 75
  spores of, 88, 91
Cyanophora, cyanelles in, 45
cyclic pathways, efficiency of, 217, 219
cysteine residues: in ferredoxins, 205–6, and in hemagglutinin of influenza virus, 304
cysts, bacterial, 87, 88, 91
cytochromes, 11, 73, 176
  of photosynthetic bacteria, 14, 177

cytoplasm: compartmentation of, in eukaryotes, 36–7
cytoplasmic streaming, in eukaryotes, 37, (not originally present), 39

dehydrogenases, family of, 226
dendrograms for bacteria, produced by rRNA cataloguing method, 5
denitrification, not found in eukaryotes, 40
3-deoxy-D-arabino-heptulosonate 7-phosphate (DAHP) synthases in system synthesizing aromatic amino acids, 231–2, 234, 242, 250–1
derepression of enzymes, 249
desiccation, resistance of spores to, 111
Desulfotomaculum, 88, 93, 94
Desulfotomaculum nigrificans, 94
Desulfovibrio desulfuricans, 182, 196
Desulfovibrio gigas, 179, 181, 182, 187–8, 189, 190
Desulfovibrio sulfuricans, 15
Desulfuromonas acetoxydans, 184
Deuteromycotina, 110
diaminopimelic acid, in peptidoglycans of sporulating Bacilli, 101
dictyosomes, in eukaryotes, 36
Dictyostelium, 47, 225
diguanosine diphosphate, in metabolic controls in *E. coli*, 252
Dinoflagellates, 63, 72, 73n
dioxygenases, 224
diphosphoguanosine diphosphate (ppGpp), in *E. coli*, 252
dipicolinic acid, in spores, 93, 103
DNA
  of Dinoflagellates and Parabasalia, containing hydroxymethyl uracil, 63
  of Eufungi, lacking in repetitive sequences, 49
  of eukaryotes: attached to nuclear membrane in interphase, 34; non-coding parts of, 65; regulatory proteins binding to, 254, 259
  of mitochondria, *see under* mitochondria
  of trypanosome groups, 329–32
DNA binding protein: in mitochondria, 62, and in prokaryotes, 61
DNA–DNA hybridization, 1–2, 3–4
DNA gyrase: in mitochondria, 62, and in prokaryotes, 60–1, 62
DNA ligase, 61
DNA polymerase, and replication of DNA containing insertion sequences, 136, 139
DNA rearrangements
  by deletions, duplication or mobilization of genes, integration of plasmids, replicon fusion, 143–4
  gross genome structure after, and evolution, 145–6
  by insertion of sequences, in *E. coli* K12, 131–2, 351; at *gal* operon, 132–9, 146; at *lac* operon, 139–42; 'mini' and 'supermini' insertions, 136, 139
  by insertion of sequences, in other organisms, 142–3
DNA replication
  in eukaryotes, 34, 38–9, 52
  in prokaryotes, 56; limitations of, 51–2
DNA–rRNA hybridization, 5 groups of Pseudomonads defined by, 16
DNA topoisomerase, 62
DNA transposase, 61, 66
Drosophila, 142, 144
Dunaliella salina, 192, 193
dynein (tubulin ATPase), restricted to eukaryotes, 37

effectors, low-molecular-weight substances binding to and modulating enzymes, 230, 254–5, 256
Eimeria schubergi, 87
electron transfer
  cyclic, in photosynthesis, 278
  by ferredoxin, 178, and small molecules containing Fe-S clusters, 205
  by hydrogenase, 282
endoplasmic reticulum, 34, 49n, 56
endospores, 88, 89, 91
  possible course of evolution of, 93, 94–7
energy-converting systems
  evolution of, 273–4, from 'primeval soup', 274, through fermentative metabolism and substrate-level phosphorylation, 274–5, and anaerobic respiration, 275–6, to energy crisis, 276–7
  proton-transporting systems for, (ATPase), 279–80, (oxido-reduction), 280–2
  vectorial enzymes for, 277–9
enteric bacteria, 14, 15, 246
enzymes
  basic, relatively non-specific, of earliest organisms, 226
  families of, 226
  membrane-bound, developed from cytoplasmic, 277–9
  metabolic control at level of preformed, 231–7; spatial organization in, (compartmentation), 239–40, (juxtaposition), 237–9; speculations on origin of controls, 244–8, and on their selective advantage, 240–4
  metabolic control at level of synthesis of, by induction or repression, 248–52, 262–3; control devices for, 254–9; speculations on reasons for mixed control devices, 259–60, and on selective advantage of controls, 252–4

INDEX

erythrocytes, superoxide dismutase of, 198, 199
Escherichia coli, 24–5
  chromosome of, 50, 52, 145; insertion of sequence into, *see under* DNA rearrangements
  citric acid cycle in, during anaerobic growth, 231
  control of RNA synthesis in, 252
  enzyme repression in, 249, 250, 252, 253, 255, 349
  enzymes of: aspartate carbamoyl transferase, 244; aspartate kinases and homoserine dehydrogenases, 235, 237, 242; DAHP synthases, 231–2, 242, 243, 250–1; superoxide dismutases, 196, 199, 200, 202; hydrogenase, 281; threonine dehydratase, 236
  ferredoxins of, 192
  genetic map of, 145
  mutants of: with asymmetric cell division, 98; oxygen-sensitive, lacking superoxide dismutase, 195
  plasmids in, 141, 152, 288, 289
  regulons in, 261
  replicons in, carrying resistance to chloramphenicol, 141
  transacetylase gene of unknown function, in *lac* operon of, 144
  tryptophan system of, 238, 255
eubacteria, 6, 22–3, 25–6
  see also Gram-negative bacteria, Gram-positive bacteria
Eubacterium, 11
Eufungi, 49–50, 72, 73
Euglena gracilis, 196
Euglenophytes, 46, 72, 73n
eukaryotes
  cell of, a genetic chimera, 6–7
  characters of prokaryotes and, 40–1
  evolution of: gradual or quantum, 75–7; prokaryotes a stage in? 113–14; suggested phylogeny of 8 major groups of, 74
  fundamental differences between prokaryotes and, 33–9
  sporulation in, 87, 88, 104–10
  superoxide dismutases of: Cu/Zn in cytosol, Mn in mitochondria, 206
  theories of origin of, 40–3; basic possibilities for first eukaryote, of which aerobic fungus feeding osmotrophically is preferred, 44–7; symbiotic theories for origin of mitochondria and plastids, not applicable to cell, 43–4
  theory of origin of, from fungus, 47–50, occurring through coincident mutations 1000 million years ago, 76, 77; this required development of coadaptation of chromosome replication, segregation, and transcription, 50–3, histones and nucleosome, 58–64, linkage revolution and transcriptional control, 67–8, mitochondria and plastids, 71–5, mitosis, 53–6, nuclear envelope and pore complex, 64–5, ribosome diversification, 68–70, and syngamy and meiosis, 56–8
evolution
  of endospore formation, postulates for, 96–7
  of eukaryotes: gradual or quantum, 75–7; requirements for adequate theory of, 33
  phenotypic and genotypic measures of, 23–4
  rate of, 24–6
exocytosis: universal in eukaryotes, unknown in prokaryotes, 36–7, 47, 77
exospores: in fungi, 89; and in prokaryotes, 88, 91, 94

fatty acids, long-chain
  metabolism of, 223–5; citric acid cycle in, 217; ferredoxin in, 180
  polyunsaturated, 40, 49
  synthesis of: enzyme complex for, in *Saccharomyces*, 243; pentose phosphate cycle in, 221
fermentation, by primitive organisms, 275
ferredoxins, 177–8
  amino-acid sequences of, 188–90, 192–3
  electron-transfer reactions catalysed by, 180–1
  EPR spectra of, 179, 183
  identification of Fe-S cluster types in, 182–3
  reduction potentials of, 178, 182, 185, 186–7
  structure of, 183, 184, 190–1
  types of: 2Fe-2S (red), 180, 190–2; 3Fe-3S, 187; 4Fe-4S, (brownish black), 181, 185–7; high potential 4Fe-4S, 186; 2(4Fe-4S), 183–5; mixed, 187–8
fish, leech-transmitted trypanosomes in bloodstream of, 332–5
flagella, present in some prokaryotes, not found in eukaryotes, 40
  phase variation in antigens of, in *Salmonella*, 352
Flavobacterium, 18
fossils: of cyanobacteria, 17, 92, early eukaryotes, 92, 105–6, and early prokaryotes, 92, 106, 175
fowl plague virus, hemagglutinin gene of, 287
fungi
  nitrogen enzymes of, repressed by ammonia, 252

fungi—*contd*
  sporulation in, 86, 87, 89, 91
  theory of origin of eukaryotes from, 47–50

β-galactosidase: hybrid of, with membrane protein, 144
gas vacuoles: in some prokaryotes, not found in eukaryotes, 40
gene dosage, gene linkage, 67, 68
genes
  clustering of those of related function, *see* operons
  duplication of, 142, 143; in evolution, 244–5
  essential, of *E. coli*, clustered round replication origin, 145
  expression of, affected by change of position in genome, 144
  introns in, *see* introns
  split, 34, 38; origin of, 65–7
  transfer of, between species of prokaryotes, 176–7; transfer in eukaryotes usually by cell fusion, 40
genetic code, common to prokaryotes and eukaryotes, 41
  of mitochondria, uses fewer tRNAs than normal code, 71, 73
genome
  gross structure of, fairly stable, 145
  size of: in eukaryotes, length of cell cycle proportional to, 51–3; larger, of eukaryotes, 40, 50, 65; smaller, of prokaryotes, 40, and Eufungi, 49
genotypic evolution, reflects rate of mutation, 24
Geodermatophilus, 11
germination, of endospores, 104
Glaucophytes, 72
β-globin of mouse, introns in gene for, 151, 167
Glossina (tsetse fly), transmitting trypanosomes to mammals, 313
  origin of association of trypanosomes with, 332
  not easily infected by most species of trypanosome, 332, 334
  trypanosomes in, 314–15, 322
  virus-like particles in, 337
glucans, α- and β-1,3-, 72, 94
glucose
  as catabolite repressor, 252
  Entner-Doudoroff pathway of degradation of, in some micro-organisms, 225
  germination response of some spores to, 225
glucose dehydrogenase, in sporulating but not vegetative cells of Bacilli, 101
glucose-6-phosphate: use and regeneration of, in pentose phosphate cycle, 220–1

glutamate, as reserve in spores, 103
glutamate synthetase, sensitive to nitrogen status of cell, 237, 252, 257, 260
γ-glutamyl cycle, transports amino acids into cell, 221
glycerol ethers, in membranes of archaebacteria, 20
α-glycerophosphate cycle, in bloodstream trypanosomes, 314
glycogen
  in ancestral Eufungi, 72
  enzyme for degradation of, specific to sporulation in *Saccharomyces*, 109
  reserves resembling, in Cryptophytes, Dinophytes, Glaucophytes, and Rhodophytes, 72
  storage cycle of, 221
glycolysis
  in bloodstream trypanosomes, 314
  regulation of, in *Dictyostelium*, 225
glyoxylate cycle, synthesizing succinate from acetyl CoA, with use and regeneration of glyoxylic acid, 219–20
Golgi cisternae, 36–7, 48, 49
Gram-negative bacteria, not a coherent group, 13–14, 16
  cell wall of, 94
  many derived from photosynthetic ancestry, 14–15
  septation in, 115
  sporulation in prokaryotes a stage in the evolution of, 113–16
  superoxide dismutases of, 206
Gram-positive bacteria
  ancestral phenotype of, 12
  cell wall of, 12, 94
  with DNA having high content of G + C, 7–12, or low content, 12–13
  origin of sporulation among progenitors of, 93
  septation in, 115
growth rate, metabolic controls relating to, 252, 253
guanine nucleotide, sporulation in relation to deprivation of, 100

Haemophilus influenzae, 334
Halobacteriaceae, 6, 19
Halobacterium halobium, 182, 191, 193
Halobacterium spp., 19, 20, 23
heat, resistance of spores to, 111
hemagglutinin of influenza virus, occurring in 3-molecular spikes attached to viral membrane, 292
  amino-acid sequences of, 286, 292–7; percentage homology in, between different subtypes of virus, 300; for 3 subtypes and an avian type, 297–300
  antibodies primarily directed against, 285;

antigenic sites on, 300–1, 302; antigenic drift in, 300–2; antigenic shift in, 302–4
fatty-acid attachment site on, 297
glycosylation site on, in different strains, 304–5
post-translational processing of, 292
RNA gene for: codon usage in, 305, 306; DNA copies of, 286, 287–8, 289; sequences of, in different strains, 287, 291
three-dimensional structure of, probably the same for all strains, 304, 306, 307
Hemiascomycetes, 73
possible resemblance of, to first eukaryotes, 48–9, 77
primitive characters of, 50, 64
hemocuprein, 193
heterokaryosis, in fungi, 112, 113
histidase, 252
histidine
as ligand for metals in Cu/Zn superoxide dismutase, 198, 199, 200
operons for degradation of (*hut*), 251, 259, 260, 261, and synthesis of (*his*), 258, 261
histones, 39, 62–3
origin of, in eukaryotes, 49, 58–64
homoserine dehydrogenases, 235
hormones, peptide, 226
hydrogen donors, for primeval organisms, 276–7
hydrogen metabolism, ferredoxins in, 180, · 181
hydrogen sulphide, hydrogen donor for primitive organisms, 276–7
hydrogenase, 194, 281, 282
β-hydroxybutyrate, induces cyst formation in *Azotobacter*, 91
Hyphochytrid fungi, suggested origin of, 46

immunoglobulins, introns in genes for, 168
induction of enzymes, 249; sequential, 251
influenza virus, showing gradual (drift) and sudden (shift) antigenic and chemical changes in surface proteins, 285, 352
changes in amino-acid sequence of hemagglutinin in drift, 300–2, and in shift, 302–4
in lower mammals and birds, 286, 287
RNA segments of (8), 286, 287–8, 289; sequences of, 288, 290–2
shifts in, perhaps due to reassortment of genes with animal strains, 305, 307
types of, since 1918, 285–6
types of, since 1977, 303–4, 307
insulin, introns in gene for, 168
introns (intervening sequences) in genes, 151
excision of, followed by ligation, 156–8
possible structure of, 158–60
iron, catalyses superoxide dismutation, 195

iron-sulphur proteins, simple and conjugated, 178
isocitrate lyase, in glyoxylate cycle, 220
isoprenoid quinones, in eubacteria having DNA with high G + C, 11

kinetochores, in eukaryotes, 35, 39
in evolution, 55, 56, 57, 59
Klebsiella, 143
Klebsiella aerogenes, 251, 252
Klebsiella pneumoniae, 334
Kurthia, 12

Lactobacillus–Bacillus–Streptococcus group, 12, 13
leeches, trypanosomes transmitted by, 332–3
Leishmania hertigi, parasite of porcupines, 337
Leptospira, 8, 17
light energy, development of organisms capable of using, 276–7
lipid bilayers of primitive cells, incorporation of protein into, 279
lipids: straight-chain in membranes of eubacteria, glycerol ethers in those of archaebacteria, 20
liposomes, self-replicating, 275
lysine
diaminopimelic acid and aminoadipic pathways of synthesis of, in phylogeny, 46, 47
in peptidoglycan, 101
as reserve in spores, 103
lysosomes, in eukaryotes, 37
lytic enzymes: extracellular, secreted by bacteria after beginning of sporulation, 100

magnesium, required for tRNA ligase of yeast, 162
malate synthase, in glyoxylate cycle, 220
manganese, catalyses superoxide dismutation, 195
Mastigocladus laminosus, 179, 192, 193
meiosis, 39, 57
origin of, in eukaryotes, 59–60; scheme for stepwise evolution of, 119–21
membrane-bound bioenergetic systems, 273, 275, 277
metabolic pathways
cyclic, 215, 226–7; citric acid cycle, 215–18; glyoxylate cycle, 218–20; photosynthesis, 220; pentose phosphate cycle, 220–1
linear reaction sequences, 221–2; long-chain fatty-acid metabolism, 223–5; propionic acid fermentation, 222–3
related enzymes involved in, 226
metabolism
controls of: at level of enzyme synthesis,

metabolism, controls of—*contd*
 248–63; at level of preformed enzymes, 230–48
 evolution of regulation of, 229–30
 intermediate, ferredoxins in, 180, 181
Methanobacteriaceae, Methanococcaceae, Methanomicrobiaceae, 6, 19
 unusual coenzymes of, 20
Methanosarcina, 19, 20
methionine cycle, for transfer of methyl groups, 221
Methylobacter, 88, 91
Methylomonas, 94
Methylosinus, 88, 91
Micrococcaceae, not a phylogenetically valid family, 21
micrococci, as degenerate forms of arthrobacteria, 9, 10, 11
Micrococcus, 9, 21
Micrococcus luteus, 24–5
Micrococcus radiodurans group (*M. radiodurans, M. radiophilus, M. roseus*), 7, 8, 18, 202
microfilaments of actin, in eukaryotes, 37
 in evolution, 49n, 55, 56, 57
microsomes, mono-oxygenase systems of, 192
microtubules of tubulin, in eukaryotes, 37
 in evolution, 49n, 55, 56, 57
 in mitosis, 35, 39, 55
 theory of origin of, from cilia, 41
mitochondria
 characteristic of eukaryotes, lost in anaerobic protozoa and fungi, 34, 39; acquired before plastids, 75
 DNA of, 38, 73, 151, 351; evolution of, 71–5; few molecules coded for, 62, 71
 DNA binding protein and DNA gyrase in, of bacterial type, 62
 ferredoxin of, 180, 192
 of plants, 6
 superoxide dismutase of, 196, 200; resembles that of prokaryotes, 201, 203, 206
 theories of origin of: autogenous, 43, 49n; endosymbiotic, 43
 in trypanosomes, in part of life cycle, 314
mitosis, in eukaryotes, 35, 38–9
 primitive form of, in Hemiascomycetes and Zygomycetes, 49–50
 theory of origin of, 50, 53, 55–8
 unique form of, in Dinoflagellates and Parabasalia, 63
mitotic spindle of eukaryotes, 37, 39, 57
 persists during interphase in some Hemiascomycetes, 50, and in *Saccharomyces*, 64
mono-oxygenase systems, containing ferredoxin, 191, 192, 224–5

morphology, not a reliable phylogenetic indicator for bacteria, 1, 11, 21, 22
motility, of cytoplasm in eukaryotes, 37
muconolactone isomerases, 247–8
Mucor, 91
Mucorales, 106–7, 109–10
muramic acid, *see* peptidoglycan
Mycobacterium, 11; *M. smegmatis*, 187
mycolic acids, 11
Mycoplasma, 12; *M. pneumoniae*, 195
mycoplasmas, 6, 12, 13
 as offshoot of Clostridia, not a separate class, 21
myosin (actin ATPase), restricted to eukaryotes, 37
Myxobacteria, 91, 94
Myxobacterium sp., 196
Myxococcus, 18

NAD: energy transduction by, common to prokaryotes and eukaryotes, 41
NAD couple, in control of citric acid cycle, 218
NADH, metabolic enzymes inhibited by, 236
$NADPH_2$, generated in pentose phosphate cycle, 220–1
Nannomonas, see Trypanosoma congolense
Neisseria gonorrhoeae, 195, 202
neuraminidase, of influenza virus, 285, 286
Neurospora, 73, 91
 enzymes of: DAHP synthases, 231–2; superoxide dismutase, 196
 synthesis in: of arginine and pyrimidine, 239; of tryptophan, 238, 243
Nitella, 201
nitrate, 99, 194
 ferredoxin in reduction of, 190
nitrite, ferredoxin in reduction of, 180, 190
nitrogen supply, of primitive organisms, 99
nitrogen fixation: in some prokaryotes, not found in eukaryotes, 40, 92, 181, 190
nitrogen metabolite repression of enzymes, 251, 252, 253
nitrogenase, of *Azotobacter*, 191
Nocardia, 9, 11
 sporulation in, 88, 91
Nostoc, 192
nuclear envelope, in eukaryotes, 34–6
 evolution of, 49n, 56, 64–5
 persists in mitosis in Hemiascomycetes and Zygomycetes, 50, 64
nucleic acids, phylogeny from nucleotide sequences in, 176
nucleolus, in eukaryotes only, 34, 38, 50, 64
nucleomorph: in plastids of Chromophyte and Cryptophyte algae, Dinoflagellates, and Euglenophyta, 72
nucleosomes, characteristic of eukaryotes, lost in free-living Dinoflagellates, 39, 60

origin of, 58–64
small in Ascomycetes, 49, 64
nucleosides (AMP, ADP, GDP), in control of catabolic enzymes, 236, 240, 241, 244

Oerskovia turbata, 9, 11
oncornaviruses, glycoprotein of trypanosome coat resembles glycoproteins of, 337, 338
Oomycetes, 46, 89
operons, 261–2
  in bacteria, especially coliforms, 245–6; speculations on origins of, 246–7; as mechanism for maintaining relative concentrations of different genes, 67
  fifty or more involved in sporulation in *B. subtilis*, 95
orang-utan, DNA difference between Sumatran and Bornean populations of, 146
ornithine carbamoyl transferase, 237, 239, 240
ovalbumin, duplicated sequence in gene for, 144
oxaloacetate, in citric acid cycle, 217, 218
oxidoreduction, 275, 280–2
β-oxoadipate *enol*-lactone hydrolases, of *Acinetobacter*, 248

Paracoccus, 21, 45, 48
Paracoccus denitrificans, 196, 200, 203
Paracoccus sp. (halophile), 186
pentose phosphate cycle, 220–1
peptidoglycan cell walls, not found in eukaryotes, 40
  change to chitin from, in development of first eukaryotes, 48, 77
  of cyanelles in *Cyanophora*, 45
  distribution of different types of, in actinomycete-coryneform group, 10, 11
  of Gram-negative bacteria, 115
  synthesis of, in endospore formation, 95, 97, 99, 100
Peptococcus aerogenes, 13
  ferredoxins of, 184, 186, 188, 189
Peptococcus glycinophilus, 13
peroxisomes, in eukaryotes, 37
phagocytosis, widespread in eukaryotes, not found in prokaryotes or first eukaryotes, 39
  evolution of, 47, 75, 77
  trypanosomes and encapsulated bacteria resistant to, 333–4
phagotrophy, advanced rather than primitive eukaryote character, 47
phenotypic evolution, sporadic and saltatory, 24
phenylalanine, controls in synthetic pathway for, 231–5

phosphate: movements of, in systems of proton-translocation by membrane-bound enzymes, 278, 280
phosphofructokinase, of *E. coli*, 231
3-phosphoglycerate, as reserve in spores, 103
phospholipids: in outer membrane of Gram-negative bacteria, and of spores, 115
phosphorylation
  oxidative, inefficient in some micro-organisms, 225
  substrate-level, of primitive organisms in anaerobic conditions, 275
Photobacterium leiognathi, symbiont in fish, 196, 200–1
photophosphorylation, ferredoxins in, 180
photosynthesis
  anoxygenic, not found in eukaryotes, 40
  as cyclic process, with use and regeneration of 1,5-ribulose bisphosphate, 220; cyclic electron flow in, 278
  ferredoxins in, 190, 192
  Mn dependence of, and Mn superoxide dismutase, 202
  as property of common ancestor of eubacteria?, 22–3; date suggested for development of, 277; lost in evolution of first eukaryote, 48
photosynthetic bacteria, 26, 185, 277
  ferredoxins of, 185, 191
phycobilins, 46, 72
phycobilisomes, 72
Phycomycetes, sporulation in, 87, 89
phylogeny, 5–7
  of archaebacteria, 18–21
  of bacteria (true), 7; of cyanobacteria, 16–17, and other Gram-negative bacteria, 17–18; of Gram-positive bacteria, 7–13; of purple photosynthetic bacteria and relatives, 13–16
  characteristics valid in, 21–2
  rRNA cataloguing method for finding, 2–5
  from sequences in proteins and nucleic acids, 2, 175–6
  suggested, for 8 major groups of eukaryotes, 74, 75, and 9 kingdoms of organisms, 46
Pilobolus, discharge of spores by, 112
pinocytosis, 39
Planococcus, questionable if a valid genus, 21
plants, 72, 73, 180
  date of divergence of, from animals, 176
plasma membrane, with phospholipid bilayer structure, common to prokaryotes and eukaryotes, 41
  double, of spores, 95, 98, 109

plasma membrane—*contd*
  hybrid proteins, introducing cytoplasmic enzymes into, 144
plasmids
  fusion of, within some cells, 141
  interspecies transfer of, 145
Plasmodium, antigenic variation in, 313
plastids, not present in earliest eukaryotes, 34, 39
  probable symbiotic origin of, 43, 45; occurring more than once, after evolution of phagocytosis?, 72, 75, 77
  see also chloroplasts
plastocyanin, 203
Plectonema boryanum, 196
Pleurotus olearius, 196
pore complex of eukaryote nuclear membrane, 34, 35, 65
Porphyra umbilicalis, 192, 193
Porphyridium cruentum, 196
prephenate dehydratase, prephenate dehydrogenase, acting also as chorismate mutase, 233, 234
'primeval soup', 274, 276
Prochloron, 17
Prochlorophytes, 45–6, 72, 75
prokaryotes
  characters of eukaryotes and, 40–1
  fundamental differences between eukaryotes and, 33–9
  inactivation of enzymes of, on entry into stationary phase or sporulation, 237
  sporulation in, 87, 88, 96; metabolic preconditions for, 100; metabolism in, 103–4; origins of, 92–4; possible course of evolution of, 94–7, (first stable intermediate in) 97–100; spore coat, 101–3; spore germination and outgrowth, 104
  usually contain Fe and Mn superoxide dismutases, 206
Propionibacteria, 9, 10, 11
Propionibacterium shermanii, 223
propionic acid fermentation of lactic acid, 222–3
proteases, family of, 226
protein synthesis, in prokaryotes and eukaryotes, 34, 70
  control of rate of, 252
  during sporulation in *B. subtilis*, 95
proteins
  amino-acid sequences in: evolutionary change in, 203–4; phylogeny from, 175–6
  of bacteria, 176; of spore coat of Bacilli, 101
  conformation of, in cytochromes, 176, and ferredoxins, 182
  differences in, correlate less well with phenotypic differences than do DNA rearrangements, 146
  hybrid, with new functions, 144
  of universal common ancestor, 27
proteolipid of mitochondrial ATPase, 73
Proteus vulgaris, 238
protons, evolution of systems for transporting
  by ATPase, 275, 279–80
  by oxidoreduction systems, 275, 278, 280–2
Protozoa, 90, 313
Pseudomonads, 16, 236, 261
Pseudomonas aeruginosa
  enzymes of: catabolite repression of, 252; constitutive and induced, 250, 251; DAHP synthase, 234, 242, 251; histidase, 236; operons for, 246; of tryptophan pathway, 238, 249–50
  strain of, not using lysine, 263
Pseudomonas diminutia, 16
Pseudomonas maltophilia, 16
Pseudomonas ovalis, 187, 189, 190, 200
Pseudomonas putida
  enzymes of: induction of, 251; muconolactone isomerases, 247–8; operons for, 246; of tryptophan pathway, 238, 249–50
  ferredoxins of, 182, 191
Puccinia graminis, dispersal of spores of, 112
purple photosynthetic bacteria, 8, 14–15
  respiratory mechanisms and cytochromes of, resemble those of eukaryotes, 44–5; stages suggested for development of eukaryotes from, 38
pyrimidine, pathway of synthesis of arginine and, 239–40, 250
pyrophosphatase, cytoplasmic: in development of proton-transporting ATPase, 279–80
pyrophosphate, metabolic enzymes inhibited by, 236
pyruvate, excreted by bloodstream trypanosomes, 314

quinones, in hydrophobic environment of cell membrane, 281

radiation
  resistance to: of group of micrococci, 7, 8, 18; of spores, 111
  superoxide radical produced by, in anaerobic conditions, 202
regulons, (scattered genes regulated by same device(s), 261
repression of enzyme synthesis, 249–52, 253
  speculations on origin of, 254–6
replicons, in eukaryote chromosomes, 35, 39

plural, necessary for larger genomes and numerous chromosomes of eukaryotes, 50, 52, 58, 68
reptiles, salivarian trypanosomes transmitted by leeches to, 332
resources
  multiple use of, 218, 219–20
  optimum use of: in citric acid cycle, 217–18; and evolutionary success, 215, 227
respiration
  aerobic, on plate-like cisternae, in development of first eukaryotes, 49n
  anaerobic, 275–6; not found in eukaryotes, 40
Rhizobium japonicum, 184
Rhizopus, fumarate produced by, 225
Rhodococcus group, 9
Rhodomicrobium, 88
Rhodophyte algae, 45–6, 72, 90
Rhodopseudomonas capsulata, 236
Rhodopseudomonas gelatinosa group, 14
Rhodopseudomonas sphaeroides, 196, 200
Rhodopseudomonas spheroides group, 14, 44–5, 48
rhodopsin, oxygen-dependent step in synthesis of, 23
Rhodospirillum rubrum, 181, 182, 190, 191
Rhodospirillum tenue, 21–2
ribonucleotides, germination response of *Bacillus* spores to, 104
ribosomes
  of eukaryotes, 34, 35, 38, diversification of, 68–70
  of prokaryotes, 35, 38; can exchange subunits with plastids, and still show some function, 68
1,5-ribulose bisphosphate: use and regeneration of, in photosynthesis, 220
RNA of eukaryotes
  introns in, 66, 67, 151
  processing of, in nucleolus, 38
  splicing of, 38, 66–7, 170
  synthesis of, in nucleus, 34, 36, 38; rate of, 38, 68, 252, 253
  transport of, from nucleus to cytoplasm, 38, 55, 167–8; suggested mechanism for, 65
mRNA of eukaryotes
  capped with 5′-methylguanosine, 35, 38, 66
  introns in, 167, 169
mRNA of prokaryotes, translated in nascent condition, 35, 36, 38
rRNA of eukaryotes, 38
  size of, in cytoplasm and mitochondria of different groups of organisms, 69, 70
rRNA of prokaryotes, 38, showing resemblance to that of plastids, 69, 70

5S, saltatory changes in, 3
16S, sequence analysis of, for measuring genealogical relations in bacteria (cataloguing approach), 2–3, 93, 350; methods, 3–5; S_{AB} values for pairs of organisms obtained by, used to make dendrograms, 5
snRNA (small nuclear RNA), involved in RNA splicing? 170
  antibodies to, inhibit splicing of adenovirus RNA, 170
tRNAs, 38
  of archaebacteria, chemically different from those of eubacteria, 20
  archetypal structure of, found in all organisms, 168
  electrophoretic separation of precursors of, 152–5
  glycine-valine family of, 350
  introns in, 168–9; role of, in expression of tRNA genes, 162–7
  splicing of, 170
  splicing of precursors of, *in vitro*, 155–6; ATP in ligation step, 156–8, 180; structure of excision product, 158–60
tRNA ligase, partial purification of, 161–2
RNA polymerases
  binding of repressor to DNA blocks access of, to promoter site, 254
  distribution of 2 types of, in archaebacteria, 18, 20
  in eukaryotes, separate for mRNA, rRNA, and tRNA, 38; origin of separate, 62
  in prokaryotes, 35; divergence of eukaryote type from, 62
  reaction of DNA-bound activator with, 256
RNases, evolved to destroy excised intron RNA? 66
Ruminococcus albus, 184
Ruminococcus bromii, 13

S_{AB} value for a pair of organisms (16S rRNA sequences in common), 5, 18, 24–5
Saccharomyces
  centrosome plaque in, 56, 57
  chromosomes of, 50, 61
  DNA of, 47, 142; mitochondrial, 73
  enzymes of: carbamoyl phosphate synthetases, 240; fatty acid synthetase complex, 243; ornithine carbamoyl transferase, 237; superoxide dismutases, 198, 199, 200
  introns in tRNA of, 168–9; role of, in expression of *tyr* ochre suppressor gene, 165–7
  mating-type switch in, 262, 352

Saccharomyces—contd
  mitotic spindle of, persistent in interphase, 64
  mutations in, affecting cell division and sporulation, 101, 121
  primitive characters in, 47, 50; resembles ancestral eukaryotes? 147
  sporulation in, 91, 98; cytology of meiosis and, 107–9
Salmonella, phase variation in, 262, 352
Salmonella typhimurium
  enzymes of: aspartokinases and homoserine dehydrogenases, 235; DAHP synthases, 231–2; induction of, 251
  genetic map of, 145
Sarcina, 12, 21
scalar reactions v. vectorial reactions, 277–9
Scenedesmus obliquus, 182
Scenedesmus quadricauda, 192
Schizosaccharomyces pombe, 73, 91
secretion, in prokaryotes and eukaryotes, 36–7
  of cell wall, stage in origin of eukaryotes, 47, 48
  transport of products of, 77; microfilaments and microtubules in, 55, 56
septation
  in Gram-negative and Gram-positive bacteria, 115
  in normal cell division and in sporulation, 97, 98–9
serodemes, of Trypanosoma clones, 317–18
Serpens flexibilis, 16, 21
Serratia, 238, 246
sexual reproduction: with genetic recombination, confers advantage of variability, 112
  evolution of: meiosis in, 57–8; sporulation in, 118–22; syngamy in 58–9
  primitive form of, in Hemiascomycetes and Zygomycetes, 49–50
  by sexual spores, in algae, 90; and in Basidiomycetes, Oomycetes, and Zygomycetes, 89
Shigella dysenteriae, 255–6
Shigella sonnei, 263
shikimate kinase, 234
simian virus (SV) 40, 167, 169, 305
species, distinguished by DNA rearrangements rather than by protein differences? 146
Spinacia oleracea, ferredoxins of, 192
spiral bacteria, not a phylogenetic group, 16, 21
Spirillum volutans, 15
Spirochaeta aurantia, 24–5
spirochaetes, 8, 17

Spirogyra, 201
Spirulina platensis
  ferredoxins of, 182, 190–1, 192
  superoxide dismutase of, 196, 200
spore coat of endospores, protecting against lysozyme and divalent cations, not involved in resistance to heat and desiccation, 95, 96, 102
  assembly of protein of, resembles assembly of capsids of bacteriophages, 102
  in Saccharomyces, 107
Sporolactobacillus, 88, 93
Sporosarcina, 93
Sporozoa, 90
sporulation: in algae, 86, 87, 90; in eukaryotes, 87, 88, 104–10; in fungi, 86, 89; in prokaryotes, 87, 88, 92–104
  as ancestral characteristic of Gram-positive bacteria? 12–13
  endospores and exospores produced in, 91
  in evolution, 85, 92, 106; evolutionary advantages of, 111–13, and disadvantages, 110–11; of Gram-positive bacteria and eukaryotes, 113–16; of sexual reproduction, 118–22
  in a filamentous segmented bacterium, 116–18
  many types of, 85–91
  mutants lacking, 110–11
  as response to starvation, 91, 95, 97, 106
Staphylococcus, not a valid genus, 12, 21
starch, in chloroplasts of green plants and Cryptophyte algae, 72
starvation, sporulation as response to, 91, 95, 97, 106
steroids, sterols, 40, 192
Streptococcus, see Clostridium–Bacillus–Streptococcus group
Streptococcus–Bacillus–Lactobacillus group, 12, 13
Streptococcus pneumoniae, 334
Streptococcus mutans, 196
Streptomyces, 87, 88, 91
Streptomycetes, 9, 10, 11
sugar catabolism and neogenesis, amphibolic pathways of, 233
Sulfolobus, 18, 19
sulphate-reducing bacteria, 185
sulphite-reducing bacteria, 180, 181, 190
superoxide dismutases, 193–5
  amino-acid sequences of, 195, 198–200, 205–6
  different types of, 195–7
  in study of evolution, 200–3; could have been produced from Fe sulphides and amino acids in primeval reducing atmosphere, 204–5

Cu/Zn, in many eukaryotes, 177, 196, 197, 198, 199
 Fe, of prokaryotes, transferred into intracellular organelles, 177, 197, 200
 Mn, of prokaryotes, transferred into intracellular organelles, 177, 199, 200; shows clear homology with Fe enzyme (but has 4 subunits instead of 2), 201
superoxide radical, respiratory intermediate of aerobic organisms, 193, 194, 195
 produced in photolysis of water in anaerobic conditions, 202
symbionts: endocellular, in eukaryotes, not originally present, 39
 theory of mitochondria as, 43
 theory of plastids as, 43, 45, 72, 75, 77
syngamy, 58

thermacidophiles (archaebacteria), 6, 20
Thermoactinomyces, 11–12, 88, 93
Thermoplasma, 18, 19, 20
Thermus aquaticus, 196
Thermus thermophilus, 182, 187, 190
Thiobacillus denitrificans, 196
thylakoids of plastids, in different groups of algae, 72
transcription, in eukaryotes, 67–8
 rate of, 38
 uncoupled from translation, 35, 36–7
translation
 apparatus for, in universal common ancestor?, 27
 coupled to transcription in prokaryotes, 35, 36–7
 hypothetical methods for regulating metabolism at level of, 262
transport cycles, 221
transposons, in bacteria, 141, 144, 146, 351
Trypanoplasma borrelli, in bloodstream of fish, 333
Trypanosoma brucei (causing sleeping sickness), 313–15
 antigenic types of, 318–20
 glycoproteins of surface coat of bloodstream forms of, associated with antigenic changes, 315, 317; amino-acid composition and isoelectric points of, 323, 324; complement fixation by, 327; homologies of, 324–5; terminal portion of, containing membrane-binding site and carbohydrate, 328
 section of bloodstream form of, 316
Trypanosoma (Nannomonas) congolense, 316, 321, 334, 335
Trypanosoma cruzi (stercorarian), 332, 333, 335–6
Trypanosoma danilewski, in bloodstream of fish, 333
Trypanosoma equiperdum, 317, 332

Trypanosoma evansi, 332
Trypanosoma vivax, 316, 321, 334
trypanosomes, salivarian, 46
 antigenic types of, crossing species borders (isotypic variant antigenic types: iso-VATS), 318, 328–9
 antigenic variations in, 317–23; biochemistry of, 323–9; hypotheses for evolution of, 339, (mobile gene), 341–2, (mobile promoter), 339–40
 classification of, 314
 culture system for, 318
 DNA sequences in different groups of, 329–32
 evolution of, 332–3; of surface coat of, 333–9, (originally a glucose-carrier protein?), 335
 life cycle of, 313–15
 pathogenic evade immune system of host by antigenic variation, 313, 352, and by resistance of surface coat to complement and phagocytes, 333
 probable sexual stage in, 318, 341
 sections of different species of, 316
tryptophan
 feedback inhibition of anthranilate synthase by, 231, 232, 234
 synthase complex for, in *E. coli*, 234, 255; attenuation and repression of, 257–8, 259
 synthetic pathway for, 231–5, 238, 250
tubulin, in all eukaryotes, 37
tyrosine, controls in synthetic pathway for, 231–5

uric acid: germination response of spores of *B. fastidiosus* to, 10

vectorial reactions *v.* scalar reactions, 277–9
Veillonella alcalescens, 184
vesicles, secretory: in eukaryotes, 36, 37, 65
vibrios, 14, 15
viruses, 336–7
 glycoproteins of, compared with glycoprotein coat of trypanosomes, 337–8

water: development of organisms capable of using, as hydrogen donor, 276–7

Xanthomonas group, 15
Xenococcus (cyanobacterium), 87

yeast, 73, 224
 genes of, 73; introns in, 152–5
 see also Saccharomyces, Schizosaccharomyces
Yersinia enterocolitica, 146

Zea mays, 142, 146
zinc, in eukaryote superoxide dismutase, may be replaced by Cd, Co, or Hg, 196
Zoological Myth, of origin of eukaryotes, 49

zoospores: process of formation of, in prokaryotes, 88
Zygomycetes, 49–50, 89, 110

RAYMOND H. FOGLER LIBRARY
DATE